Springer Series in Solid-State Sciences

Volume 185

The Springer Series in Solid-State Sciences consists of fundamental scientific books prepared by leading researchers in the field. They strive to communicate, in a systematic and comprehensive way, the basic principles as well as new developments in theoretical and experimental solid-state physics.

More information about this series at http://www.springer.com/series/682

Sergey I. Bozhevolnyi · Luis Martin-Moreno
Francisco Garcia-Vidal
Editors

Quantum Plasmonics

Springer

Editors
Sergey I. Bozhevolnyi
Centre for Nano Optics
University of Southern Denmark
Odense M
Denmark

Francisco Garcia-Vidal
Condensed Matter Theory
Universidad Autonoma de Madrid
Madrid
Spain

Luis Martin-Moreno
Theory and Simulation of Materials
Instituto de Ciencia de Materiales de Aragón
Zaragoza
Spain

ISSN 0171-1873 ISSN 2197-4179 (electronic)
Springer Series in Solid-State Sciences
ISBN 978-3-319-45819-9 ISBN 978-3-319-45820-5 (eBook)
DOI 10.1007/978-3-319-45820-5

Library of Congress Control Number: 2016951692

Printed on acid-free paper

This Springer imprint is published by Springer Nature
The registered company is Springer International Publishing AG
The registered company address is: Gewerbestrasse 11, 6330 Cham, Switzerland

Preface

Quantum plasmonics is a very rapidly developing field that emerged recently at the border of two fields, both rich in fundamental physics and highly innovative in technology: quantum optics and plasmonics (the latter can also be viewed as nanophotonics of metal structures).

Nanophotonics concerns with the interaction of nanostructures with light, thereby aiming at providing photonic capabilities at smaller length scales and lower energy requirements. But even more importantly, nanophotonics also aims at engineering the light–matter interaction at unprecedented high strengths and/or subwavelength spatial resolutions. The latter usually involves the use of metals as these support surface electromagnetic modes (known as surface plasmons), which are confined to metal surfaces within subwavelength distances. In the last fifteen years, studies in what became known as "plasmonics" have been concentrated on plasmonic circuits (composed of subwavelength-sized waveguides and waveguide components), optical antennas (as efficient transducers between the far- and near-field wave components, squeezing in volume and boosting up in strength local fields), and surface-enhanced spectroscopic techniques (as "surface plasmon resonance sensing" and "surface-enhanced Raman spectroscopy"), with implications to diverse fields including photonics, optoelectronics, material science, bio-imaging, medicine, and energy.

These research directions continue to flourish but, additionally, plasmonics has now acquired a level of maturity that paves the way toward new venues, notably those involving quantum effects that arise from the interaction between plasmons (understood as quanta of localized or propagating surface plasmon excitations) and quantum systems characterized by a few discrete energy levels, such as molecules or quantum dots. This interaction opens up many different possibilities. Plasmons can be exchanged between quantum emitters, modifying their effective interaction and thus their physical properties. Alternatively, few-level systems can be used to induce efficient plasmon–plasmon interactions, leading eventually to strong nonlinear optical properties at the single-photon level. Additionally, when the interaction between plasmons and matter is strong enough, the combined system may acquire completely different (to those of the constituents) properties, opening new possibilities for the design of materials with novel functionalities. Another important area of research in quantum plasmonics refers not so much to the quantum features of plasmons, but to the influence of electron tunneling in the optical response of plasmonic nanostructures. This aspect, usually termed as "quantum effects in plasmonics," has a paramount importance in the properties of metal structures containing nanometer-sized gaps.

After a rapid initial evolution of quantum plasmonics [1], it became clear in 2014 that there is a need for a monographic workshop that would bring together researchers, belonging to different communities and covering various aspects of this nascent field. In 2015, we undertook this task and launched a workshop in the "Centro de Ciencias" in the beautiful village of Benasque, located in the heart of the Spanish Pyrenees. The success of the workshop, being reflected both in the quality of presentations and in the spirit of scientific discussions, indicated the aptness and importance of putting together the present state of the art in an easy-to-access manner. The same motivation has also been the origin of this book that, although not being a book of conference proceedings, is a compilation of the research done by several of the more representative groups attended the 2015 Benasque meeting.

This book addresses the following aspects:

(i) Quantum optics in the few-emitter and few-plasmon limit (Chaps. 1–4).
(ii) Single-photon sources and nano-lasers based in metal structures (Chaps. 5 and 8).
(iii) Polariton condensation and collective strong coupling between organic molecules and nanophotonic structures (Chaps. 6 and 7).
(iv) Plasmon-enhanced effects in Schottky and tunnel junctions (Chaps. 9 and 10).
(v) Non-local effects in metamaterials and metal nanostructures (Chaps. 11–13).

Understanding of quite complicated topics covered by these chapters requires certain knowledge of the fundamentals that can be refreshed by making use of recent introductory textbooks [2–6].

Odense M, Denmark — Sergey I. Bozhevolnyi
Zaragoza, Spain — Luis Martin-Moreno
Madrid, Spain — Francisco Garcia-Vidal

References

1. M.S. Tame, K.R. McEnery, S.K. Özdemir, J. Lee, S.A. Maier, M.S. Kim, Nat. Phys. **9**, 329 (2013)
2. M. Fox, *Quantum Optics: An Introduction* (Oxford University Press, 2006).
3. S.A. Maier, *Plasmonics: Fundamentals and Applications* (Springer, New York, 2007).

4. D. Sarid, W. Challener, *Modern Introduction to Surface Plasmons: Theory, Mathematica Modeling, and Applications* (Cambridge University Press, New York, 2010).
5. L. Novotny, B. Hecht, *Principles of Nano-Optics*, 2nd edn. (Cambridge University Press, New York, 2012).
6. M. Pelton, G.W. Bryant, *Introduction to Metal-Nanoparticle Plasmonics* (Willey, Hoboken, 2013).

Contents

Contributors

Mario Agio Laboratory of Nano-Optics, Universität Siegen, Siegen, Germany

A. Alù Department of Electrical and Computer Engineering, The University of Texas at Austin, UTA 7.215, Austin, TX, USA; Center for Nanophotonics, FOM Institute AMOLF, Amsterdam, The Netherlands

Pierre Berini Department of Physics, School of Electrical Engineering and Computer Science, Centre for Research in Photonics, University of Ottawa, Ottawa, ON, Canada

Esteban Bermúdez-Ureña ICFO-Institut de Ciencies Fotoniques, Barcelona Institute of Science and Technology, Castelldefels, Barcelona, Spain

Palash Bharadwaj Photonics Laboratory, ETH Zürich, Zürich, Switzerland

S. Bogdanov School of Electrical and Computer Engineering, Birck Nanotechnology Center, and Purdue Quantum Center, Purdue University, West Lafayette, IN, USA

A. Boltasseva School of Electrical and Computer Engineering, Birck Nanotechnology Center, and Purdue Quantum Center, Purdue University, West Lafayette, IN, USA

Sergey I. Bozhevolnyi Centre for Nano Optics, University of Southern Denmark, Odense M, Denmark

R. Chandrasekar School of Electrical and Computer Engineering, Birck Nano technology Center, and Purdue Quantum Center, Purdue University, West Lafayette, IN, USA

Konstantinos S. Daskalakis Department of Applied Physics, COMP Centre of Excellence, Aalto University, Aalto, Finland

Shanhui Fan Ginzton Laboratory, Department of Electrical Engineering, Stanford University, Stanford, CA, USA

Michael Geiselmann École Polytechnique Fédérale de Lausanne (EPFL), Lausanne, Switzerland

Pavel Ginzburg Department of Electrical Engineering Physical Electronics, Tel Aviv University, Tel Aviv, Israel

Harald R. Haakh Nano-Optics Division, Max-Planck-Institute for the Science of Light, Erlangen, Germany

Y. Hadad Department of Electrical and Computer Engineering, The University of Texas at Austin, UTA 7.215, Austin, TX, USA

Tommi K. Hakala COMP Centre of Excellence, Department of Applied Physics, Aalto University, Aalto, Finland

Jacob B. Khurgin Department of Electrical and Computer Engineering, Johns Hopkins University, Baltimore, MD, USA

A.V. Kildishev School of Electrical and Computer Engineering, Birck Nanotechnology Center, and Purdue Quantum Center, Purdue University, West Lafayette, IN, USA

A.F. Koenderink Center for Nanophotonics, FOM Institute AMOLF, Amsterdam, The Netherlands

Shailesh Kumar Centre for Nano Optics, University of Southern Denmark, Odense M, Denmark

Stéphane Kéna-Cohen Department of Engineering Physics, Polytechnique Montréal, Montreal, Canada

A. Lagutchev School of Electrical and Computer Engineering, Birck Nanotechnology Center, and Purdue Quantum Center, Purdue University, West Lafayette, IN, USA

Stefan A. Maier Department of Physics, Imperial College London, London, UK

O.A. Makarova School of Electrical and Computer Engineering, Birck Nanotechnology Center, and Purdue Quantum Center, Purdue University, West Lafayette, IN, USA

Jani-Petri Martikainen COMP Centre of Excellence, Department of Applied Physics, Aalto University, Aalto, Finland

Diego Martín-Cano Nano-Optics Division, Max-Planck-Institute for the Science of Light, Erlangen, Germany

X. Meng School of Electrical and Computer Engineering, Birck Nanotechnology Center, and Purdue Quantum Center, Purdue University, West Lafayette, IN, USA

Robert J. Moerland Department of Imaging Physics, Delft University of Technology, Delft, The Netherlands

N. Asger Mortensen Department of Photonics Engineering and Center for Nanostructured Graphene, Technical University of Denmark, Kongens Lyngby, Denmark

Lukas Novotny Photonics Laboratory, ETH Zürich, Zürich, Switzerland

Markus Parzefall Photonics Laboratory, ETH Zürich, Zürich, Switzerland

Viktor A. Podolskiy Department of Physics and Applied Physics, University of Massachusetts Lowell, Lowell, MA, USA

Romain Quidant ICFO-Institut de Ciencies Fotoniques, Barcelona Institute of Science and Technology, Castelldefels, Barcelona, Spain; ICREA-Instució Catalana de Recerca i Estudis Avançats, Barcelona, Spain

Heikki T. Rekola COMP Centre of Excellence, Department of Applied Physics, Aalto University, Aalto, Finland

A.H. Schokker Center for Nanophotonics, FOM Institute AMOLF, Amsterdam, The Netherlands

V.M. Shalaev School of Electrical and Computer Engineering, Birck Nanotechnology Center, and Purdue Quantum Center, Purdue University, West Lafayette, IN, USA

M.Y. Shalaginov School of Electrical and Computer Engineering, Birck Nanotechnology Center, and Purdue Quantum Center, Purdue University, West Lafayette, IN, USA

Greg Sun Department of Physics, University of Massachusetts at Boston, Boston, MA, USA

Päivi Törmä COMP Centre of Excellence, Department of Applied Physics, Aalto University, Aalto, Finland

F. van Riggelen Center for Nanophotonics, FOM Institute AMOLF, Amsterdam, The Netherlands

Aaro I. Väkeväinen COMP Centre of Excellence, Department of Applied Physics, Aalto University, Aalto, Finland

Z. Wang School of Electrical and Computer Engineering, Birck Nanotechnology Center, and Purdue Quantum Center, Purdue University, West Lafayette, IN, USA

Brian Wells Department of Physics, University of Hartford, West Hartford, CT, USA

Martijn Wubs Department of Photonics Engineering and Center for Nanostructured Graphene, Technical University of Denmark, Kongens Lyngby, Denmark

Shanshan Xu Department of Physics, Stanford University, Stanford, CA, USA

Anatoly V. Zayats Department of Physics, King's College London, London, UK

Chapter 1
Input-Output Formalism for Few-Photon Transport

Shanshan Xu and Shanhui Fan

Abstract We extend the input-output formalism of quantum optics to analyze few-photon transport in waveguide quantum electrodynamics (QED) systems. We provide explicit analytical derivations for one- and two-photon scattering matrix elements based on the quantum causality relation. The computation scheme can be generalized to N-photon scattering systematically.

1.1 Introduction

The capability to create strong photon-photon interaction at a few-photon level in integrated photonic systems is of central importance for quantum information processing. To achieve such a capability, an important approach is to use the waveguide quantum electrodynamics (QED) system, which consists of a waveguide that is strongly coupled to a local quantum system. Experimentally, the waveguides that have been used for this purpose include optical fibers [1], metallic plasmonic nanowires [2], photonic crystal waveguides [3], and microwave transmission line [4]. The local quantum system typically incorporates a variety of quantum multi-level systems such as actual atoms [1], quantum dots [2, 3], or microwave qubits [4], where the strong nonlinearity of these multi-level systems forms the basis for strong photon-photon interactions. These multi-level systems moreover can be embedded in cavity structures to further control their nonlinear properties [5–10].

The rapid experimental developments, in turn, have motivated significant theoretical efforts. From a fundamental physics perspective, the photon-photon interaction is characterized by the multi-photon scattering matrix (S matrix). Therefore, a

S. Xu (✉)
Department of Physics, Stanford University, Stanford, CA 94305, USA
e-mail: xuss@stanford.edu

S. Fan
Ginzton Laboratory, Department of Electrical Engineering,
Stanford University, Stanford, CA 94305, USA
e-mail: shanhui@stanford.edu

S.I. Bozhevolnyi et al. (eds.), *Quantum Plasmonics*,
Springer Series in Solid-State Sciences 185,
DOI 10.1007/978-3-319-45820-5_1

natural objective for theoretical works is to compute such multi-photon S matrix. Moreover, from an engineering perspective, the systems considered here are envisioned as devices that process quantum states. To describe these systems as a device one naturally have to specify its input-output relation. The S matrix, which relates the input and output states, therefore provides a natural basis for device engineering as well.

Motivated by both the physics and engineering considerations as discussed above, a large body of theoretical works have been therefore devoted to the computation the S matrix of various waveguide QED systems [11–37]. In this chapter, we extend the input-output formalism [38, 39] of quantum optics—a Heisenberg picture approach originally introduced to analyze the interaction between an atom in a cavity and a continuous set of electromagnetic states outside of the atom-cavity system—to analyze the transport of few-photon states in waveguide QED system [17, 20, 25, 32, 35–37].

The chapter is organized as follows. In Sect. 1.2 we introduce the Hamiltonian of the system and present the input-output formalism. In Sect. 1.3 we discuss the quantum causality condition and prove certain time-ordering relations, which are the key to compute the S matrix of waveguide photons. In Sect. 1.4 we build the link between the scattering theory and the input-output formalism and continue in Sect. 1.5 with the derivation of the one-photon transport properties. In Sect. 1.6 we show how to extend the calculations to the two-photon case. In Sect. 1.7 as an example of the application of this formalism, we calculate the exact single- and two-photon S matrix for a waveguide coupled a cavity containing a medium with Kerr nonlinearity. As a check, we calculate the same example in Sect. 1.8 in the wavefunction approach. We conclude in Sect. 1.9.

1.2 Hamiltonian and Input-Output Formalism

To illustrate the formalism, as a concrete example, we consider a cavity coupled to a single polarization, single-mode waveguide [40] and treat the transport properties of few-photon states in such a system (Fig. 1.1). The Hamiltonian $\tilde{H}$ is defined as ($\hbar = 1$)

$$\tilde{H} = \tilde{H}_0 + \tilde{H}_1.$$

Here, $\tilde{H}_0$ describes a chiral (i.e. one-way) waveguide where photons propagate in only one direction:

$$\tilde{H}_0 = \int_0^{\infty} d\beta \tilde{\omega}(\beta) \tilde{c}_\beta^\dagger \tilde{c}_\beta,$$

where $\tilde{c}_\beta$ and $\tilde{c}_\beta^\dagger$ are the respective annihilation and creation operators for the photons with wave vector β that obey the commutation relation $\left[\tilde{c}_\beta, \tilde{c}_{\beta'}^\dagger\right] = \delta(\beta - \beta')$. $\tilde{H}_1$ describes the cavity as well as the waveguide-cavity interaction:

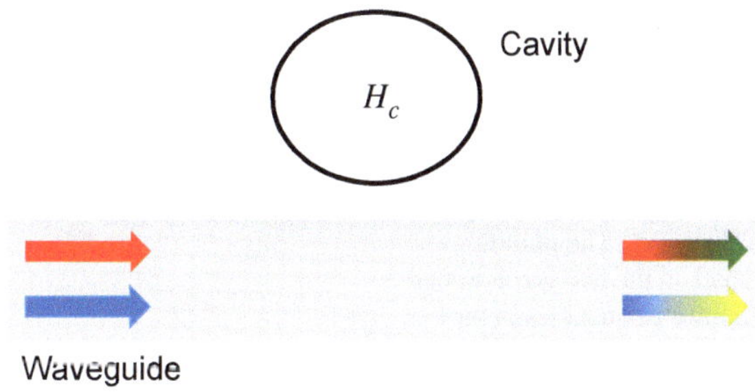

Fig. 1.1 Schematic representation of two photons in a waveguide moving to the right toward a side-coupled cavity

$$\tilde{H}_1 = \tilde{H}_c + V \int_0^\infty d\beta \left(\tilde{c}_\beta^\dagger a + a^\dagger \tilde{c}_\beta \right).$$

$\tilde{H}_c$ is the cavity Hamiltonian and a $\left(a^\dagger\right)$ is its annihilation (creation) operator satisfying the commutation relation $\left[a, a^\dagger\right] = 1$. V denotes the coupling strength between the cavity modes and waveguide modes.

We assume that the cavity system, in the absence of the waveguide, conserves the total number of excitations inside the cavity, i.e. there exists a conserved excitation number operator N_c for the total number of excitations, satisfying

$$\left[N_c, \tilde{H}_c\right] = 0.$$

The operator N_c takes non-negative integer as its eigenvalues. Removing a cavity photon should reduce the total number of excitations in the cavity system by unity, and hence $\left[N_c, a\right] = -a$. A natural form of the number operator N_c is therefore

$$N_c = a^\dagger a + O, \tag{1.1}$$

where O consists of other degrees of freedom in the cavity with $[a, O] = 0$. In our form of the Hamiltonian $\tilde{H}_1$, only the cavity operator a couples to the waveguide, whereas these other degrees of freedom do not couple with the waveguide directly.

It will be useful to label the waveguide photon operators in terms of the frequencies rather than their wave vectors; therefore, we linearize the waveguide dispersion around $\left(\beta_0, \omega_0\right)$ as $\tilde{\omega}(\beta) = \omega_0 + v_g(\beta - \beta_0)$ (see Fig. 1.2). Notice that the total excitation operator of the whole system

$$N_E = \int_0^\infty d\beta \tilde{c}_\beta^\dagger \tilde{c}_\beta + N_c$$

commutes with $\tilde{H}$ (i.e. $\left[\tilde{H}, N_E\right] = 0$). We could thus equivalently solve a system described by

$$H = \tilde{H} - \omega_0 N_E = H_0 + H_1, \tag{1.2}$$

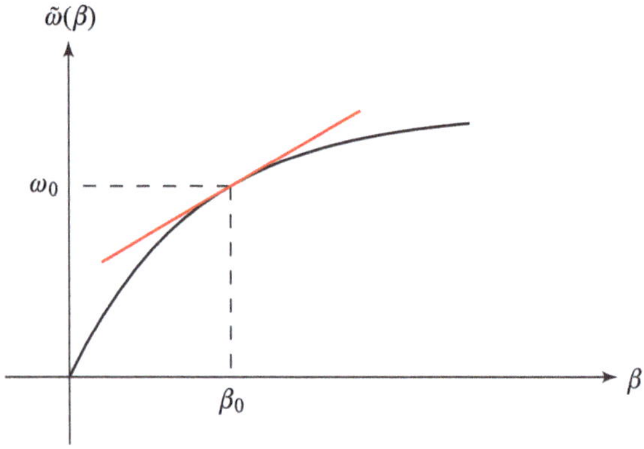

Fig. 1.2 Linearization of a surface-plasmon-like waveguide dispersion relation $\tilde{\omega}(\beta)$ around a wave vector β_0. The *slope* of the line is equal to the group velocity v_g. The photon states in the text are assumed to have frequencies in the vicinity of ω_0 so that the linearization is justified

where

$$H_0 = \int_{-\infty}^{+\infty} d\beta v_g \left(\beta - \beta_0\right) \tilde{c}_\beta^\dagger \tilde{c}_\beta,$$

$$H_1 = H_c + V \int_{-\infty}^{\infty} d\beta \left(\tilde{c}_\beta^\dagger a + a^\dagger \tilde{c}_\beta\right).$$

Here $H_c = \tilde{H}_c - \omega_0 N_c$, and we also extended the lower limit of integration to $-\infty$ so that we can define the Fourier transform of operators later in this Chapter. Since we will be dealing with states with wave vectors around β_0, the extension of the integration limit is well justified. With these transformations, we can now label the waveguide photon operators in terms of frequency $\omega \equiv v_g \beta$ with the definition $c_\omega \equiv \tilde{c}_{\beta+\beta_0} / \sqrt{v_g}$, which satisfies the commutation relation $\left[c_\omega, c_{\omega'}^\dagger\right] = \delta(\omega - \omega')$. From now on, besides ω, the labels for photon degrees of freedom, for example k and p, also refer to photon frequency. As a result of all these changes, we have

$$H_0 = \int dk\, k\, c_k^\dagger c_k, \tag{1.3}$$

$$H_1 = H_c + \frac{V}{\sqrt{v_g}} \int dk \left(c_k^\dagger a + a^\dagger c_k\right). \tag{1.4}$$

Following [17, 38], we now define the input and output operators as

$$c_{\text{in}}(t) \equiv \int \frac{dk}{\sqrt{2\pi}}\, c_k(t_0)\, e^{-ik(t-t_0)}, \tag{1.5}$$

$$c_{\text{out}}(t) \equiv \int \frac{dk}{\sqrt{2\pi}}\, c_k(t_1)\, e^{-ik(t-t_1)}, \tag{1.6}$$

with $t_0 \to -\infty$ and $t_1 \to +\infty$. We note that $c_{in}(t)$ and $c_{out}(t)$ consist of Heisenberg operators of waveguide photons at time $-\infty$ and $+\infty$, respectively. They satisfy the commutation relations

$$\left[c_{in}(t), c_{in}(t')\right] = \left[c_{out}(t), c_{out}(t')\right] = 0,$$
$$\left[c_{in}(t), c_{in}^{\dagger}(t')\right] = \left[c_{out}(t), c_{out}^{\dagger}(t')\right] = \delta(t - t'). \tag{1.7}$$

As shown in the Appendix, for the system described by the Hamiltonian (1.2)–(1.4), one can develop the standard input-output formalism [38] that relates $c_{in}(t)$, $c_{out}(t)$ and a as:

$$c_{out}(t) = c_{in}(t) - i\sqrt{\gamma}\, a(t), \tag{1.8}$$
$$\frac{da}{dt} = -i\left[a, H_c\right] - \frac{\gamma}{2}a - i\sqrt{\gamma}\, c_{in} \tag{1.9}$$
$$= -i\left[a, H_c\right] + \frac{\gamma}{2}a - i\sqrt{\gamma}\, c_{out}, \tag{1.10}$$

where $\gamma \equiv 2\pi V^2/v_g$, and $a(t) \equiv e^{iHt}a(0)e^{-iHt}$ is the cavity photon operator in the Heisenberg picture.

In the standard quantum optics literature, the input-output formalism have been applied widely on computing properties related to an input state that is a coherent state, a thermal state, or a squeezed state. Here we will extend it to computations for Fock state input, since in general the transport property of Fock states is qualitatively different from that of the coherent state.

1.3 Quantum Causality Relation

Integrating (1.9) and (1.10) from $t = -\infty$ and $t = \infty$, respectively, result in:

$$a(t) = a(-\infty) - i\int_{-\infty}^{t} d\tau\left[a, H_c\right] - \frac{\gamma}{2}\int_{-\infty}^{t} d\tau\, a - i\sqrt{\gamma}\int_{-\infty}^{t} d\tau\, c_{in}, \tag{1.11}$$

$$a(t) = a(+\infty) - i\int_{+\infty}^{t} d\tau\left[a, H_c\right] + \frac{\gamma}{2}\int_{+\infty}^{t} d\tau\, a - i\sqrt{\gamma}\int_{+\infty}^{t} d\tau\, c_{out}, \tag{1.12}$$

where the integrands are operators at time τ. Equations (1.11) and (1.12) can be used to prove a quantum causality relation. When using (1.11) to evaluate $a(t)$ or $a^{\dagger}(t)$, the integral should result in an expression that involves only $c_{in}(\tau)$ and $c_{in}^{\dagger}(\tau)$ with $\tau < t$. Therefore, by the commutation relation (1.7), one concludes from (1.11) that for $t < t'$,

$$\left[a(t), I(t')\right] = \left[a(-\infty), I(t')\right], \quad \left[a^{\dagger}(t), I(t')\right] = \left[a^{\dagger}(-\infty), I(t')\right],$$

where $I(t')$ is a shorthand notation for the input operators that represent either $c_{\text{in}}(t')$ or $c^{\dagger}_{\text{in}}(t')$. On the other hand, the operator I is really a Heisenberg operator at time $-\infty$ for the waveguide photon as can be seen in (1.5) above, and hence commute with the cavity operator $a(-\infty)$ and $a^{\dagger}(-\infty)$. Therefore, we have

$$\left[a(t), I(t')\right] = \left[a^{\dagger}(t), I(t')\right] = 0, \quad \text{for } t < t'. \tag{1.13}$$

Similarly, one can prove

$$\left[a(t), O(t')\right] = \left[a^{\dagger}(t), O(t')\right] = 0, \quad \text{for } t > t', \tag{1.14}$$

where $O(t')$ is a shorthand notation for the output operators that represent either $c_{\text{out}}(t')$ or $c^{\dagger}_{\text{out}}(t')$, by utilizing (1.12) and the fact that the output operators are really Heisenberg operators for waveguide photons at time $+\infty$. Following [38], we refer to (1.13) and (1.14) as the *quantum causality condition*. The operator $a(t)$, which characterizes the physical field in the local system, depends only on the input field $c_{\text{in}}(\tau)$ with $\tau \leq t$, and generate only output field $c_{\text{out}}(\tau)$ with $\tau \geq t$.

With quantum causality condition, the commutator $\left[a(t), c^{\dagger}_{\text{in}}(t')\right]$ for $t > t'$ can then be computed as:

$$\left[a(t), c^{\dagger}_{\text{in}}(t')\right] = \left[a(t), c^{\dagger}_{\text{out}}(t') - i\sqrt{\gamma}\, a^{\dagger}(t')\right] = -i\sqrt{\gamma}\left[a(t), a^{\dagger}(t')\right],$$

which, in combination with (1.13), leads to the relation

$$\left[a(t), c^{\dagger}_{\text{in}}(t')\right] = -i\sqrt{\gamma}\left[a(t), a^{\dagger}(t')\right]\theta(t - t'), \tag{1.15}$$

where

$$\theta(t) \equiv \begin{cases} 1 & t > 0 \\ 1/2 & t = 0 \\ 0 & t < 0 \end{cases}$$

is the Heaviside step function. Similarly, we can derive

$$\left[a(t), c^{\dagger}_{\text{out}}(t')\right] = i\sqrt{\gamma}\left[a(t), a^{\dagger}(t')\right]\theta(t' - t). \tag{1.16}$$

To study the few-photon transport, we will need to consider some properties of a time-ordered product involving a and the input or output operators. Here, we define the time-ordered product of operators $A(t)$ and $B(t')$ as

$$\mathscr{T} A(t)B(t') \equiv \begin{cases} A(t)B(t') & t > t' \\ \frac{1}{2}\left[A(t)B(t') + B(t')A(t)\right] & t = t' \\ B(t')A(t) & t < t' \end{cases}.$$

With the proved commutation relations (1.15) and (1.16), we can prove that

$$\mathscr{T} a(t) c_{\text{in}}^{\dagger}(t') = a(t) c_{\text{in}}^{\dagger}(t') + \frac{i}{4}\sqrt{\gamma}\,\delta_{t,t'}, \tag{1.17}$$

$$\mathscr{T} a(t) c_{\text{out}}^{\dagger}(t') = c_{\text{out}}^{\dagger}(t') a(t) + \frac{i}{4}\sqrt{\gamma}\,\delta_{t,t'}, \tag{1.18}$$

where $\delta_{t,t'} = 1$ for $t = t'$ and 0 when $t \neq t'$. We emphasize that $\delta_{t,t'}$ is not the Dirac δ-function $\delta(t - t')$. Take (1.17) as an example. By definition, $\mathscr{T} a(t) c_{\text{in}}^{\dagger}(t') = a(t) c_{\text{in}}^{\dagger}(t')$ for $t > t'$; When $t < t'$, by the commutation relation (1.15), $\mathscr{T} a(t) c_{\text{in}}^{\dagger}(t') = c_{\text{in}}^{\dagger}(t') a(t) = a(t) c_{\text{in}}^{\dagger}(t')$; When $t = t'$, $\mathscr{T} a(t) c_{\text{in}}^{\dagger}(t) = \frac{1}{2} a(t) c_{\text{in}}^{\dagger}(t) + \frac{1}{2} c_{\text{in}}^{\dagger}(t) a(t)$ $= a(t) c_{\text{in}}^{\dagger}(t) - \frac{1}{2}\left[a(t), c_{\text{in}}^{\dagger}(t')\right] = a(t) c_{\text{in}}^{\dagger}(t) + \frac{i}{4}\sqrt{\gamma}$, completing the proof. Equation (1.18) can be proved similarly.

In this chapter, our objective is to compute the few-photon scattering matrix in the frequency domain. For this purpose, we will perform Fourier transformations to the time-ordered products such as those in (1.17) and (1.18). Since the $\delta_{t,t'}$ term vanishes upon Fourier transformation, for our purpose, it can be safely ignored. With this consideration in mind, we rewrite the relations (1.17) and (1.18) as

$$\mathscr{T} a(t) I(t') = a(t) I(t'), \tag{1.19}$$

$$\mathscr{T} a(t) O(t') = O(t') a(t). \tag{1.20}$$

More generally, from (1.19) and (1.20), we have the following relation regarding the time-ordered product:

$$\mathscr{T} \prod_{i,j} a(t_i) I(t_j') = \left[\mathscr{T} \prod_i a(t_i)\right] \cdot \left[\mathscr{T} \prod_j I(t_j')\right], \tag{1.21}$$

$$\mathscr{T} \prod_{i,j} a(t_i) O(t_j') = \left[\mathscr{T} \prod_j O(t_j')\right] \cdot \left[\mathscr{T} \prod_i a(t_i)\right], \tag{1.22}$$

where bracket is used to indicate the range over which the time-ordering is being applied. Again, in (1.21) and (1.22), the equality is to be understood in the frequency domain, i.e., the equality holds after Fourier transformations to all the time variables are performed.

Equations (1.21) and (1.22) can be proved in a similar way. Here we show only the proof of (1.21). The proof of (1.21) can be constructed from induction with respect to the number of operators. The base case is already proved in (1.19). Now suppose (1.21) holds for all cases involving a total number of N operators of a and I. Consider a time-ordered product involving $N + 1$ operators, if the operator with the largest time label is $a(t_{\max})$,

$$\mathscr{T}\prod_{I,j} a(t_I)I(t_j') = a(t_{\max})\left[\mathscr{T}\prod_{i,j} a(t_i)I(t_j')\right] = a(t_{\max})\left[\mathscr{T}\prod_{i} a(t_i)\right]\cdot\left[\mathscr{T}\prod_{j} I(t_j')\right]$$
$$= \left[\mathscr{T}\prod_{I} a(t_I)\right]\cdot\left[\mathscr{T}\prod_{j} I(t_j')\right],$$

where the definition of time-ordered product is used in the first and last steps, and the induction hypothesis is used in the second step. On the other hand, if the operator with the largest time label is $I(t_{\max})$, we have

$$\mathscr{T}\prod_{i,J} a(t_i)I(t_J') = I(t_{\max})\left[\mathscr{T}\prod_{i,j} a(t_i)I(t_j')\right] = I(t_{\max})\left[\mathscr{T}\prod_{i} a(t_i)\right]\cdot\left[\mathscr{T}\prod_{j} I(t_j')\right]$$
$$= \left[\mathscr{T}\prod_{i} a(t_i)\right]\cdot I(t_{\max})\left[\mathscr{T}\prod_{j} I(t_j')\right] = \left[\mathscr{T}\prod_{i} a(t_i)\right]\cdot\left[\mathscr{T}\prod_{J} I(t_J')\right],$$

where we use the induction hypothesis in the second step and the commutation relation (1.13) in the third step. Therefore, (1.21) holds for $N + 1$ operators, completing the proof.

1.4 Connection to Scattering Theory

In a typical scattering experiment, various input states are prepared and sent toward a scattering region. After the scattering takes place, the outgoing states of the experiment are observed, and information about the interaction is deduced. Here as an example we consider two-particle scattering. This process is commonly described using the scattering matrix with elements of the form

$$S_{p_1p_2k_1k_2} = \langle p_1p_2|S|k_1k_2\rangle,$$

where $|k_1k_2\rangle$ denotes an input state—here given as a two-particle state with frequencies k_1 and k_2, and $|p_1p_2\rangle$ denotes an outgoing state. The S operator is equal to the evolution operator U_I in the interaction picture from time $-\infty$ to $+\infty$:

$$S = \lim_{\substack{t_0\to-\infty \\ t_1\to+\infty}} U_I(t_1, t_0) = \lim_{\substack{t_0\to-\infty \\ t_1\to+\infty}} e^{iH_0t_1}e^{-iH(t_1-t_0)}e^{-iH_0t_0},$$

where H_0 is the noninteracting part of the Hamiltonian, and $H = H_0 + H_I$ is the total Hamiltonian. In order to have a more compact notation, in this section whenever we use t_0 or t_1, we imply a limiting process of $t_0 \to -\infty$ and $t_1 \to +\infty$, respectively, and we drop the limit notation.

An equivalent way to describe the scattering is in terms of the scattering eigenstates $|k_1 k_2^+\rangle$ and $|k_1 k_2^-\rangle$ defined as

$$|k_1 k_2^+\rangle \equiv U_I(0, t_0)|k_1 k_2\rangle = e^{iHt_0} e^{-iH_0 t_0} |k_1 k_2\rangle,$$
$$|k_1 k_2^-\rangle \equiv U_I(0, t_1)|k_1 k_2\rangle = e^{iHt_1} e^{-iH_0 t_1} |k_1 k_2\rangle.$$

We can then write the scattering matrix elements as

$$\langle p_1 p_2 | S | k_1 k_2 \rangle = \langle p_1 p_2^- | k_1 k_2^+ \rangle.$$

It is possible to denote the scattering matrix elements by an appropriate definition of input and output operators such that

$$\langle p_1 p_2^- | k_1 k_2^+ \rangle = \langle 0 | c_{\text{out}}(p_1) c_{\text{out}}(p_1) c_{\text{in}}^\dagger(k_1) c_{\text{in}}^\dagger(k_2) | 0 \rangle, \tag{1.23}$$

where

$$c_{\text{in}}^\dagger(k) \equiv e^{iHt_0} e^{-iH_0 t_0} c_k^\dagger e^{iH_0 t_0} e^{-iHt_0}, \tag{1.24}$$
$$c_{\text{out}}^\dagger(k) \equiv e^{iHt_1} e^{-iH_0 t_1} c_k^\dagger e^{iH_0 t_1} e^{-iHt_1} \tag{1.25}$$

create input and output scattering eigenstates from vacuum, i.e.,

$$c_{\text{in}}^\dagger(k)\, |0\rangle = |k^+\rangle,$$
$$c_{\text{out}}^\dagger(p)\, |0\rangle = |p^-\rangle,$$

and satisfy the commutation relations

$$\left[c_{\text{in}}(k), c_{\text{in}}^\dagger(p)\right] = \left[c_{\text{out}}(k), c_{\text{out}}^\dagger(p)\right] = \delta(k - p).$$

We now relate the scattering theory, as briefly sketched above, to the input-output formalism [38, 39] of quantum optics. To do so, we start from the definition of the input field operator $c_{\text{in}}(t)$ (1.5), in which $c_k(t_0) \equiv e^{iHt_0} c_k e^{-iHt_0}$. The relationship between $c_{\text{in}}(t)$—which is defined in the input-output formalism—and $c_{\text{in}}(k)$—which is defined above in (1.24) as a part of the scattering theory—can then be determined by noting that

$$\begin{aligned} c_{\text{in}}(t) &= \frac{1}{\sqrt{2\pi}} \int dk\, e^{iHt_0} c_k e^{-iHt_0} e^{-ik(t-t_0)} \\ &= \frac{1}{\sqrt{2\pi}} \int dk\, e^{iHt_0} e^{-iH_0t_0} c_k e^{iH_0t_0} e^{-iHt_0} e^{-ikt} \\ &= \frac{1}{\sqrt{2\pi}} \int dk\, c_{\text{in}}(k) e^{-ikt}, \end{aligned} \tag{1.26}$$

where in the second line we used the fact that $\left[H_0,\, c_k\right] = -k\, c_k$ to convert the $c_k e^{ikt_0}$ term into $e^{-iH_0t_0} c_k e^{iH_0t_0}$. As a result, $c_{\text{in}}(k)$ provides the spectral representation of $c_{\text{in}}(t)$. Similarly, the output field operator (1.6) in the input-output formalism is related to $c_{\text{out}}(k)$ in the scattering theory through

$$c_{\text{out}}(t) = \frac{1}{\sqrt{2\pi}} \int dk\, c_{\text{out}}(k) e^{-ikt}. \tag{1.27}$$

We have thus established a direct connection between the input-output formalism and the scattering theory.

1.5 Single-Photon Transport

Having established the relationship between the input-output formalism and the scattering theory, we now calculate the S matrix elements $\langle p|S|k\rangle$ between two single-photon states $|k\rangle$ and $|p\rangle$. The single-photon S matrix is related to the input and output operators by

$$S_{pk} = \langle p|S|k\rangle = \langle 0|c_{\text{out}}(p)c^{\dagger}_{\text{in}}(k)|0\rangle = \int \frac{dt'}{\sqrt{2\pi}} e^{ipt'} \int \frac{dt}{\sqrt{2\pi}} e^{-ikt}\, \langle 0|c_{\text{out}}(t')c^{\dagger}_{\text{in}}(t)|0\rangle,$$

where we used (1.26) and (1.27) to write $c^{\dagger}_{\text{in}}(k)$, $c_{\text{out}}(p)$ in terms of $c^{\dagger}_{\text{in}}(t)$, $c_{\text{out}}(t')$, respectively. It is therefore sufficient to first calculate $\langle 0|c_{\text{out}}(t')c^{\dagger}_{\text{in}}(t)|0\rangle$ and then perform Fourier transformation to determine the single-photon S matrix. As the starting point, we use the input-output formalism (1.8) to obtain

$$\langle 0|c_{\text{out}}(t')c^{\dagger}_{\text{in}}(t)|0\rangle = \langle 0|c_{\text{in}}(t')c^{\dagger}_{\text{in}}(t)|0\rangle - i\sqrt{\gamma}\, \langle 0|a(t')c^{\dagger}_{\text{in}}(t)|0\rangle.$$

The first term is simply $\delta(t'-t)$ following the commutation relation (1.7). The second term can be computed as

$$\begin{aligned}\langle 0|a(t')c_{\text{in}}^{\dagger}(t)|0\rangle &= \langle 0|\mathscr{T}a(t')c_{\text{in}}^{\dagger}(t)|0\rangle - \frac{i}{4}\sqrt{\gamma}\,\delta_{t,t'} \\ &= \langle 0|\mathscr{T}a(t')c_{\text{out}}^{\dagger}(t)|0\rangle - i\sqrt{\gamma}\,\langle 0|\mathscr{T}a(t')a^{\dagger}(t)|0\rangle - \frac{i}{4}\sqrt{\gamma}\,\delta_{t,t'} \\ &= \langle 0|c_{\text{out}}^{\dagger}(t)a(t')|0\rangle - i\sqrt{\gamma}\,\langle 0|\mathscr{T}a(t')a^{\dagger}(t)|0\rangle \\ &= -i\sqrt{\gamma}\,\langle 0|\mathscr{T}a(t')a^{\dagger}(t)|0\rangle, \end{aligned} \tag{1.28}$$

where we first utilized the time-ordered relation (1.17) to introduce the time-ordered operation, then inserted the input-output formalism (1.8) in the second line to transform $c_{\text{in}}^{\dagger}$ to $c_{\text{out}}^{\dagger}$, and finally utilized the time-ordered relation (1.18) to move $c_{\text{out}}^{\dagger}$ to the leftmost. We note that as indicated above in the discussions associated with (1.17) and (1.18), the $\delta_{t,t'}$ terms in (1.19) and (1.20) do not contribute after Fourier transformation and can be safely dropped in the calculation. Therefore, for the rest of the chapter, we will directly use (1.21) and (1.22).

We define the two-point Green function of the cavity as

$$G(t';t) \equiv -\gamma\,\langle 0|\mathscr{T}a(t')a^{\dagger}(t)|0\rangle. \tag{1.29}$$

and relate single-photon S matrix to the cavity's two-point Green function as

$$S(t';t) \equiv \langle 0|c_{\text{out}}(t')c_{\text{in}}^{\dagger}(t)|0\rangle = \delta(t'-t) + G(t';t), \tag{1.30}$$

or

$$S_{pk} = \delta(p-k) + G(p;k), \tag{1.31}$$

where $G(p;k)$ is the Fourier transformation of (1.29) in the frequency domain. As shown in the Appendix, the two-point Green function (1.29) can be computed using an effective Hamiltonian

$$H_{\text{eff}} = H_c - i\frac{\gamma}{2}a^{\dagger}a \tag{1.32}$$

without involving the waveguide degrees of freedoms. As a result, we only need to solve a cavity system which has a finite, and typically small, number of degrees of freedom.

For cavities with total excitation number as defined in (1.1), we have $\left[H_{\text{eff}}, N_c\right] = 0$ because of $\left[H_c, N_c\right] = 0$ and $[a, O] = 0$. As a result, H_{eff} can be block-diagonalized as

$$H_{\text{eff}}\,|\mathscr{E}_n\rangle = \mathscr{E}_n|\mathscr{E}_n\rangle, \quad N_c\,|\mathscr{E}_n\rangle = n\,|\mathscr{E}_n\rangle,$$

where $|\mathscr{E}_n\rangle$ is a right eigenstate of H_{eff} with eigenvalue $\mathscr{E}_n$. Especially, $|\mathscr{E}_0\rangle$ is the ground state $|0\rangle$ and $\mathscr{E}_0 = 0$. Since H_{eff} is not Hermitian, $\mathscr{E}_n$ is in general complex. By inserting a complete set of biorthogonal basis states and noting that only the single excitation state contributes to the summation, we compute the two-point Green function (1.29) as

$$\begin{aligned}G(t';t) &= -\gamma\,\langle 0|a(t')a^\dagger(t)|0\rangle\theta(t'-t) = -\gamma\langle 0|a\,e^{-iH_{\text{eff}}(t'-t)}a^\dagger|0\rangle\theta(t'-t)\\ &= -\gamma\sum_{\mathscr{E}_1}\langle 0|a\,|\mathscr{E}_1\rangle\langle\bar{\mathscr{E}}_1|e^{-iH_{\text{eff}}(t'-t)}a^\dagger|0\rangle\theta(t'-t)\\ &= -\gamma\sum_{\mathscr{E}_1}\langle 0|a\,|\mathscr{E}_1\rangle\langle\bar{\mathscr{E}}_1|a^\dagger|0\rangle\,e^{-i\mathscr{E}_1(t'-t)}\theta(t'-t).\end{aligned}$$

and thus

$$G(p;k) = -i\gamma\sum_{\mathscr{E}_1}\frac{\langle 0|a\,|\mathscr{E}_1\rangle\langle\bar{\mathscr{E}}_1|a^\dagger|0\rangle}{k-\mathscr{E}_1}\delta(p-k).$$

The single-photon S matrix is then

$$S_{pk} = \left[1 - i\gamma\sum_{\mathscr{E}_1}\frac{\langle 0|a\,|\mathscr{E}_1\rangle\langle\bar{\mathscr{E}}_1|a^\dagger|0\rangle}{k-\mathscr{E}_1}\right]\delta(p-k). \tag{1.33}$$

1.6 Two-Photon Transport

Our aim in this section is to calculate the two-photon S matrix based on the quantum causality condition, building upon the results we obtained for the single-photon case. The two-photon S matrix element (1.23) is related to the input and output operators by

$$S_{p_1p_2k_1k_2} = \left(\prod_{l=1,2}\int\frac{dt'_l}{\sqrt{2\pi}}e^{ip_lt'_l}\int\frac{dt_l}{\sqrt{2\pi}}e^{-ik_lt_l}\right)\langle 0|c_{\text{out}}(t'_1)c_{\text{out}}(t'_2)c^\dagger_{\text{in}}(t_1)c^\dagger_{\text{in}}(t_2)|0\rangle.$$

We begin by computing the two-photon S matrix element in the time domain. We first insert the input-output formalism (1.8) as

$$\begin{aligned}S(t'_1t'_2;t_1t_2) &\equiv \langle 0|c_{\text{out}}(t'_1)c_{\text{out}}(t'_2)c^\dagger_{\text{in}}(t_1)c^\dagger_{\text{in}}(t_2)|0\rangle\\ &= \langle 0|\left[\mathscr{T}c_{\text{out}}(t'_1)c_{\text{out}}(t'_2)\right]c^\dagger_{\text{in}}(t_1)c^\dagger_{\text{in}}(t_2)|0\rangle\\ &= \langle 0|\left[\mathscr{T}\left(c_{\text{in}}(t'_1)-i\sqrt{\gamma}a(t'_1)\right)\left(c_{\text{in}}(t'_2)-i\sqrt{\gamma}a(t'_2)\right)\right]c^\dagger_{\text{in}}(t_1)c^\dagger_{\text{in}}(t_2)|0\rangle\\ &= \langle 0|\left[\mathscr{T}c_{\text{in}}(t'_1)c_{\text{in}}(t'_2)\right]c^\dagger_{\text{in}}(t_1)c^\dagger_{\text{in}}(t_2)|0\rangle\\ &\quad -i\sqrt{\gamma}\,\langle 0|\left[\mathscr{T}c_{\text{in}}(t'_1)a(t'_2)\right]c^\dagger_{\text{in}}(t_1)c^\dagger_{\text{in}}(t_2)|0\rangle\\ &\quad -i\sqrt{\gamma}\,\langle 0|\left[\mathscr{T}a(t'_1)c_{\text{in}}(t'_2)\right]c^\dagger_{\text{in}}(t_1)c^\dagger_{\text{in}}(t_2)|0\rangle\\ &\quad -\gamma\,\langle 0|\left[\mathscr{T}a(t'_1)a(t'_2)\right]c^\dagger_{\text{in}}(t_1)c^\dagger_{\text{in}}(t_2)|0\rangle.\end{aligned}$$

The first term in the final equality above is simply $\langle 0|c_{\text{in}}(t'_1)c_{\text{in}}(t'_2)c^\dagger_{\text{in}}(t_1)c^\dagger_{\text{in}}(t_2)|0\rangle = \delta(t'_1-t_1)\delta(t'_2-t_2)+\delta(t'_1-t_2)\delta(t'_2-t_1)$. By the time-ordered relation (1.19) and the result in the single-photon case (1.28), the second term can be computed as

$$\begin{aligned}
&\langle 0| \left[\mathscr{T} c_{\text{in}}(t_1')a(t_2')\right] c_{\text{in}}^{\dagger}(t_1)c_{\text{in}}^{\dagger}(t_2)|0\rangle = \langle 0|a(t_2')c_{\text{in}}(t_1')c_{\text{in}}^{\dagger}(t_1)c_{\text{in}}^{\dagger}(t_2)|0\rangle \\
&= \langle 0|a(t_2')c_{\text{in}}^{\dagger}(t_2)|0\rangle\delta(t_1'-t_1) + \langle 0|a(t_2')c_{\text{in}}^{\dagger}(t_1)|0\rangle\delta(t_1'-t_2) \\
&= -i\sqrt{\gamma}\langle 0|\mathscr{T} a(t_2')a^{\dagger}(t_2)|0\rangle\delta(t_1'-t_1) - i\sqrt{\gamma}\langle 0|\mathscr{T} a(t_2')a^{\dagger}(t_1)|0\rangle\delta(t_1'-t_2).
\end{aligned}$$

Similarly, the evaluation of the third term gives $-i\sqrt{\gamma}\langle 0|\mathscr{T} a(t_1')a^{\dagger}(t_2)|0\rangle\delta(t_2'-t_1) - i\sqrt{\gamma}\langle 0|\mathscr{T} a(t_1')a^{\dagger}(t_1)|0\rangle\delta(t_2'-t_2)$. For the last term, we have

$$\begin{aligned}
&\langle 0| \left[\mathscr{T} a(t_1')a(t_2')\right] c_{\text{in}}^{\dagger}(t_1)c_{\text{in}}^{\dagger}(t_2)|0\rangle \\
&= \langle 0| \left[\mathscr{T} a(t_1')a(t_2')\right] \left[\mathscr{T} c_{\text{in}}^{\dagger}(t_1)c_{\text{in}}^{\dagger}(t_2)\right] |0\rangle \\
&= \langle 0|\mathscr{T} a(t_1')a(t_2')c_{\text{in}}^{\dagger}(t_1)c_{\text{in}}^{\dagger}(t_2)|0\rangle \\
&= \langle 0|\mathscr{T} a(t_1')a(t_2') \left(c_{\text{out}}^{\dagger}(t_1) - i\sqrt{\gamma}\, a^{\dagger}(t_1)\right) \left(c_{\text{out}}^{\dagger}(t_2) - i\sqrt{\gamma}\, a^{\dagger}(t_2)\right) |0\rangle \\
&= -\gamma\, \langle 0|\mathscr{T} a(t_1')a(t_2')a^{\dagger}(t_1)a^{\dagger}(t_2)|0\rangle + \langle 0|c_{\text{out}}^{\dagger}(t_1)c_{\text{out}}^{\dagger}(t_2)a(t_1')a(t_2')|0\rangle \\
&\quad - i\sqrt{\gamma}\, \langle 0|c_{\text{out}}^{\dagger}(t_1) \left[\mathscr{T} a(t_1')a(t_2')a^{\dagger}(t_2)\right] |0\rangle - i\sqrt{\gamma}\, \langle 0|c_{\text{out}}^{\dagger}(t_2) \left[\mathscr{T} a(t_1')a(t_2')a^{\dagger}(t_1)\right] |0\rangle \\
&= -\gamma\, \langle 0|\mathscr{T} a(t_1')a(t_2')a^{\dagger}(t_1)a^{\dagger}(t_2)|0\rangle.
\end{aligned}$$

where we utilized time-ordered relation (1.21) in the second line and inserted the input-output formalism (1.8) in the third line. In the fourth line, we applied the time-ordered relation (1.22) to move the output operators to the leftmost to annihilate the $\langle 0|$ state.

Combing all the computations leads to the result of two-photon S matrix in the time domain. To make further simplification, we define the four-point Green function of the cavity system as

$$G(t_1', t_2'; t_1, t_2) \equiv \gamma^2\, \langle 0|\mathscr{T} a(t_1')a(t_2')a^{\dagger}(t_1)a^{\dagger}(t_2)|0\rangle. \tag{1.34}$$

In general, its Fourier component $G(p_1, p_2; k_1, k_2)$ can be expressed as a sum of various terms containing products of several δ-functions [41]. Among all these terms, we define the term that contains only $\delta(p_1 + p_2 - k_1 - k_2)$ and no other δ functions as the *connected* Green function, $G_C(p_1, p_2; k_1, k_2)$. We thus have the decomposition

$$G(p_1, p_2; k_1, k_2) = G(p_1; k_1)G(p_2; k_2) + G(p_1; k_2)G(p_2; k_1) + G_C(p_1, p_2; k_1, k_2),$$

or in the time domain,

$$G(t_1', t_2'; t_1, t_2) = G(t_1'; t_1)G(t_2'; t_2) + G(t_1'; t_2)G(t_2'; t_1) + G_C(t_1', t_2'; t_1, t_2).$$

Consequently, the two-photon S matrix can be written in a compact form [35]

$$S(t_1't_2'; t_1t_2) = S(t_1', t_1)S(t_2', t_2) + S(t_2', t_1)S(t_1', t_2) + G_C(t_1', t_2'; t_1, t_2)$$

with single-photon S matrix (1.30) and the connected four-point Green function. Or in the frequency domain, we have

$$S_{p_1p_2k_1k_2} = S_{p_1k_1}S_{p_2k_2} + S_{p_2k_1}S_{p_1k_2} + G_C(p_1, p_2; k_1, k_2), \tag{1.35}$$

where S_{pk} is computed in (1.31) or (1.33). The form of this decomposition satisfies the cluster decomposition principle [28, 42, 43] in quantum field theory. From the computational perspective, the result here reduces the computation of the two-photon S matrix to the evaluation of the connected four-point Green function of the cavity system, which can again be computed using the effective Hamiltonian (1.32).

In general, for N-photon S matrix, similar decompositions hold and our computational scheme based on the input-output formalism and quantum causality condition can be generalized systematically [35].

1.7 Example: A Waveguide Coupled to a Kerr-Nonlinear Cavity

As an example of the application of the formalism developed above, we compute the S matrix of two-photon transport in a single-mode waveguide that is side-coupled to a ring resonator incorporating Kerr nonlinear media. The full Hamiltonian has the same form as (1.2) with the specific form of H_c:

$$H_c = \omega_c\, a^\dagger a + \frac{\chi}{2}\, a^\dagger a^\dagger aa.$$

Let $\alpha \equiv \omega_c - i\frac{\gamma}{2}$, the effective Hamiltonian (1.32) in this case is

$$H_{\text{eff}} = \alpha\, a^\dagger a + \frac{\chi}{2}\, a^\dagger a^\dagger aa,$$

and can be diagonalized as

$$H_{\text{eff}}|n\rangle = \left[\alpha n + \frac{\chi}{2} n(n-1)\right]|n\rangle, \tag{1.36}$$

with the eigenvalues $\mathscr{E}_n \equiv \alpha n + \frac{\chi}{2} n(n-1)$ and the eigenstates $|\mathscr{E}_n\rangle \equiv |n\rangle$.

The single-photon S matrix can be computed directly by (1.33) as

$$\begin{aligned} S_{pk} &= \left[1 - i\gamma \frac{\langle 0|a\,|1\rangle\langle 1|a^\dagger\,|0\rangle}{k-\alpha}\right]\delta(p-k) = \left[1 - i\gamma\,\frac{1}{k-\omega_c + i\gamma/2}\right]\delta(p-k) \\ &= \frac{k-\omega_c - i\gamma/2}{k-\omega_c + i\gamma/2}\delta(p-k) = \left[1 + s_k\right]\delta(p-k), \end{aligned} \tag{1.37}$$

where, for later convenience, we defined

$$s_k \equiv -\gamma \frac{i}{k-\alpha}.$$

For two-photon S matrix, from (1.35), we only need to compute the connected four-point Green function. We start by computing the four-point Green function $G(t_1', t_2'; t_1, t_2)$. Depending on the values of the four time labels, the time ordering operation would give rise non-zero terms that can be classified into two types: $\langle aa^\dagger aa^\dagger \rangle$ and $\langle aaa^\dagger a^\dagger \rangle$, i.e.

$$G(t_1', t_2'; t_1, t_2) \equiv \sum_{j=1}^{2} G^{(j)}(t_1', t_2'; t_1, t_2),$$

with

$$G^{(1)}(t_1', t_2'; t_1, t_2) = (-\gamma)^2 \sum_{P,Q} \langle 0|a(t'_{Q_1})a^\dagger(t_{P_1})a(t'_{Q_2})a^\dagger(t_{P_2})|0\rangle \theta(t'_{Q_1} - t_{P_1})\theta(t_{P_1} - t'_{Q_2})\theta(t'_{Q_2} - t_{P_2}),$$
$$G^{(2)}(t_1', t_2'; t_1, t_2) = (-\gamma)^2 \sum_{P,Q} \langle 0|a(t'_{Q_1})a(t'_{Q_2})a^\dagger(t_{P_1})a^\dagger(t_{P_2})|0\rangle \theta(t'_{Q_1} - t'_{Q_2})\theta(t'_{Q_2} - t_{P_1})\theta(t_{P_1} - t_{P_2}),$$

where both P and Q are permutations over indices $\{1, 2\}$. We calculate each term by inserting the complete sets of eigenstates, which results in:

$$\begin{aligned}
\langle 0|a(t_1')a^\dagger(t_1)a(t_2')a^\dagger(t_2)|0\rangle &= \sum_{m,n,l} \langle 0|a(t_1')|m\rangle\langle m|a^\dagger(t_1)|n\rangle\langle n|a(t_2')|l\rangle\langle l|a^\dagger(t_2)|0\rangle \\
&= \langle 0|a(t_1')|1\rangle\langle 1|a^\dagger(t_1)|0\rangle\langle 0|a(t_2')|1\rangle\langle 1|a^\dagger(t_2)|0\rangle \\
&= e^{-i\alpha(t_1'-t_1)}e^{-i\alpha(t_2'-t_2)},
\end{aligned}$$

and

$$\begin{aligned}
\langle 0|a(t_1')a(t_2')a^\dagger(t_1)a^\dagger(t_2)|0\rangle &= \langle 0|a(t_1')|1\rangle\langle 1|a(t_2')|2\rangle\langle 2|a^\dagger(t_1)|1\rangle\langle 1|a^\dagger(t_2)|0\rangle \\
&= 2e^{-i\alpha(t_1'-t_2)}e^{-i(\alpha+\chi)(t_2'-t_1)}.
\end{aligned}$$

Their Fourier transformations are

$$G^{(1)}(p_1, p_2; k_1, k_2) = -\frac{i}{2\pi} \sum_{P,Q} s_{p_{Q_1}} s_{k_{P_2}} \frac{1}{p_{Q_2} - k_{P_2} - i\varepsilon} \delta(p_1 + p_2 - k_1 - k_2),$$
$$G^{(2)}(p_1, p_2; k_1, k_2) = \frac{i}{\pi} \sum_{P,Q} s_{p_{Q_1}} s_{k_{P_2}} \frac{1}{k_1 + k_2 - 2\alpha - \chi} \delta(p_1 + p_2 - k_1 - k_2),$$

where an infinitesimal imaginary part in the denominator arises due to the Fourier transform of the θ function. Moreover, we note that

$$\frac{1}{p-k-i\varepsilon} = \frac{\mathscr{P}}{p-k} + i\pi\delta(p-k).$$

On the other hand, since the connected four-point Green function contains only a single δ function, only the principal part contributes to $G_C(p_1, p_2; k_1, k_2)$. Therefore, we have

$$G_C(p_1, p_2; k_1, k_2) = \sum_{j=1}^{2} i\mathcal{M}^{(j)}_{p_1 p_2; k_1 k_2}\delta(p_1 + p_2 - k_1 - k_2),$$

with

$$\begin{aligned} i\mathcal{M}^{(1)}_{p_1 p_2; k_1 k_2} &= -\frac{i}{2\pi}\sum_{P,Q} s_{p_{Q_1}} s_{k_{P_2}} \frac{\mathscr{P}}{p_{Q_2} - k_{P_2}}, \\ i\mathcal{M}^{(2)}_{p_1 p_2; k_1 k_2} &= \frac{i}{\pi}\sum_{P,Q} s_{p_{Q_1}} s_{k_{P_2}} \frac{1}{k_1 + k_2 - 2\alpha - \chi}. \end{aligned}$$

Finally, we can sum over all the permutation terms and obtain a compact form:

$$G_C(p_1, p_2; k_1, k_2) = -\frac{\chi}{\pi\gamma} s_{k_1} s_{k_2} \left(s_{p_1} + s_{p_2}\right) \frac{1}{k_1 + k_2 - 2\alpha - \chi}\delta(p_1 + p_2 - k_1 - k_2). \tag{1.38}$$

The summation of all principal parts vanishes. The final result (1.38) indeed has the exact analytical structure constrained by the cluster decomposition principle [28]. The only singularities are isolated poles corresponding to one and two-photon excitations in the cavity.

1.8 Wavefunction Approach

As a check on the input-output formalism, we recalculate the example problem in the last section using the wavefunction approach [12, 13, 15, 16, 19, 22, 23, 26, 29, 31, 33]. We start by rewriting the system's Hamiltonian in $H = H_0 + H_{\text{int}}$ in the coordinate space

$$\begin{aligned} H_0 &= \int dx\, c^\dagger(x)\left(-i\frac{d}{dx}\right)c(x) + \omega_c a^\dagger a + \frac{\chi}{2}a^\dagger a^\dagger aa, \\ H_{\text{int}} &= +\sqrt{\gamma}\int dx\, \delta(x)\left[c^\dagger(x)a + a^\dagger c(x)\right], \end{aligned}$$

where $c^\dagger(x)$ is the creation operator for a right-going photon at position x satisfying the commutation relation $\left[c(x), c^\dagger(x')\right] = \delta(x - x')$. For an input state of one-photon Fock state, the most general time-independent interacting eigenstate for this Hamiltonian is

$$|k^+\rangle = \int dx\, \phi_k(x) c^\dagger(x)|0\rangle + e_a\, a^\dagger|0\rangle.$$

The time-independent Schrodinger equation $H|k^+\rangle = k|k^+\rangle$ yields the following equations of motion:

$$-i\frac{d}{dx}\phi_k(x) + \sqrt{\gamma}\, e_a\, \delta(x) = k\, \phi_k(x), \tag{1.39}$$

$$\omega_c\, e_a + \sqrt{\gamma}\, \phi(0) = k\, e_a. \tag{1.40}$$

Our aim is to solve the transmission amplitude for an incident photon. For this purpose, we take

$$\phi_k(x) = \frac{e^{ikx}}{\sqrt{2\pi}}\left[\theta(-x) + t_k\theta(x)\right],$$

where t_k is the transmission amplitude. Submitting the above ansatz into the equations of motion (1.39) and (1.40) gives the solution

$$e_a = \frac{1}{\sqrt{2\pi}}\frac{\sqrt{\gamma}}{k - \omega_c + i\gamma/2}, \qquad t_k = \frac{k - \omega_c - i\gamma/2}{k - \omega_c + i\gamma/2}. \tag{1.41}$$

For two-photon transport, consider the two-excitation eigenstate

$$|E_2\rangle = \int dx_1 dx_2\, g(x_1, x_2) c^\dagger(x_1) c^\dagger(x_2)|0\rangle + f_a \frac{1}{\sqrt{2}} a^\dagger a^\dagger|0\rangle + \int dx\, h(x) c^\dagger(x) a^\dagger|0\rangle.$$

From $H|E_2\rangle = E_2|E_2\rangle$, by equating the coefficients of the terms $c^\dagger c^\dagger|0\rangle$, $c^\dagger a^\dagger|0\rangle$, $a^\dagger a^\dagger|0\rangle$, respectively, we obtain the equations of motion

$$\left(-i\frac{\partial}{\partial x_1} - i\frac{\partial}{\partial x_2} - E_2\right) g(x_1, x_2) + \frac{\sqrt{\gamma}}{2}\left[\delta(x_1)h(x_2) + h(x_1)\delta(x_2)\right] = 0,$$

$$\left(-i\frac{d}{dx} + \omega_c - E_2\right) h(x) + \sqrt{\gamma}\left[g(x, 0) + g(0, x)\right] + \sqrt{2\gamma} f_a\, \delta(x) = 0,$$

$$(2\omega_c + \chi - E_2) f_a + \sqrt{2\gamma}\, h(0) = 0,$$

where $g(x, 0) \equiv [g(x, 0^+) + g(x, 0^-)]/2 = g(0, x) \equiv [g(0^+, x) + g(0^-, x)]/2$ and $h(0) \equiv [h(0^+) + h(0^-)]/2$. We eliminate $h(x)$ and f_a from the preceding equations and obtain equations on $g(x_1, x_2)$:

$$\left(-i\frac{\partial}{\partial x_1}-i\frac{\partial}{\partial x_2}-E_2\right)g(x_1,x_2)=0, \tag{1.42}$$

$$\left(-i\frac{d}{dx}-E_2+\omega_c-i\frac{\gamma}{2}\right)g(0^+,x)=\left(-i\frac{d}{dx}-E_2+\omega_c+i\frac{\gamma}{2}\right)g(0^-,x), \tag{1.43}$$

$$\frac{g(0^+,0^+)-g(0^-,0^+)}{g(0^+,0^-)-g(0^-,0^-)}=\frac{E-2\omega_c-\chi-i\gamma}{E-2\omega_c-\chi+i\gamma}. \tag{1.44}$$

From (1.42), the general solution of $g(x_1,x_2)$ thus has the form of

$$g(x_1,x_2)=\tilde{g}(x_1-x_2)e^{iE_2x_2}. \tag{1.45}$$

Because of the exchange symmetry of $g(x_1,x_2)$, we can restrict to the half space $x_1 \le x_2$. Following [13, 16], we divide this half space into three regions (a) $x_1 \le x_2 < 0$, (b) $x_1 < 0 < x_2$, (c) $0 < x_1 \le x_2$ and denote $g^{(a)}(x_1,x_2)$, $g^{(b)}(x_1,x_2)$, $g^{(c)}(x_1,x_2)$ as the function $g(x_1,x_2)$ in each respective region. Using (1.43) and (1.44), we have

$$\left(-i\frac{d}{dx}-E_2+\omega_c-i\frac{\gamma}{2}\right)g^{(b)}(x,0^+)=\left(-i\frac{d}{dx}-E_2+\omega_c+i\frac{\gamma}{2}\right)g^{(a)}(x,0^-), \tag{1.46}$$

$$\left(-i\frac{d}{dx}-E_2+\omega_c-i\frac{\gamma}{2}\right)g^{(c)}(0^+,x)=\left(-i\frac{d}{dx}-E_2+\omega_c+i\frac{\gamma}{2}\right)g^{(b)}(0^-,x), \tag{1.47}$$

$$\frac{g^{(c)}(0^+,0^+)-g^{(b)}(0^-,0^+)}{g^{(b)}(0^-,0^+)-g^{(a)}(0^-,0^-)}=\frac{E_2-2\omega_c-\chi-i\gamma}{E_2-2\omega_c-\chi+i\gamma}. \tag{1.48}$$

In order to construct the S matrix, one interprets $g(x_1,x_2)$ in the third quadrant $(x_1,x_2<0)$ as the in-state, and in the first quadrant $(x_1,x_2>0)$ as the out-state. We consider an incident two-photon planewave state comprised of two photons with frequencies k_1 and k_2, i.e.,

$$g^{(a)}(x_1,x_2)=\frac{1}{2\sqrt{2\pi}}\left(e^{ik_1x_1}e^{ik_2x_2}+e^{ik_1x_2}e^{ik_2x_1}\right), \quad \text{for } x_1,x_2<0,$$

which gives $E_2=k_1+k_2$ and $g^{(a)}(x,0^-)=\frac{1}{2\sqrt{2\pi}}\left(e^{ik_1x}+e^{ik_2x}\right)$ directly. We can then obtain $g^{(b)}(x,0^+)$ by solving (1.46) as

$$g^{(b)}(x,0^+)=\frac{1}{2\sqrt{2\pi}}\left(t_{k_2}e^{ik_1x}+t_{k_1}e^{ik_2x}\right)+A\,e^{i\left(k_1+k_2-\omega_c+i\frac{\gamma}{2}\right)x},$$

where t_k is the single-photon transmission amplitude solved in (1.41) and A is a constant to be determined. Using (1.45), we can determine $g^{(b)}(x_1,x_2)$ from $g^{(b)}(x_1,0^+)$ as

$$g^{(b)}(x_1,x_2)=\frac{1}{2\sqrt{2\pi}}\left[t_{k_2}e^{i(k_1x_1+k_2x_2)}+t_{k_1}e^{i(k_1x_2+k_2x_1)}\right]+A\,e^{i\left(\omega_c-i\frac{\gamma}{2}\right)(x_2-x_1)}e^{i(k_1+k_2)x_1}.$$

To avoid the divergence when $x_2 - x_1 \to +\infty$, we have to set $A = 0$. As a result, the final solution of $g^{(b)}(x_1, x_2)$ is

$$g^{(b)}(x_1,x_2) = \frac{1}{2\sqrt{2}\pi}\left[t_{k_2} e^{i(k_1x_1+k_2x_2)} + t_{k_1} e^{i(k_1x_2+k_2x_1)}\right].$$

Similarly, with $g^{(b)}(0^-, x_2)$, we can solve $g^{(c)}(0^+, x_2)$ from (1.47) and then determine $g^{(c)}(x_1, x_2)$ as

$$g^{(c)}(x_1,x_2) = \frac{1}{2\sqrt{2}\pi} t_{k_1} t_{k_2}\left[e^{i(k_1x_1+k_2x_2)} + e^{i(k_1x_2+k_2x_1)}\right] + B\, e^{-i\left(\omega_c - i\frac{\gamma}{2}\right)(x_2-x_1)} e^{i(k_1+k_2)x_2}.$$

The constant B can be determined by (1.48) as

$$B = -\frac{\chi}{\sqrt{2}\pi}\frac{\gamma^2}{\left(k_1 - \omega_c + i\frac{\gamma}{2}\right)\left(k_2 - \omega_c + i\frac{\gamma}{2}\right)\left(k_1 + k_2 - 2\omega_c - \chi + i\gamma\right)}.$$

Finally, to solve the two-photon S matrix, we compute the initial and final wavefunctions, $\psi_i(x_1, x_2)$ and $\psi_f(x_1, x_2)$, respectively, by the Lippmann-Schwinger equations [13]. That is,

$$\begin{aligned}
\psi_i(x_1,x_2) &= \langle x_1,x_2|E_2\rangle - \langle x_1,x_2|\frac{1}{k_1+k_2-H_0+i0^+}H_{\text{int}}|E_2\rangle \\
&= g^{(a)}(x_1,x_2)\theta(x_2-x_1) + g^{(a)}(x_2,x_1)\theta(x_1-x_2), \\
\psi_f(x_1,x_2) &= \langle x_1,x_2|E_2\rangle - \langle x_1,x_2|\frac{1}{k_1+k_2-H_0-i0^+}H_{\text{int}}|E_2\rangle \\
&= g^{(c)}(x_1,x_2)\theta(x_2-x_1) + g^{(c)}(x_2,x_1)\theta(x_1-x_2).
\end{aligned}$$

$\psi_i(x_1, x_2)$ is the plane wave consisting of two photons with frequencies k_1 and k_2, and forms a complete basis for the two-photon Hilbert space. Therefore, the two-photon S matrix can be computed by

$$S_{p_1p_2k_1k_2} = \int dx_1 dx_2 \frac{1}{2\sqrt{2}\pi}\left[e^{-i(p_1x_1+p_2x_2)} + e^{-i(p_1x_2+p_2x_1)}\right]\psi_f(x_1,x_2),$$

which leads to the exactly the same result as (1.35) and (1.38).

1.9 Conclusion

In this chapter, we extend the input-output formalism of quantum optics to analyze few-photon scattering in waveguide QED systems. We develop the relationship between the input-output operators and the scattering theory, which in turn enables us to analytically calculate the photon scattering matrix elements. With the quantum causality condition, We relate the S matrix to the cavity's Green function without the need of knowing the specific details of the cavity Hamiltonian. The approach is therefore generally applicable for a large number of waveguide QED systems with different local quantum systems. Finally, the computation is systematic and can be generalized to N-photon scattering case.

Acknowledgements This work is supported by the AFOSR-MURI program, Grant No. FA9550-12-1-0488.

Appendix

Derivation of Input-Output Formalism

The Heisenberg equations associated with the Hamiltonian (1.2)–(1.4) are

$$\frac{d}{dt}c_k = -i\,k\,c_k - i\,\frac{V}{\sqrt{v_g}}\,a, \tag{1.49}$$

$$\frac{d}{dt}a = -i\,\left[a, H_c\right] - i\,\frac{V}{\sqrt{v_g}}\int dk\,c_k\,. \tag{1.50}$$

We let

$$\Phi(t) \equiv \int \frac{dk}{\sqrt{2\pi}}\,c_k(t),$$

and define the input and output operators

$$c_{\text{in}}(t) = \int \frac{dk}{\sqrt{2\pi}}\,c_k(t_0)\,e^{-ik(t-t_0)},$$

$$c_{\text{out}}(t) = \int \frac{dk}{\sqrt{2\pi}}\,c_k(t_1)\,e^{-ik(t-t_1)},$$

with $t_0 \to -\infty, t_1 \to +\infty$.

After multiplying (1.49) by the factor $\exp(ikt)$, we integrate it from an initial time t_0 to get

$$c_k(t) = c_k(t_0)e^{-ik(t-t_0)} - i\,\frac{V}{\sqrt{v_g}}\int_{t_0}^{t} d\tau\,a(\tau)e^{-ik(t-\tau)},$$

and then integrate it with respect to k to get

$$\Phi(t) = c_{\text{in}}(t) - i\frac{V}{\sqrt{v_g}}\frac{1}{2}\sqrt{2\pi}\,a(t) = c_{\text{in}}(t) - i\frac{\sqrt{\gamma}}{2}\,a(t). \tag{1.51}$$

Here, notice that we integrate over half the delta function, which results in a factor of $1/2$ and γ is defined as $\gamma \equiv 2\pi V^2/v_g$.

Furthermore, plugging (1.51) into (1.50) results in

$$\frac{d}{dt}a = -i\left[a, H_{\mathrm{c}}\right] - \frac{\gamma}{2}a - i\sqrt{\gamma}\, c_{\mathrm{in}}(t).$$

Similarly, we integrate (1.49) up to a final time $t_1 > t$ and obtain

$$\Phi(t) = c_{\mathrm{out}}(t) + i\frac{\sqrt{\gamma}}{2}\, a(t). \tag{1.52}$$

Combining (1.51) and (1.52), we finally obtain

$$c_{\mathrm{out}}(t) = c_{\mathrm{in}}(t) - i\sqrt{\gamma}\, a(t). \tag{1.53}$$

Derivation of Effective Hamiltonian

We first prove that the propagator of the cavity can be computed using the effective Hamiltonian (1.32). That is, when $H_{\mathrm{c}} = H_{\mathrm{c}}^{(0)} \equiv \omega_c a^\dagger a$,

$$G^{(0)}(t', t) = \widetilde{G}^{(0)}(t', t),$$

where

$$G^{(0)}(t', t) \equiv \langle 0|\mathscr{T}\, a(t')a^\dagger(t)|0\rangle,$$
$$\widetilde{G}^{(0)}(t', t) \equiv \langle 0|\mathscr{T}\, \widetilde{a}(t')\widetilde{a^\dagger}(t)|0\rangle.$$

$a(t)$ and $a^\dagger(t)$ are Heisenberg operators in the input-output formalism (1.8)–(1.10). $\widetilde{a}(t)$ and $\widetilde{a^\dagger}(t)$ are defined as

$$\widetilde{a}(t) = e^{iH_{\mathrm{eff}}t}\, a\, e^{-iH_{\mathrm{eff}}t}, \qquad \widetilde{a^\dagger}(t) = e^{iH_{\mathrm{eff}}t}\, a^\dagger\, e^{-iH_{\mathrm{eff}}t}, \tag{1.54}$$

whose evolution is controlled by the effective Hamiltonian (1.32).

The proof is as follows. When $t' > t$,

$$\begin{aligned}
\frac{\partial}{\partial t'}G^{(0)}(t', t) &= \langle 0|\, \frac{da(t')}{dt'}\, a^\dagger(t)\, |0\rangle \\
&= -i\left(\omega_c - i\frac{\gamma}{2}\right)\langle 0|a(t')\, a^\dagger(t)|0\rangle - i\sqrt{\gamma}\,\langle 0|c_{\mathrm{in}}(t')a^\dagger(t)|0\rangle \\
&= -i\left(\omega_c - i\frac{\gamma}{2}\right) G^{(0)}(t', t),
\end{aligned}$$

$$
\begin{aligned}
\frac{\partial}{\partial t} G^{(0)}(t',t) &= \langle 0|\, a(t') \frac{da^\dagger(t)}{dt}\, |0\rangle \\
&= i\left(\omega_c - i\frac{\gamma}{2}\right) \langle 0|a(t')\, a^\dagger(t)|0\rangle + i\sqrt{\gamma}\, \langle 0|a(t')\, c^\dagger_{\text{out}}(t)|0\rangle \\
&= i\left(\omega_c - i\frac{\gamma}{2}\right) G^{(0)}(t',t),
\end{aligned}
$$

where we plug in the input-output formalism (1.9) and (1.10) and use the respective quantum casuality (1.15) and (1.16) so that $\langle 0|c_{\text{in}}(t')a^\dagger(t)|0\rangle = \langle 0|a^\dagger(t)\, c_{\text{in}}(t')|0\rangle = 0$ and $\langle 0|a(t')\, c^\dagger_{\text{out}}(t)|0\rangle = \langle 0|c^\dagger_{\text{out}}(t)\, a(t')|0\rangle = 0$ when $t' > t$. On the other hand, by (1.54), one can compute

$$
\begin{aligned}
\frac{\partial}{\partial t'} \widetilde{G}^{(0)}(t',t) &= \langle 0|\, \frac{d\widetilde{a}(t')}{dt'}\, \widetilde{a}^\dagger(t)\, |0\rangle = -i\, \langle 0|\, [\widetilde{a}, H_{\text{eff}}](t')\, \widetilde{a}^\dagger(t)\, |0\rangle \\
&= -i\left(\omega_c - i\frac{\gamma}{2}\right) \langle 0|\widetilde{a}(t')\, \widetilde{a}^\dagger(t)|0\rangle = -i\left(\omega_c - i\frac{\gamma}{2}\right) \widetilde{G}^{(0)}(t',t), \\
\frac{\partial}{\partial t} \widetilde{G}^{(0)}(t',t) &= \langle 0|\widetilde{a}(t') \frac{d\widetilde{a}^\dagger(t)}{dt}\, |0\rangle = -i\, \langle 0|\, \widetilde{a}(t')\, [\widetilde{a}^\dagger, H_{\text{eff}}](t)\, |0\rangle \\
&= i\left(\omega_c - i\frac{\gamma}{2}\right) \langle 0|\widetilde{a}(t')\, \widetilde{a}^\dagger(t)|0\rangle = i\left(\omega_c - i\frac{\gamma}{2}\right) \widetilde{G}^{(0)}(t',t).
\end{aligned}
$$

So $G^{(0)}(t',t)$ and $\widetilde{G}^{(0)}(t',t)$ satisfy exactly the same differential equations when $t' > t$. They also have the same initial values at $t' = t$. Therefore, by the uniqueness theorem for differential equations, we can conclude that $G^{(0)}(t',t) = \widetilde{G}^{(0)}(t',t)$.

Now for a general Hamiltonian $H_c = H_c^{(0)} + V$, according to the perturbation theory in quantum field theory, all Green functions in principle are completely determined by the propagator $G^{(0)}$and the interaction vertices. The vertices only rely on the form of the interaction term V and is independent of the waveguide photons. As a result, all Green functions, including higher-order ones, can be computed by the effective Hamiltonian (1.32).

In [18, 35], the effective Hamiltonian is obtained in the path integral formalism by integrating out the waveguide degrees of freedom in the full Hamiltonian. The derivation here only relies on the input-output formalism (1.8)–(1.10) and the resulting quantum causality (1.15) and (1.16).

References

1. T. Aoki, B. Dayan, E. Wilcut, W.P. Bowen, A.S. Parkins, T.J. Kippenberg, K.J. Vahala, H.J. Kimble, Nature **443**, 671–674 (2006)
2. A.V. Akimov, A. Mukherjee, C.L. Yu, D.E. Chang, A.S. Zibrov, P.R. Hemmer, H. Park, M.D. Lukin, Nature **450**, 402–406 (2007)
3. T. Lund-Hansen, S. Stobbe, B. Julsgaard, H. Thyrrestrup, T. Sünner, M. Kamp, A. Forchel, P. Lodahl, Phys. Rev. Lett. **101**, 113903 (2008)
4. A. Wallraff, D.I. Schuster, A. Blais, L. Frunzio, R.-S. Huang, J. Majer, S. Kumar, S.M. Girvin, R.J. Schoelkopf, Nature **431**, 162–167 (2004)

5. K.M. Birnbaum, A. Boca, R. Miller, A.D. Boozer, T.E. Northup, H.J. Kimble, Nature **436**, 87–90 (2005)
6. B. Dayan, A.S. Parkins, T. Aoki, E.P. Ostby, K.J. Vahala, H.J. Kimble, Science **319**, 1062 (2008)
7. K. Srinivasan, O. Painter, Nature **450**, 862–865 (2007)
8. D. Englund, A. Majumdar, A. Faraon, M. Toishi, N. Stoltz, P. Petroff, J. Vučković, Phys. Rev. Lett. **104**, 073904 (2010)
9. K. Hennessy, A. Badolato, M. Winger, D. Gerace, M. Atatüre, S. Gulde, S. Fält, E.L. Hu, A. Imamoğlu, Nature **445**, 896–899 (2007)
10. C. Lang, D. Bozyigit, C. Eichler, L. Steffen, J.M. Fink, A.A. Abdumalikov Jr., M. Baur, S. Filipp, M.P. da Silva, A. Blais, A. Wallraff, Phys. Rev. Lett. **106**, 243601 (2011)
11. D.E. Chang, A.S. Sørensen, E.A. Demler, M.D. Lukin, Nat. Phys. **3**, 807–812 (2007)
12. J.T. Shen, S. Fan, Phys. Rev. Lett. **98**, 153003 (2007)
13. J.T. Shen, S. Fan, Phys. Rev. A **76**, 062709 (2007)
14. T. Shi, C.P. Sun, Phys. Rev. B **79**, 205111 (2009)
15. J.Q. Liao, C.K. Law, Phys. Rev. A **82**, 053836 (2010)
16. H. Zheng, D.J. Gauthier, H.U. Baranger, Phys. Rev. A **82**, 063816 (2010)
17. S. Fan, S.E. Kocabas, J.T. Shen, Phys. Rev. A **82**, 063821 (2010)
18. T. Shi, S. Fan, C.P. Sun, Phys. Rev. A **84**, 063803 (2011)
19. P. Longo, P. Schmitteckert, K. Busch, Phys. Rev. A **83**, 063828 (2011)
20. E. Rephaeli, S.E. Kocabas, S. Fan, Phys. Rev. A **84**, 063832 (2011)
21. A. Gonzalez-Tudela, D. Martin-Cano, E. Moreno, L. Martin-Moreno, C. Tejedor, F.J. Garcia-Vidal, Phys. Rev. Lett. **106**, 020501 (2011)
22. D. Roy, Phys. Rev. Lett. **106**, 053601 (2011)
23. P. Kolchin, R.F. Oulton, X. Zhang, Phys. Rev. Lett. **106**, 113601 (2011)
24. H. Zheng, D.J. Gauthier, H.U. Baranger, Phys. Rev. A **85**, 043832 (2012)
25. E. Rephaeli, S. Fan, IEEE J. Sel. Top. Quantum Electron. **18** (2012)
26. Z. Ji, S. Gao, Opt. Commun. **285**, 1302 (2012)
27. H. Zheng, H.U. Baranger, Phys. Rev. Lett. **110**, 113601 (2013)
28. S. Xu, E. Rephaeli, S. Fan, Phys. Rev. Lett. **111**, 223602 (2013)
29. J.Q. Liao, C.K. Law, Phys. Rev. A **87**, 043809 (2013)
30. T. Shi, S. Fan, Phys. Rev. A **87**, 063818 (2013)
31. D. Roy, Phys. Rev. A **87**, 063819 (2013)
32. E. Rephaeli, S. Fan, Photonics Res. **1**, 110 (2013)
33. E. Snchez-Burillo, D. Zueco, J. Garca-Ripoll, L. Martn-Moreno. arXiv:1406.5779. [quant-ph]
34. T.Y. Li, J.F. Huang, C.K. Law, Phys. Rev. A **91**, 043834 (2015)
35. S. Xu, S. Fan, Phys. Rev. A **91**, 043845 (2015)
36. T. Caneva, M.T. Manzoni, T. Shi, J.S. Douglas, J.I. Cirac, D.E. Chang, New J. Phys. **17**(11), 113001 (2015)
37. T. Shi, D.E. Chang, J.I. Cirac, Phys. Rev. A **92**, 053834 (2015)
38. C.W. Gardiner, M.J. Collett, Phys. Rev. A **31**, 3761 (1985)
39. D.F. Walls, G.J. Milburn, *Quantum Optics*, 2nd edn. (Springer, New York, 2008), See Chap. 7 for the input-output formalism in cavities, and Sec. 10.5 for resonance fluorescence
40. J.T. Shen, S. Fan, Opt. Lett. **30**, 2001 (2005)
41. V.P. Nair, *Quantum Field Theory: A Modern Perspective* (Springer, 2005), Chapter 8.7
42. E. Wichmann, J. Crichton, Phys. Rev. **132**, 2788 (1963)
43. S. Weinberg, *The Quantum Field of Fields, Volume I Foundations* (Cambridge University Press, 2005), Chapter 4.3

Chapter 2
Quadrature-Squeezed Light from Emitters in Optical Nanostructures

Diego Martín-Cano, Harald R. Haakh and Mario Agio

Abstract We discuss the concept of quadrature-squeezed light and give an overview of its applications and its relevance for quantum nanophotonics. The state-of-the-art in its detection and generation is briefly reviewed, placing special emphasis on resonance fluorescence from single emitters. We show how nanostructures allow tailoring the generation of squeezed light from a two-level emitter and from two emitters that interact through the nanoarchitecture. Finally, we comment on the future prospects of nanostructured environments for improved sources of squeezed light.

2.1 Introduction

Quadrature-squeezed light corresponds to a class of nonclassical states that are identified by their ability to reduce the electromagnetic field fluctuations below shot noise [1]. The latter arises because photons emitted by classical sources arrive at the detector in an uncorrelated way. In contrast, squeezed states of light contain nonclassical photon correlations that arise from the coherences between different photon numbers and can 'squeeze', i.e. suppress, the photodetection noise. As an important consequence of the quantum coherence involved in quadrature squeezing, the noise suppression is strongly phase dependent and thus it displays one of the few nonclassical aspects of the wave-nature of light.

The generation of quadrature-squeezed light relies intrinsically on the presence of optical nonlinearities in matter. Since most solid-state emitters have sizes on the

D. Martín-Cano (✉) · H.R. Haakh
Nano-Optics Division, Max-Planck-Institute for the Science of Light,
91058 Erlangen, Germany
e-mail: diego-martin.cano@mpl.mpg.de

H.R. Haakh
e-mail: harald.haakh@mpl.mpg.de

M. Agio
Laboratory of Nano-Optics, Universität Siegen, 57072 Siegen, Germany
e-mail: mario.agio@uni-siegen.de

S.I. Bozhevolnyi et al. (eds.), *Quantum Plasmonics*,
Springer Series in Solid-State Sciences 185,
DOI 10.1007/978-3-319-45820-5_2

nanometer scale, modified electromagnetic fields in nanostructured environments can strongly alter the inherent nonlinearities in the emitter. In fact, the key feature of optical nanostructures consists in their ability to control and concentrate the electromagnetic fields on length scales comparable or smaller than the optical wavelength [2], which allows to boost the interaction between light and matter to extreme values [3]. Therefore, optical nanostructures open new avenues to reduce the size of squeezed light sources at the nanoscale and to increase the rates and efficiency for squeezed light generation.

Enhanced single-photon generation from solid-state emitters has been demonstrated experimentally with the assistance of different nanostructures [4, 5]. Due to this nonclassical emission, time-correlation measurements between two photodetectors have shown suppressed coincidences at zero time delay, or antibunching, which reflects the impossibility to detect two photons from a single-emitter excitation at the same time. On the other hand, quadrature-squeezed light from single emitters consists of few-photons superpositions [6, 7] and its nonclassical fluctuations remain to be generated and observed in nanophotonic environments. Its phase-dependent capability to suppress the noise in photodetection makes it attractive for interferometric measurements [8] with enhancing nanostructures, e.g. for outperforming classical sensing [9], imaging [10] or information processing [11]. Testing nonclassical wave generation in nanostructured environments would also provide an improved understanding of integrated squeezed light sources with an unprecedented emission enhancement. Moreover, as this nonclassicality arises naturally in networks of emitters that are coherently connected by nanostructures, a demonstration of squeezed emission would provide a fundamental support to the studies that aim at predicting the behavior of large condensed-matter systems [12], a formidable task outside the scope of a classical computer.

In the rest of this Introduction we provide a more precise definition of squeezed light, and present the most common detection techniques and sources, placing special emphasis on resonance fluorescence from single emitters. Different observables can show nonclassical light fluctuations leading to squeezing, e.g. polarization squeezing, or amplitude squeezing. However in this chapter we will focus on quadrature-squeezed light. In Sect. 2.2 we give a short review of the theoretical description of quadrature squeezing in resonance fluorescence and in Sect. 2.3.1 we study the ability of nanostructures to tailor the generation of squeezed light from a single emitter [13, 14]. Moreover, because the generated squeezed light can interact with other emitters, they can also induce nonclassical correlations between them, providing the basis for quantum information processing [15], i.e. entanglement. In Sect. 2.3.2, we show light correlations that can only be generated from entangled emitters and how nanostructures facilitate these emitter interactions [16].

2.1.1 Quadrature-Squeezed Light

The electromagnetic field operator can be written in terms of quadratures, analogously to the canonical operators of position and momentum in a quantum harmonic oscillator [17]. Accordingly, the role of the destruction (creation) operator in the electric field quadrature is played by the positive (negative) frequency electric field operators $\hat{E}^{(+)}(\mathbf{r},t)$ $[\hat{E}^{(-)}(\mathbf{r},t)]$ at a position $\mathbf{r}$ and time t. These can be superimposed with an angle Θ to define an observable, the electric field quadrature

$$\hat{E}_k(\Theta,\mathbf{r},t) = e^{\mathrm{i}\Theta}\hat{E}_k^{(+)}(\mathbf{r},t) + e^{-\mathrm{i}\Theta}\hat{E}_k^{(-)}(\mathbf{r},t), \tag{2.1}$$

corresponding to its vectorial component along the unitary vector $\mathbf{e}_k$ in a given coordinate system. The ambiguity in the phase Θ is removed by the addition of an electric field in a coherent state, the so-called local oscillator (LO), which provides a reference phase and enables the control of the quadrature phase by its optical path.

The variance of the photodetection current carries the statistics of the quantum field fluctuations that identifies quadrature-squeezed light. On the theory side, this reduces to the calculation of the square of the variance associated to the total electric field quadrature, which is proportional to

$$(\Delta\mathcal{E})^2 \equiv \langle : [\Delta\hat{E}_k(\Theta,t)]^2 : \rangle = \langle : [\hat{E}_k(\Theta,t) - \langle\hat{E}_k(\Theta,t)\rangle]^2 : \rangle, \tag{2.2}$$

and whose negative values identify quadrature-squeezed light. Here, $\langle : \quad : \rangle$ denotes the normally-ordered expectation value, which is equivalent to subtracting from the operator mean value the vacuum expectation value associated with shot noise [17].

2.1.2 Detection Schemes

The reduced quantum fluctuations of the electromagnetic field can be measured by means of homodyne techniques. We briefly review the two main schemes that are relevant for the detection of squeezing from a single emitter. For more details on homodyne detection, we refer the reader to the textbook [17].

Balanced homodyning is the most common experimental approach for the detection of squeezed light. Within this scheme, the scattered field is mixed with a high intensity LO by means of a 50:50 beam splitter and collected by two photodetectors as illustrated in Fig. 2.1a. The photocurrents generated in the two detectors are then subtracted electronically and the resulting signal is proportional to the product of the field-quadrature mean value and the LO coherent amplitude. Interestingly, after the calculation of the joint-event probability, this differential signal is shown to preserve the probability distribution of a efficiently-collected scattered field and therefore allows for evaluating the field variance that identifies squeezing [17]. Due to the availability of high-numerical aperture collection systems and highly efficient

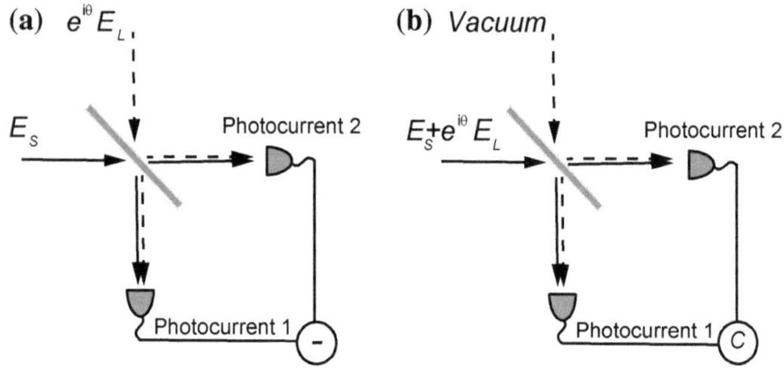

Fig. 2.1 **a** General scheme for balanced homodyne detection and **b** intensity homodyne correlations with a local oscillator field E_L and a squeezed field E_S. Adapted with permission from [14]. Copyrighted by IOP Publishing

photodetectors that prevent the degradation of the probability distribution, this technique has enabled in the detection of squeezing from a single two-level emitter (TLE) assisted by a high-Q microcavity [18] and the measurement of its small cross-section in free space [19].

The second scheme, proposed by Werner Vogel [20, 21], correlates the homodyne intensity signals arising from the configuration shown in Fig. 2.1b. Here, the scattered field comes already mixed with a LO of very low intensity. In this manner, the superimposed photodetected signals can give access to the second order intensity correlation in a Hanbury Brown and Twiss configuration. Interestingly, though the signal consists of several contributions, the one reflecting the quadrature fluctuations can be isolated due to its characteristic dependence on the phase and the LO amplitude [21]. The weak LO scheme has two advantages: firstly, its classical noise, though unbalanced, is largely suppressed and secondly, it is independent of the collection efficiency. Therefore, this homodyne correlation technique has proven particularly useful for the detection of squeezing from a single emitter in free space [22].

2.1.3 Squeezed Light sources

In the following, we briefly review the available sources of quadrature-squeezed light [11, 23], which can be summarized as atomic clouds, nonlinear crystals, and individual emitters.

Quadrature-squeezed light was first demonstrated using atomic vapors and balanced homodyning [24], which provided a modest reduction of 0.3 dB ($\approx$7 %) below the shot-noise level. The generation is based on four-wave mixing, which is commonly provided by a Λ-shaped level scheme available in alkali atoms. The two transitions can be either degenerate or nondegenerate and are both pumped, providing access to two modes, the so-called Stokes and Anti-Stokes mode. The combination of both provides reduced field fluctuations that allow quadrature squeezing when a LO is located between both mode frequencies. While atoms have a higher third-order nonlinearity as compared to crystals, the presence of fluorescence and Raman scattering have prevented high degrees of quadrature squeezing until recent advances have achieved a reduction close to 90 % [25].

Nonlinear crystals constitute the light sources with the highest degree of squeezing reported so far. Two main approaches can be identified, based on either fibers or cavities. Most fiber sources consist of amorphous materials with third-order nonlinearities, which induce a dependence of the waveguide refractive index on the incoming light intensity. For an initial coherent state, this nonlinearity induces a rotation of the quadrature phase that depends on the field amplitude and creates the phase-dependent fluctuation asymmetry characteristic of squeezed states. Because of the small nonlinearities, long fibers or pulsed coherent states have been used to increase the effective nonlinearity with a reduction in the field fluctuations close to 6 dB below the shot-noise level [11, 23]. Alternatively, most cavity-based sources use the second-order nonlinearity inherent in parametric down conversion. In this process, a pump photon with frequency ω_p incident in a material with a second-order susceptibility is split in two photons with frequencies ω_s and ω_i (signal and idler, respectively) which fulfill $\omega_p = \omega_s + \omega_i$. Because of the intrinsically low amount of squeezing due to the small nonlinearity of this process, the effect is generally enhanced by placing the nonlinear crystal in a Fabry-Pérot cavity in resonance with the idler and signal modes. The fluctuations arising from the combination of both modes enabled the first demonstration of 3 dB squeezing [26] and further technical improvements have allowed reaching the reduction of 12.7 dB below shot noise [27].

While the above sources of squeezed light are macroscopic and require high intensities, even a single two-level emitter can generate quadrature-squeezed light by means of its resonance fluorescence [6] at the few-photon level. Squeezing in resonance fluorescence arises from the coherent superposition of zero and one or two photons [7], which is intrinsic in the processes of quantum nonlinear optics. The small degree of squeezing of this process, together with technical phase noise that affects trapped emitters, has prevented its experimental demonstration until very recently [18, 22]. The first pioneering experiment [18] relied on the ability of a high-Q microcavity to increase the resonant nonlinearity provided by an atom, whereas the second [22] took advantage of the larger dipole moments found in quantum dots as compared to atoms, which facilitates both photon detection and the implementation of homodyne intensity correlation measurements in free space. Although this source is only able to produce 0.57 dB squeezing in free space (1.2 dB in a cavity [28]), it has an intrinsically much higher degree of squeezing per emitter.

2.2 Theoretical Description

2.2.1 Macroscopic Quantum Electrodynamics

We now outline the framework of macroscopic quantum electrodynamics [17, 29], which provides powerful tools to describe the TLE dynamics in arbitrary nanostructured environments. Here, the positive frequency component of the total field operator is written as

$$\hat{\mathbf{E}}^{(+)}_{\text{tot}}(\mathbf{r},t) = \hat{\mathbf{E}}^{(+)}_{\text{free}}(\mathbf{r},t) + \hat{\mathbf{E}}^{(+)}(\mathbf{r},t), \tag{2.3}$$

with the first term describing the freely evolving driving field, and the second term arising from the response of scatterers. Since the fundamental solution of the Helmholtz equation is provided by the dyadic Green's tensor

$$\nabla \times \nabla \times \mathbf{G}(\mathbf{r},\mathbf{r}_E,\omega) - \frac{\omega^2}{c^2}\varepsilon(\mathbf{r},\omega)\mathbf{G}(\mathbf{r},\mathbf{r}_E,\omega) = \mathbb{1}\delta(\mathbf{r}-\mathbf{r}_E)\,, \tag{2.4}$$

a classical dipole $\mathbf{p}$, placed in an inhomogeneous dielectric $\varepsilon(\mathbf{r},\omega)$ at position $\mathbf{r}_E$, and oscillating at a frequency ω generates the (classical) source field [2]

$$\mathbf{E}^{(+)}(\mathbf{r},\omega) = \frac{\omega^2}{\varepsilon_0 c^2}\mathbf{G}(\mathbf{r},\mathbf{r}_E,\omega)\mathbf{p}. \tag{2.5}$$

In the following, we consider a TLE with ground state $|g\rangle$ and excited state $|e\rangle$, separated by a dipole transition at a frequency $\omega_E = 2\pi c/\lambda_E$. The transition is described by lowering and raising operators $\hat{\sigma} = |g\rangle\langle e|, \hat{\sigma}^\dagger = |e\rangle\langle g|$ that anticommute $\{\hat{\sigma}^\dagger, \hat{\sigma}\} = \mathrm{II}$, whereas the commutation relation $[\hat{\sigma}^\dagger, \hat{\sigma}] = \hat{\sigma}_z$ defines the inversion operator. The transition dipole matrix element is $\mathbf{d}$. For near-resonant driving, the rotating wave approximation can be imposed and the scattered quantum field is described by [29]

$$\hat{\mathbf{E}}^{(+)}(\mathbf{r},t) \approx \mathrm{i}\int_0^\infty \frac{\mathrm{d}\omega}{\pi}\frac{\omega^2}{c^2\varepsilon_0}\mathrm{Im}\{\mathbf{G}(\mathbf{r},\mathbf{r}_\mathrm{E},\omega)\}\cdot\mathbf{d}\,K(\omega,t). \tag{2.6}$$

Near the nanostructure, correlation times are typically short and the memory kernel is evaluated in the Markov approximation, where $K(\omega,t) \approx \sigma(t)[\mathrm{i}\mathcal{P}\frac{1}{\omega_E-\omega} + \pi\delta(\omega_E - \omega)]$. We evaluate the principal value integral by complex contour integration [13, 30] and obtain the expression

$$\hat{\mathbf{E}}^{(+)}(\mathbf{r},t) = \frac{\omega_E^2}{c^2\varepsilon_0}\left[\mathbf{G}(\mathbf{r},\mathbf{r}_E,\omega_E) + \delta\mathbf{G}(\mathbf{r},\mathbf{r}_E,\omega_E)\right]\cdot\mathbf{d}\,\hat{\sigma}(t), \tag{2.7}$$

which closely resembles the previous (2.5) for the classical dipole except for the operator-valued source and a quantum correction [31]

$$\delta\mathbf{G}(\mathbf{r},\mathbf{r}_E,\omega_E)=\int_0^\infty\frac{\mathrm{d}\xi}{\pi}\frac{\xi^2}{c^2\varepsilon_0}\frac{\mathbf{G}(\mathbf{r},\mathbf{r}_E,\mathrm{i}\xi)\omega_E}{\xi^2+\omega_E^2}\,. \tag{2.8}$$

The latter arises from non-resonant quantum fluctuations and is expressed in the complex frequency plane for improved convergence. It describes dispersion interactions that shift the TLE transition frequency [13, 30, 31]. Other than that, the impact on the scattered source field distribution is only relevant at subwavelength distances. Finally, we rewrite the scattered field operator in terms of the phase and amplitude of the response function for the vector component along the direction $\mathbf{e}_k$

$$\hat{E}_k^{(+)}(\mathbf{r},t)=|g_k(\mathbf{r})|e^{\mathrm{i}\phi_k(\mathbf{r})}\hat{\sigma}(t). \tag{2.9}$$

In most complex nanoarchitectures, the specific values $g_k(\mathbf{r})$ and its complex argument $\phi_k(\mathbf{r})$ depend on the detector position $\mathbf{r}$ and need to be evaluated numerically, but closed forms of the Green's function are available in several important cases such as free space, for a plane interface, or for nanospheres [2, 32, 33]. In the following, the detector position is typically fixed and we do not write the dependence on $\mathbf{r}$.

2.2.2 The Optical Bloch Equations

We now evaluate the evolution of a quantum dipole in an external classical driving field, described by the Rabi frequency $\Omega=2\mathbf{d}\cdot\langle\hat{\mathbf{E}}_{\text{free}}^{(+)}(\mathbf{r}_\mathrm{E})\rangle/\hbar=|\Omega|e^{\mathrm{i}\phi_\mathrm{L}}$ that includes the local driving field modification due to the nanostructured environment [34]. Evaluating the Heisenberg equations of motion for the atomic operators in the Wigner-Weisskopf limit, electric-dipole and rotating-wave approximation, and considering slowly varying operators, e.g. $\langle\hat{\tilde{\sigma}}(t)\rangle=\langle\hat{\sigma}(t)\rangle e^{\mathrm{i}\omega_L t}$, in the co-rotating frame of the driving field, one finds the optical Bloch equations [29, 35]

$$\frac{\mathrm{d}}{\mathrm{d}t}\langle\hat{\tilde{\sigma}}(t)\rangle=\left(-\frac{\gamma}{2}-\gamma^*+\mathrm{i}\delta_\mathrm{L}\right)\langle\hat{\tilde{\sigma}}(t)\rangle-\mathrm{i}\frac{\Omega}{2}\langle\hat{\sigma}_z(t)\rangle, \tag{2.10}$$

$$\frac{\mathrm{d}}{\mathrm{d}t}\langle\hat{\sigma}_z(t)\rangle=\mathrm{i}\left(\Omega\langle\hat{\tilde{\sigma}}^\dagger(t)\rangle-\Omega^*\langle\hat{\tilde{\sigma}}(t)\rangle\right)-\gamma(1+\langle\hat{\sigma}_z(t)\rangle)\,. \tag{2.11}$$

This includes spontaneous decay at a rate $\gamma=\frac{2\omega_\mathrm{E}^2}{\hbar\varepsilon_0c^2}\mathbf{d}\cdot\mathrm{Im}[\mathbf{G}(\mathbf{r}_\mathrm{E},\mathbf{r}_\mathrm{E},\omega_\mathrm{E})]\cdot\mathbf{d}$ according to Fermi's golden rule. $\delta_\mathrm{L}=\omega_\mathrm{L}-\omega_\mathrm{E}-\frac{\omega_\mathrm{E}^2}{c^2\hbar\varepsilon_0}\mathbf{d}\cdot\mathrm{Re}[\mathbf{G}(\mathbf{r}_E,\mathbf{r}_E,\omega_E)]\cdot\mathbf{d}-\mathbf{d}\cdot\delta\mathbf{G}(\mathbf{r}_E,\mathbf{r}_E,\omega_E)\cdot\mathbf{d}/\hbar$ is the laser detuning with respect to the TLE transition frequency. Note that both quantities may differ from their value in a homogeneous environment, as the TLE close to a nanostructure is affected by modified local density of states [2] and dispersion potentials, respectively. The additional loss of phase coherence, e.g. due to phononic excitations, strongly affects solid-state emitters [36] and we allow for additional pure dephasing at a rate γ^* to account for such phenomena.

Equations (2.10) and (2.11) give rise to a stationary expectation value of the slowly varying coherence $\langle \hat{\tilde{\sigma}} \rangle_s$ and population $\langle \hat{\sigma}_z \rangle_s$, respectively,

$$\langle \hat{\tilde{\sigma}} \rangle_s = \frac{-\Omega[2\delta_L - i(\gamma + 2\gamma^*)]}{4\delta_L^2 + 2|\Omega|^2(1 + \frac{2\gamma^*}{\gamma}) + (\gamma + 2\gamma^*)^2}, \tag{2.12}$$

$$\langle \hat{\sigma}_z \rangle_s = -1 - 2\,\mathrm{Im}\left[\frac{\Omega}{\gamma}\langle \hat{\tilde{\sigma}}^\dagger \rangle_s\right], \tag{2.13}$$

which we will use below to study the properties of the emitted light.

2.2.3 *Squeezed Resonance Fluorescence*

Resonance fluorescence, the re-emission of light by an emitter at the driving frequency, is one of the key concepts in quantum optics. The scattered field fluctuations can be related to the source field using (2.9) in (2.2)

$$[\Delta\hat{\mathcal{E}}_k(\Theta, t)]^2 = 2|g_k|^2\left(\left[\langle \hat{\sigma}^\dagger(t)\hat{\sigma}(t)\rangle - |\langle \hat{\sigma}(t)\rangle|^2\right] - \mathrm{Re}\left[e^{2i(\phi_k+\Theta)}\langle \hat{\sigma}(t)\rangle^2\right]\right). \tag{2.14}$$

From (2.14), we see that $(\Delta\hat{\mathcal{E}}_k)^2$ is governed by the TLE's optical coherence $\hat{\sigma}$ and upper-state population $\hat{\sigma}^\dagger\hat{\sigma}$ [6]. The fluctuations $\langle \hat{\sigma}^\dagger(t)\hat{\sigma}(t)\rangle - |\langle \hat{\sigma}(t)\rangle|^2$ are always positive, and, hence, tend to destroy squeezing, but they approach zero the weaker the excitation. Since we deal with one TLE, a small detection efficiency ($\propto |g_k(\mathbf{r})|^2$ at the detection position) results in low photon count rates, which has made the detection of squeezing challenging in free space [22] and even in a cavity [18]. It is the last term in (2.14) that originates from phase-dependent quantum fluctuations in the optical coherence $\hat{\sigma}$, and it is responsible for the negative values associated to quadrature-squeezed light [1, 6].

We now use the stationary solution from (2.12) and (2.13) to express the normally-ordered field variance in a compact form

$$[\Delta\hat{\mathcal{E}}_k(\Theta, t)]^2 \underset{\text{stat.}}{=} |g_k|^2 \frac{z^2}{1 + \delta^2 + z^2}\left(1 - \frac{(1 + \delta^2)(1 + \cos[2\Phi + 2\Theta])}{(1 + x)(1 + \delta^2 + z^2)}\right), \tag{2.15}$$

where we abbreviated the normalized dephasing rate $x = 2\gamma^*/\gamma$, the normalized detuning $\delta = 2\delta_L/(\gamma + 2\gamma^*)$, and the normalized Rabi frequency $z = \sqrt{2}|\Omega|/\sqrt{\gamma(\gamma + 2\gamma^*)}$. Φ collects the phases connected with the scattering and the driving laser. The quadrature angle Θ can be controlled in a homodyne detection scheme and thus we set the cosine equal to unity without loss of generality [13]. Under such optimal conditions, one finds that $[\Delta\hat{\mathcal{E}}_k(\Theta, t)]^2 \geq -|g_k|^2/8$ provides a bound to the steady-state squeezing amplitude generated in resonance fluorescence in a Markovian (weak-coupling) environment. Besides, it can take negative values only when

$$z^2 < (1 + \delta^2)\frac{1 - x}{1 + x}, \tag{2.16}$$

which provides an upper limit to the Rabi frequency and shows that additional pure dephasing at a rate $\gamma^* \geq \gamma/2$ prevents squeezing.

The above treatment is general in terms of the parameters γ, γ^*, and Ω so that at sufficiently weak driving, even single emitters in free space will generate squeezed light in resonance fluorescence [6, 22], as mentioned before. We will assess in the following how this can be modified by a controlled coupling to an optical nanostructure.

2.3 Quadrature Squeezing Assisted by Nanostructures

2.3.1 A Single Emitter Coupled to a Nanostructure

If a TLE is placed near a nanostructure, the dynamics that generate squeezing are fundamentally changed [13]. The Green's tensor becomes $\mathbf{G} = \mathbf{G}^{(0)} + \mathbf{G}^{(\mathrm{sc})}$, where in addition to the free-space background, the electromagnetic field scattered by the nanostructure modifies the dipole emission pattern, so that both the amplitude $|g_k(\mathbf{r})|$ and phase $\phi_k(\mathbf{r})$ differ from their values in free space, $|g_k^{(0)}(\mathbf{r})|$ and $\phi_k^{(0)}(\mathbf{r})$, respectively. Hence, although the field intensity scattered by the TLE—and consequently the detection efficiency—increase [37], its quantum fluctuations can be comparatively reduced with respect to shot noise, with a squeezing amplitude $|g_k(\mathbf{r})|$. Second, since $\hat{\sigma}$ is affected by the enhancement of the driving field and the change in γ, both induced by the nanostructure [37], control of these quantities can be used to reduce the electromagnetic field fluctuations, while increasing the photon-count rate.

For a quantitative analysis, we exemplify in the following the nanostructure with a gold nanosphere (GNS) which couples to a TLE, as sketched in Fig. 2.2a. In this case, the Green's tensor is known analytically [32, 33]. Apart from the geometry, the material properties of gold enter in terms of a dielectric permittivity, which we model by interpolating tabulated optical constants of gold [38] by a Drude-Lorentz dispersion model [39].

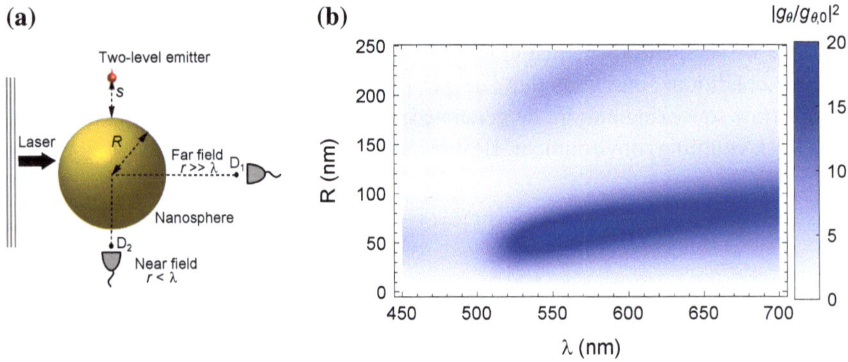

Fig. 2.2 **a** A TLE is placed at a distance s from a GNS of radius R. D_1 and D_2 are detection points in the far and near fields, respectively. D_1 is on the x-axis at $10^5\lambda_E$ from the GNS center, while D_2 is along the z-axis, 10 nm from the GNS surface. **b** Far-field squeezing amplitude $|g_\theta|$ versus R and λ for a GNS at $\lambda_E = 550$ nm and $s = 10$ nm. The amplitude values are normalized to the case in free space. Figure adapted with permission from [13]. Copyrighted by the American Physical Society

2.3.1.1 Enhanced Amplitude, Bandwidth, and Driving

The field scattered by the emitter-GNS system near a detector in the far field ($|\mathbf{r} - \mathbf{r}_E| \gg \lambda_E$) is transverse and the dominant component points along $\mathbf{e}_\theta$ in polar coordinates. We therefore restrict our analysis to this component and suppress the vector index for brevity unless stated differently. As the quantum fluctuations in (2.7) are negligible in this case, $g = |g|e^{i\phi} \approx \mathbf{e}_\theta \cdot \frac{\omega_E^2}{\varepsilon_0 c^2}\mathbf{G}(\mathbf{r}, \mathbf{r}_E, \omega_E) \cdot \mathbf{d}$ provides an excellent approximation of the amplitude and phase in (2.14).

Figure 2.2b shows the squeezing amplitude $|g|^2$ at a detection point in the far field (D_1 in Fig. 2.2a) and features several local maxima that correspond to plasmon-polariton resonances [40], which depend on the GNS radius R and on the TLE emission wavelength λ_E. The strongest one originates from the dipole resonance. Squeezing is enhanced by up to a factor of 20 due to the presence of the GNS. Comparing to a typical cavity environment, the limited transmission out of the cavity is estimated to provide a normalized squeezing amplitude slightly lower than the GNS value [13]. The maxima at larger R are associated with higher-order resonances and have a different spatial emission pattern so that nanostructures can also control the directionality of squeezed light in the far field. This effect can be optimized by careful design [41].

The presence of a nanostructure also strongly modifies the conditions under which squeezed light can be generated. This is possible because $(\Delta\hat{\mathcal{E}})^2$ depends on the frequency detuning δ_L, the Rabi frequency Ω (*i.e.* the driving field) and the TLE's spontaneous decay rate γ, which differ from their values in free space $\delta_L^{(0)}$, $\Omega^{(0)}$, and $\gamma^{(0)}$, respectively [37]. In practice, the boundaries for the generation of squeezing depend only on the ratios Ω/γ and δ_L/γ (see (2.15)). For a given configuration, these limits are shown in Fig. 2.3a as a function of the rescaled detuning and driving

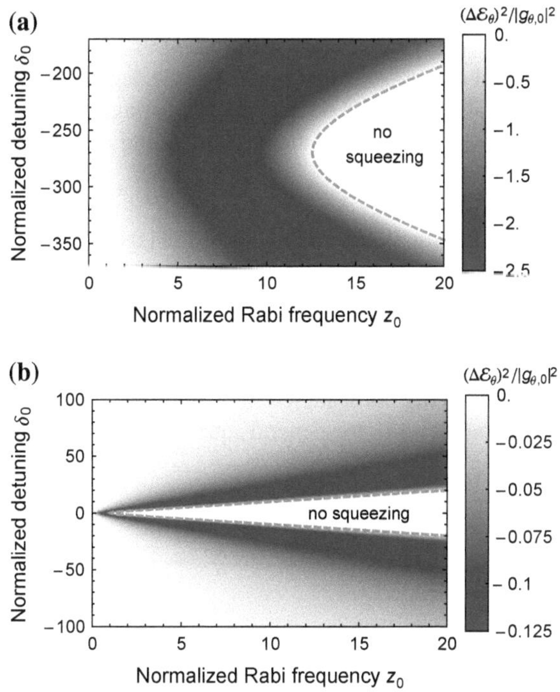

Fig. 2.3 Electric field fluctuations versus normalized detuning and Rabi frequency with (**a**) and without GNS (**b**). The parameters are $s = 10$ nm, $R = 60$ nm, $\lambda_E = 550$ nm and fields are detected at D_1 (see Fig. 2.2a). The amplitude values are normalized to the case in free space. The lower bound of the *color scale* displays the different minimum value in each panel. Their ratio emphasizes the 20-fold enhancement of squeezing due to the GNS as compared to free space. Figure adapted with permission from [13]. Copyrighted by the American Physical Society

field ($\delta_0 = 2\delta_L^{(0)}/\gamma^{(0)}$ and $z^{(0)} = \sqrt{2}\Omega^{(0)}/\gamma^{(0)}$, respectively). Importantly, we find that the detuning range with sizable squeezing has increased by two orders of magnitude with respect to free space, as displayed in Fig. 2.3b. This is directly related to the fact that the enhancement of the TLE spontaneous decay rate is enhanced by the GNS ($\gamma/\gamma^{(0)} \sim 60$), which also leads to a shift in the resonance frequency. Moreover, squeezing occurs over a much wider range of laser intensities as compared to free space, *c.f.* Fig. 2.3b at zero detuning. The reason is that the GNS has a larger impact on γ than on the field enhancement with respect to free space ($\Omega/\Omega^{(0)} \sim 4.9$ for this case), so that the ratio Ω/γ provides a weaker excitation level at the same incident power ($\propto z^{(0)}\Omega/\gamma$).

2.3.1.2 Spectrum of Squeezing

So far we determined the dependence on the driving frequency. For completeness, we now consider driving at resonance but perform the Fourier transformation of the photocurrents, which allows to obtain the spectrum of squeezing [21, 42–44]

$$S(\Theta, \omega) = \int_{-\infty}^{\infty} \mathrm{d}\tau \; e^{\mathrm{i}\omega\tau} \langle : \mathcal{T} \; \Delta\hat{\mathcal{E}}(\Theta, 0) \Delta\hat{\mathcal{E}}(\Theta, \tau) : \rangle. \tag{2.17}$$

This involves the time- and normally ordered average $\langle : \mathcal{T} \; : \rangle$ necessary to obtain the physically observable spectrum [43]. This quantity gives information about the spectral components that possess some degree of squeezing [42], indicated by negative values, and can be relevant for distinguishing different kinds of squeezed light sources [43], including the onset of strong coupling [18]. Notice that by integrating the full frequency spectrum $[\Delta\hat{\mathcal{E}}(\Theta,0)]^2 = \int_{-\infty}^{\infty} \frac{d\omega}{2\pi}\, S(\Theta,\omega)$, we recover the electric field fluctuations at zero time delay [21] used above to identify squeezed light. In the present section, the spatial dependence of the amplitude is intentionally omitted in the electric field, since the spectrum will be normalized with respect to the relative flux ratio $|g|^2/\gamma$ as in references [42, 44], which provides an absolute measure of the degree of squeezing of the field.

Introducing the scattered field quadrature in (2.17) together with the proper time ordering [17, 44], the general squeezing spectrum is written as

$$S(\Theta,\omega) = |g|^2 \int_0^{\infty} d\tau\, (e^{i\omega\tau} + e^{-i\omega\tau})\big[\langle\sigma^\dagger(0), \sigma(\tau)\rangle + e^{2i(\Theta+\phi)}\langle\sigma(\tau), \sigma(0)\rangle + c.c.\big], \tag{2.18}$$

where $\langle A, B\rangle = \langle AB\rangle - \langle A\rangle\langle B\rangle$. An elegant way of evaluating the time-integrals uses the Laplace transforms $\int_0^{\infty} d\tau\, e^{\zeta\tau}\langle\sigma(\tau), \sigma(0)\rangle$ and $\int_0^{\infty} d\tau\, e^{\zeta\tau}\langle\sigma^\dagger(0), \sigma(\tau)\rangle$ and the quantum regression theorem [17] in the limit of a narrow-band detector. Thus the expectation values of the TLE operators fulfill the previous Bloch equations [13, 17], which after the Laplace transform provide a simple linear system of equations that is easily solved analytically in the case of driving at the TLE resonance. By modifying the phase Θ via the LO, it is then possible to maximize the values of the squeezing spectrum in (2.18) and obtain [14] the squeezing spectrum of resonance fluorescence for the quadratures in-phase $[S_1(\omega) = S(-\phi,\omega)]$ and out of phase $[S_2(\omega) = S(\pi/2-\phi,\omega)]$, respectively,

$$\frac{S_1(\omega)}{|g|^2} = \frac{2\Omega^2}{\gamma(\gamma^* + \gamma/2) + \Omega^2}\,\frac{\gamma^* + \gamma/2}{(\gamma^* + \gamma/2)^2 + \omega^2}, \tag{2.19}$$

$$\frac{S_2(\omega)}{|g|^2} = \frac{2\Omega^2}{\gamma(\gamma^* + \gamma/2) + \Omega^2}\,\frac{2\gamma\Omega^2 - (\gamma/2 - \gamma^*)(\gamma^2 + \omega^2)}{\left[\gamma(\gamma^* + \gamma/2) - \omega^2 + \Omega^2\right]^2 + (\gamma^* + 3\gamma/2)^2\,\omega^2}. \tag{2.20}$$

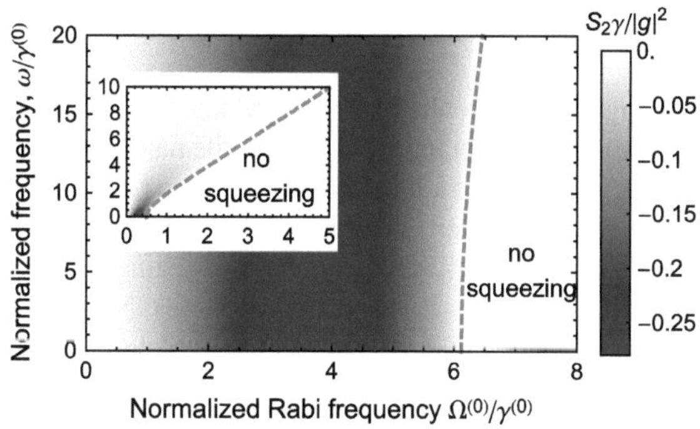

Fig. 2.4 **a** Squeezing spectrum of the out-of-phase quadrature (S_2) as a function of $\omega/\gamma^{(0)}$ and $\Omega^{(0)}/\gamma^{(0)}$, respectively. The spectrum is normalized by $|g|^2/\gamma$. Same geometrical parameters as in Fig. 2.3. The pure dephasing rate γ^* is neglected. Inset: analogous representation in the absence of GNS. Adapted from [14].

These expressions generalize previous results[1] [42–44] to the case with additional phase decoherence ($\gamma^* \neq 0$).

Notice that the in-phase quadrature spectrum S_1 shows no signature of squeezing due to the positive character of (2.19), which is consistent with the on-resonant excitation in time domain [6]. On the other hand, the out-of-phase quadrature can show negative values that originate in the coherence term $\langle\sigma(t), \sigma(0)\rangle$ in (2.18) that dominates with respect to $\langle\sigma^\dagger(0), \sigma(t)\rangle$ at low driving, similarly to the time-domain excitation spectra. Figure 2.4 shows the normalized spectrum for the out-of-phase component (2.20) as a function of the free-space driving amplitude $\Omega^{(0)}/\gamma^{(0)}$ and the spectral frequency $\omega/\gamma^{(0)}$, both normalized with respect the decay rate without nanostructure. For comparison, the spectrum in free space is represented in the inset ($\gamma/\gamma^{(0)} = 1$ and $\Omega/\Omega^{(0)} = 1$). Negative values appear at weak driving below saturation ($\Omega/\gamma < 0.5$) and reach a minimum value ($S_2 = -0.28$) at $\Omega/\gamma \approx 0.3$. We can also see that the GNS allows for squeezing over a broader range of frequencies for generating squeezing due to the broadening of the emission linewidth ($\gamma/\gamma^{(0)} > 1$). Also the driving parameter range is increased due to a smaller effective saturation parameter as compared to free space [$\Omega/\gamma = (\Omega/\Omega^{(0)})/(\gamma/\gamma^{(0)}) \approx 0.08$]. Besides the reduction of the squeezing at larger detuning, as a result of the contribution from the resonant field, and the variation of the overall absolute values, we can conclude that the spectrum displays similar qualitative information of squeezing than the time-domain case near resonant excitation.

[1]Note that the field quadrature prefactor needs to be chosen consistently and we use the convention and notation adopted in our previous work [13]. A factor 1/2 has been commonly introduced in the literature that will provide a value 1/4 of the spectrum calculated in this work. Also, the expressions are corrected with respect to those written in the text of the seminal work [42].

2.3.1.3 Robustness Against Decoherence

Pure dephasing strongly affects realistic TLEs [36] and can preclude the generation of squeezing in free space. To gain intuition on how the nanostructure may overcome this difficulty, we show in Fig. 2.5a the field fluctuations $(\Delta\hat{\mathcal{E}})^2$ as a function of the distance to the GNS surface, at zero detuning, fixed Ω and assuming an additional rate of pure dephasing, $\gamma^* = \gamma^{(0)}/2$ that excludes squeezing in free space (cf. gray dashed curve and (2.16)). In contrast, the presence of the GNS (black curves) allows for quantum squeezing over a range of distances that depend on Ω, γ^* and γ. For example, for $\Omega = 5\gamma^{(0)}$, negative values of $(\Delta\hat{\mathcal{E}})^2$ occur below $s = 35$ nm and its minimum is reached at $s = 23$ nm. This overall behavior is general, as highlighted by the other curves in Fig. 2.5a corresponding to a larger Ω. Importantly, it is the increase in the decay rate γ with respect to the free-space value $\gamma^{(0)}$ and the dephasing rate γ^* that allows to meet the condition set by (2.16). Moreover, the resulting squeezing is robust against the positioning of the TLE. In fact, squeezing can be achieved over the whole range of distances given in Fig. 2.5a by adjusting Ω (as inferred from (2.16). A significant enhancement of the squeezing amplitude is feasible at distances s of a few tens of nanometers for the case considered. As the TLE moves towards the GNS surface, optimal squeezing requires increasingly stronger Ω, especially once s falls below 10 nm, where absorption by real metals provides a dominating nonradiative decay channel for the TLE [45]. This is reflected in the growing deviation of $\gamma/\gamma^{(0)}$ from $|g/g^{(0)}|^2$ (see Fig. 2.5b). Nevertheless, this ratio can be modified by optimized nanostructures [45] and quantum squeezing may, in principle, be enhanced without considerably raising Ω to compensate for the nonradiative losses.

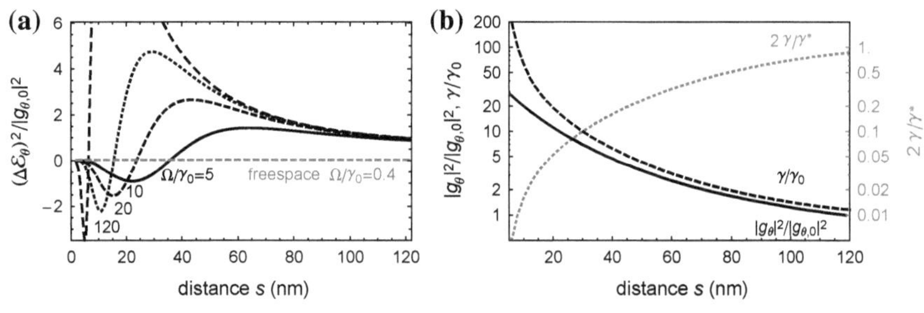

Fig. 2.5 **a** Normalized electric field fluctuations as a function of the distance s between the TLE and the GNS surface, considering pure dephasing at a rate $\gamma^* = \gamma^{(0)}/2$ and parameters as in Fig. 2.3a. The amplitude values are normalized to the case in free space. are normalized to the case in free space. The *curves* corresponds to different Rabi frequencies Ω. The result without GNS and $\Omega = 0.4\gamma^{(0)}$ is represented by a *grey dashed line*. **b** Normalized total decay rate $\gamma/\gamma^{(0)}$ (*dashed black curve*) and the field intensity enhancement factor $|g_\theta/g_\theta^{(0)}|^2$ (*solid black curve*) as a function of s, along with the ratio $2\gamma^*/\gamma$ (*dotted grey curve, right axis*). Figure adapted with permission from [13]. Copyrighted by the American Physical Society

2.3.1.4 Near-Field Squeezing and Propagation

Further enhancement of squeezing can be achieved for detection in the near field $|\mathbf{r} - \mathbf{r}_E| < \lambda_E$, where evanescent modes become relevant. Even in free space, the squeezing amplitude close to a TLE is orders of magnitude higher than in the far field, due to the spatial behavior of its dipolar field. To estimate the ability of nanostructures to manipulate squeezed light in the near field, we consider a detection point D_2 in the near field (see Fig. 2.2a). This is the regime, where the quantum correction in (2.7) and (2.8) modifies the field distribution and needs to be considered [13]. As the radial field component along $\mathbf{e}_r$ in polar coordinates dominates in the near-field region, we restrict this analysis to the corresponding amplitude $|g_r|^2$.

For better intuition, we give in Fig. 2.6a the near-field squeezing pattern for a GNS of $R = 200$ nm. At the point opposite side from the emitter 10 nm below the GNS, the GNS provides enhancement by up to a factor of 2000 as compared to free space [13]. The pattern shows two lateral lobes, which stem from the excitation of higher-order plasmon-polariton resonances. These are superimposed with the dipolar contribution indicated by the presence of the top and bottom lobes, more clearly visible in Fig. 2.6b in the case of a smaller GNS ($R = 60$ nm). Note that despite the huge enhancements found for large GNSs, the small one improves the squeezing amplitude, e.g. by a factor 30 at the point opposite from the emitter. This is the result of detection closer to the TLE combined with a stronger near-field enhancement. It turns out that the squeezing enhancement remains huge up to very large radii [13], when eventually it is limited by propagation losses. These results, together with significant experimental evidence [46], suggest that squeezed light generated in such a hybrid system could also be efficiently transferred over a considerable distance by nanoscale waveguides (see Chap. 4).

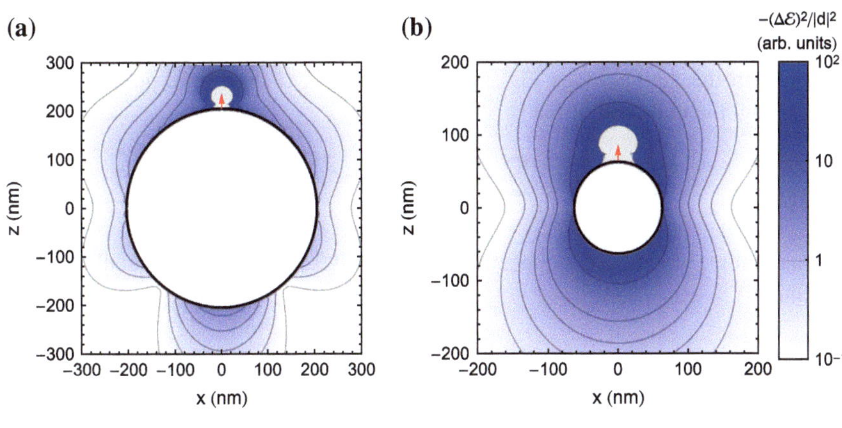

Fig. 2.6 a-b Near field contour maps of the negative field fluctuations for the radial component for $R = 200$ nm (**a**) and $R = 60$ nm (**b**), with $s = 10$ nm and $\lambda_E = 550$ nm. The values are normalized to the square modulus of the dipole moment $|\mathbf{d}|^2$ to be independent of a specific TLE. The emitter and the GNS are represented by a *black arrow* and a disk in the *xz*-plane, respectively. $-(\Delta\mathcal{E})^2/|d|^2$ is given in arbitrary units. Figure adapted with permission from [13]. Copyrighted by the American Physical Society

2.3.2 Cooperative Quadrature Squeezing

2.3.2.1 Nonclassical Correlations in Coupled Emitters

We now turn to collective effects in a system of two TLEs and establish a connection of the steady-state entanglement between the emitters and the reduced light fluctuations in cooperative resonance fluorescence assisted by nanostructures [16]. For a theoretical description of more than one emitter, the optical Bloch equations are replaced by the master equation that describes the evolution of the density matrix [47, 48]

$$\dot{\rho} = -\frac{\mathrm{i}}{\hbar}[H,\rho] - \sum_{m,n} \frac{\gamma_{mn}}{2}\left(\sigma_m^\dagger \sigma_n \rho + \rho \sigma_m^\dagger \sigma_n - 2\sigma_m \rho \sigma_n^\dagger\right), \tag{2.21}$$

$$H = \sum_m \hbar(\omega_m - \omega_L)\sigma_m^\dagger \sigma_m - \sum_m \left(\frac{\hbar\Omega_m}{2}\sigma_m^+ + h.c.\right) - \sum_{m\neq n} \Omega_{12}\sigma_m^\dagger \sigma_n, \tag{2.22}$$

where the summations run over the TLEs identified by indices $m, n = 1, 2$. As before, the expressions hold in general Markovian environments [48] and the impact of the specific nanostructure enters via the coupling constants. This includes the dipole-dipole coupling Ω_{12}, the incoherent coupling rate γ_{12}, as well as the transition frequencies ω_m, decay rate $\gamma_{mm} = \gamma_m$ and Rabi frequency Ω_m for each emitter. The corresponding values in free space used for comparison are again indicated by a superscript $^{(0)}$. As in the case of a single emitter, the modified response speeds up the dynamics, increases the bandwidth of the interaction, and requires lower laser powers.

The interaction generates quantum correlations between the emitters [47, 49–51] that strongly affect the statistical properties of the scattered light. As before, we consider far-field detection and restrict the analysis to the transverse vector component along $\mathbf{e}_\theta$ in the far field (see Fig. 2.7a). The general electric field quadrature now arises from the emission of the two emitters

$$\hat{E}(\Theta, t) = \sum_{m=1,2} [e^{\mathrm{i}\Theta}\hat{E}_m^+(\Theta, t) + e^{-\mathrm{i}\Theta}\hat{E}_m^-(\Theta, t)], \tag{2.23}$$

and as before, we identify squeezing by a negative value of the normally ordered variance

$$(\Delta\mathcal{E})^2 = \langle :(\hat{E} - \langle\hat{E}\rangle)^2 : \rangle = (\Delta\mathcal{E}_1)^2 + (\Delta\mathcal{E}_2)^2 + (\Delta\mathcal{E}_{12})^2. \tag{2.24}$$

Here, the first two terms recover fluctuations from single emitters in the form of (2.14) with the respective operators $\hat{\sigma}_m$ and scattering functions $g_m = |g_m|e^{i\phi_m}$, where the subscript now indicates the emitter. The interaction modifies each of these terms and gives rise to the additional cross-correlations

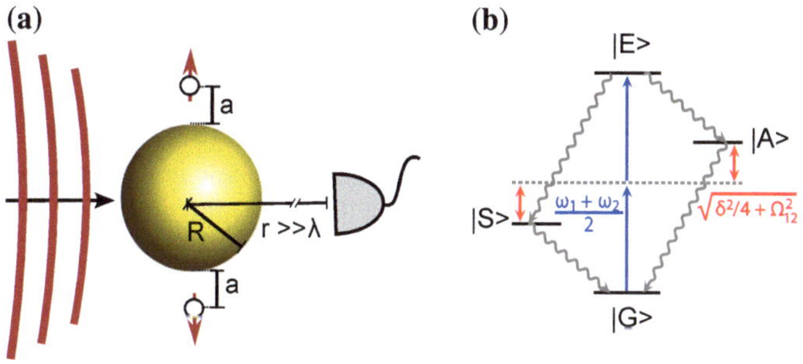

Fig. 2.7 **a** Two TLEs coupled to a nanoantenna and illuminated by a coherent driving field. The observation takes place in the far field. The emitters are placed at symmetrical distances $a =$ 25 nm from a $R = 40$ nm GNS ($\lambda = 780\,\text{nm}, \delta = 400\gamma_0, \Omega_{12} = -6.4\gamma_0, \gamma_{12} = -2.6\gamma_0, \gamma = 2.9\gamma_0$, $\Omega \approx 2\Omega_0$). **b** Level scheme of the coupled system, consisting of the fundamental state $|G\rangle$ and doubly excited state $|E\rangle$ connected by a nonlinear two-photon process (*blue arrow*). Due to the detuning δ and coupling Ω_{12}, the states $|A\rangle$ and $|S\rangle$ are nondegenerate. Reprinted with permission from [16]. Copyright 2015 American Chemical Society

$$\frac{(\Delta\mathcal{E}_{12})^2}{2|g_1 g_2|} = 2\text{Re}\left[e^{\mathrm{i}(\phi_1-\phi_2)}\left(\langle\hat{\sigma}_2^\dagger\hat{\sigma}_1\rangle - \langle\hat{\sigma}_2^\dagger\rangle\langle\hat{\sigma}_1\rangle\right) + e^{\mathrm{i}(2\Theta+\phi_1+\phi_2)}\left(\langle\hat{\sigma}_2\hat{\sigma}_1\rangle - \langle\hat{\sigma}_2\rangle\langle\hat{\sigma}_1\rangle\right)\right] \tag{2.25}$$

that clearly vanish for uncorrelated emitters. Therefore, negative values arising from the cross-terms denote reduced fluctuations that can identify bipartite entanglement [16].

2.3.2.2 Two-Photon Excitation

To exemplify the connection between entanglement and squeezing, we consider a particular driving scheme based on a two-photon process in a coupled pair of TLEs with transition frequencies detuned by many linewidths ($\omega_2 - \omega_1 = \delta \gg \gamma$) (see Fig. 2.7b). In broad-band environments, the single-emitter levels can still be hybridized due to the dipole-dipole coupling potential $\hbar\Omega_{12}$, which arises from the scattering of photons between the emitters and is present even at off-resonant driving as long as $\omega_1, \omega_2 \gg \delta$ and the frequency response of the environment varies slowly over δ [47, 48]. Here, they form a collective four-level system depicted in Fig. 2.7b, consisting of the fundamental state $|G\rangle = |g_1 g_2\rangle$, the doubly excited state $|E\rangle = |e_1 e_2\rangle$ and two states $|A\rangle, |S\rangle$ that involve superpositions of single-excitation states. Two-photon transitions $|G\rangle \rightarrow |E\rangle$ at the midfrequency $(\omega_1 + \omega_2)/2$ (blue arrows) become allowed as a result of the dipole-dipole coupling. For emitters coupled by the extreme near field in a bulk medium [52], the incoherent regime of such nonlinear transitions has been verified experimentally [53].

In contrast, we exploit the coherent regime of the transition far below saturation and find that both squeezed light and entanglement can be generated when Ω_{12}

becomes comparable to the emitters' decay rate γ and when the driving fields, characterized by the Rabi frequency Ω and the laser detuning $\Delta = \omega_L - (\omega_1 + \omega_2)/2$, are weak. The coherence, however, degrades in stronger fields, when the doubly excited state is populated via collective transitions ($\propto \Omega^4\Omega_{12}$) or—to a lesser extent—single-photon transitions ($\propto \Omega^4\Delta$) [52]. We show in the following that the nanostructure-assisted nonlinearity allows for the generation of entanglement and the collective emission of squeezed light, which exceeds the performance of uncoupled emitters [16].

2.3.2.3 Entanglement and Cooperative Quadrature Squeezing

We quantify bipartite entanglement in the following by nonzero values of the concurrence $C = \max[\{0, \lambda_1 - \lambda_2 - \lambda_3 - \lambda_4\}]$, where the λ_n^2 are the eigenvalues of the operator $\rho(\sigma_{y,1} \otimes \sigma_{y,2})\rho^\star(\sigma_{y,1} \otimes \sigma_{y,2})$ ordered by descending magnitude, where $\sigma_{y,m}$ denotes the y-component of the Pauli matrix of the mth emitter and * denotes the complex conjugate. Importantly, this magnitude ranges between 0 for fully separable states and 1 for maximally entangled states [54]. Figure 2.8 shows the steady-state concurrence (red curve) in direct comparison with the optimum degree of squeezing $(\Delta\mathcal{E})^2/2|g|^2$ (black curve), obtained by optimizing the phases Θ, ϕ_m, as a function of the laser detuning. The driving $\Omega^{(0)} = 60\gamma^{(0)}$ was chosen well below the two-photon saturation and we focus on a favorable symmetric detection configuration. A distinct minimum of the fluctuations arises close to the two-photon resonance $\Delta = 0$ and mirrors a maximum in the concurrence.

To provide additional insight into the main mechanism underlying the present scheme, we consider a simplified model for weak driving on the two-photon reso-

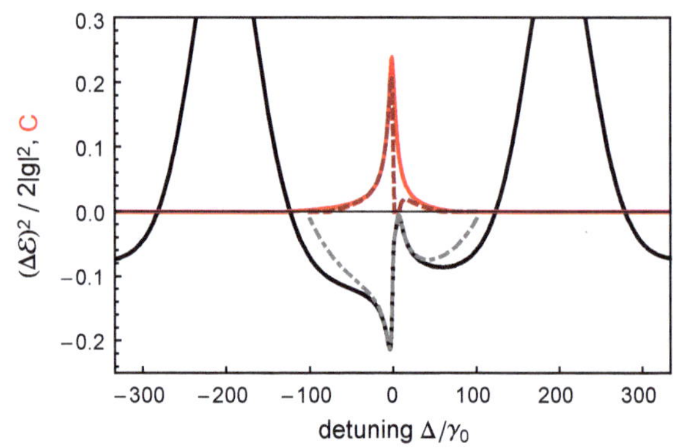

Fig. 2.8 Electric field fluctuations (*black curve*) optimized by the detection quadrature angle and steady state concurrence (*red curve*) versus laser detuning at $\Omega^{(0)} = 60\gamma^{(0)}$. Parameters as in Fig. 2.7. Negative values indicate squeezed light. The *dash-dotted* and *dashed lines* correspond to the two-level approximation on the $|E\rangle - |G\rangle$-transition for squeezing, and concurrence, respectively. Reprinted with permission from [16]. Copyright 2015 American Chemical Society

nance. An effective two-level picture arises, because the single-photon transitions are far detuned and the dipole-dipole coupling mainly induces a coherence ρ_{EG} between the fundamental and doubly excited state, so that the density matrix is dominated by this element and the populations. From these and (2.14) and (2.25) we find the normalized degree of squeezing

$$\frac{(\Delta\mathcal{E})^2}{2|g|^2} \approx 1 - \rho_{GG} + \rho_{EE} + \mathrm{Re}\left[e^{\mathrm{i}[\Theta+\phi_1+\phi_2]}\rho_{EG}\right], \tag{2.26}$$

which is optimized when the last term equals $-|\rho_{EG}|$ and it originates from the cross-correlation term (see (2.25)). The approximation (gray dash-dotted curve) agrees excellently with the full squeezing amplitude (black curve). Interestingly, (2.26) is closely related to the spin squeezing [55, 56] that reinforces the link to entanglement. A similar treatment of the concurrence near the resonance (red dashed line) is equally dominated by $|\rho_{EG}|$ [16]. Altogether this identifies the two-photon coherence as the source of both squeezing and entanglement.

For a full picture, Fig. 2.9a shows the steady-state concurrence as a function of the driving field detuning and amplitude for the previous configuration. Significant entanglement can be reached close to the two-photon resonance. As Ω is increased, the onset of saturation in the collective transition shifts this maximum in frequency

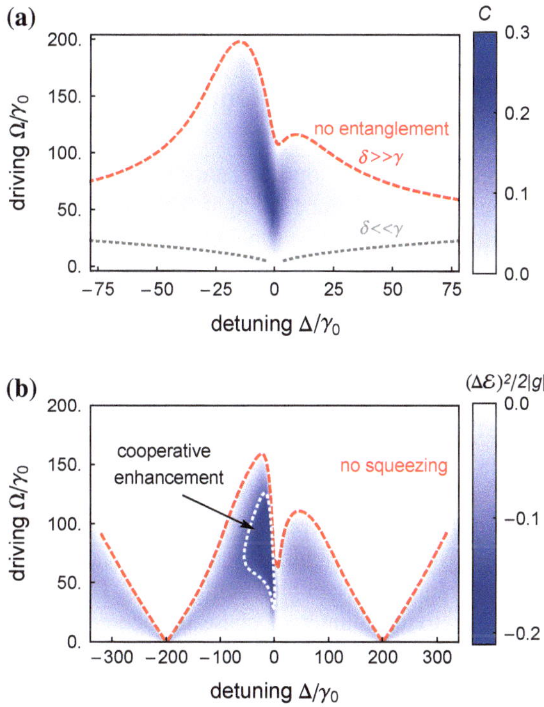

Fig. 2.9 **a** Steady state concurrence vs the driving field amplitude and detuning. Parameters as in Fig. 2.7. Bipartite entanglement is excluded in the *white area* for detuned emitters ($\delta > \Omega_{12}$) and is restricted to the region below the *grey dotted curve* for identical emitters ($\delta \ll \Omega_{12}$). **b** Electric field fluctuations vs the driving field detuning and amplitude. The detection quadrature angle is optimized. No squeezing can be observed above the *red dashed* limit. Inside the *white* contour, cooperative squeezing exceeds the limit feasible by two independent emitters. Reprinted with permission from [16]. Copyright 2015 American Chemical Society

(asymmetrically due to unequal populations of the entangled states $|A\rangle$ and $|S\rangle$), and finally suppresses entanglement. This behavior differs strongly from the case of identical emitters ($\delta \ll \Omega, \gamma$), for which single-emitter saturation suppresses the generation of entanglement at comparable driving Ω (in the region above the dotted gray curve in Fig. 2.9a). Moreover, the maximum concurrence $C \approx 0.3$ is comparable to typical values for identical emitters coupled by nanowaveguides via a single photon process [57], although concurrence values close to unity could be achieved by engineered nanostructures with negligible quenching and $\gamma_{12}/\gamma \approx 1$ [58]. These features bear striking similarity to the squeezing signal shown in Fig. 2.9b as a function of the detuning and Rabi frequency. At the chosen parameters, the strongest squeezing is encountered at $\Omega^{(0)} \approx 75\gamma^{(0)}$ at $\Delta \approx 0$. It closely follows the regime that allows for maximum bipartite entanglement in Fig. 2.9a and it is similarly suppressed by saturation of the two-photon transition as Ω is increased. As the collective interaction also affects the single-body terms, c.f. (2.14), the boundaries of squeezed light near $\Delta = \pm 200\gamma^{(0)}$ are deformed with respect to the symmetric hyperbolas of uncoupled emitters, c.f. (2.16) and Fig. 2.3. The strongest squeezing with values $(\Delta\mathcal{E})^2/(2|g|^2) \approx -0.21$ significantly exceeds the universal threshold $-1/8$ that limits the squeezed resonance fluorescence per independent emitter (2.15) and is indicated by the white contour in Fig. 2.9b. Thus, a strong collective enhancement of the squeezing per emitter [59] can be achieved despite their large detuning. This is a general result for pairs of TLEs coupled by means of strong dipole-dipole interactions and can be used to identify entanglement.

2.4 Conclusions and Outlook

We have shown how optical nanostructures can improve the conditions for generating quadrature-squeezed light from a TLE in terms of bandwidth and driving amplitude and how the near fields can control its direction and scattering amplitude. These prospects are promising for demonstrating the first generation of quadrature-squeezed light in nanostructured environments and thus creating novel integrated sources of nonclassical light. This proof would show the generation of nonclassical coherence, i.e. superpositions of few-photon excitations, that is inherently used in quantum nonlinear nanophotonics and thus to give evidence of the underlying mechanism that allows to outperform classical schemes [12]. In particular, we have shown that the squeezed light emission assisted by a nanostructure can generate entanglement between TLEs and at the same time be a witness of their nonclassical interaction. This system surpasses the limiting squeezing level of independent emitters and can overcome a significant amount of additional pure dephasing, even beyond the constraints on uncoupled emitters [13, 16]. As both nanophotonic enhancement and collectivity counteract phase decoherence—a fundamental limit in setups at nonzero temperatures—nanostructures are promising to scale up the operating temperatures of solid-state emitters. The results obtained from a GNS can be transferred to other broad-band environments such as optical antennas, subwavelength cavities, or nanowaveguides, to push further the limits for generating squeezed light and analyze its nonclassical wave behavior. Experimentally, these sources can be realized by promising emerging techniques based on the deterministic positioning of quantum dots close to nanophotonic structures [41] or by scanning a nanostructured probe over a spin-coated layer of emitters [53]. We expect that collective effects may be scaled up to higher numbers of emitters [60–62] or more emitter levels, a study of which can provide a better understanding of squeezing in structured nonlinear media and additional microscopic nonlinearities [1] that can provide reduced fluctuations of almost 90 % below the shot noise.

Acknowledgements Financial support from the Max Planck Society and the EC Seventh Framework Programme (284584) are gratefully acknowledged. The authors wish to thank the support of Vahid Sandoghdar in this work and to acknowledge helpful and stimulating discussions with Karim Murr.

References

1. D.F. Walls, Nature **306**, 141 (1983)
2. L. Novotny, B. Hecht, *Principles of Nano-Optics* (Cambridge University Press, Cambridge, UK, 2006)
3. M. Agio, A. Alù (eds.), *Optical Antennas* (Cambridge University Press, Cambridge, UK, 2013)
4. M. Tame, K. McEnery, Ş. Özdemir, J. Lee, S. Maier, M. Kim, Nat. Phys. **9**, 329 (2013)
5. P. Lodahl, S. Mahmoodian, S. Stobbe, Rev. Mod. Phys. **87**, 347 (2015)
6. D.F. Walls, P. Zoller, Phys. Rev. Lett. **47**, 709 (1981)
7. K. Wodkiewicz, P. Knight, S. Buckle, S. Barnett, Phys. Rev. A **35**, 2567 (1987)
8. C.M. Caves, Phys. Rev. D **23**, 1693 (1981)
9. R.C. Pooser, B. Lawrie, ACS Photon. **3**, 8 (2016)
10. M.A. Taylor, W.P. Bowen, Phys. Rep. **615**, 1 (2016)
11. A. Lvovsky, *Photonics, Volume 1: Fundamentals of Photonics and Physics*, chap. Squeezed light, (John Wiley & Sons, 2015), pp. 121–164
12. D.E. Chang, V. Vuletić, M.D. Lukin, Nat. Photon. **8**, 685 (2014)
13. D. Martín-Cano, H.R. Haakh, K. Murr, M. Agio, Phys. Rev. Lett. **113**, 263605 (2014)
14. D. Martín-Cano, H.R. Haakh, M. Agio, J. Opt. **18**, 024010 (2016)
15. M. Bremner, C. Dawson, J. Dodd, A. Gilchrist, A. Harrow, D. Mortimer, M. Nielsen, T. Osborne, Phys. Rev. Lett. **89**, 247902 (2002)
16. H.R. Haakh, D. Martín-Cano, ACS Photon. **2**, 1686 (2015)
17. W. Vogel, D.G. Welsch, *Quantum Optics* (WILEY-VCH Verlag, Weinheim, 2006)
18. A. Ourjoumtsev, A. Kubanek, M. Koch, C. Sames, P.W. Pinkse, G. Rempe, K. Murr, Nature **474**, 623 (2011)
19. P. Kukura, M. Celebrano, A. Renn, V. Sandoghdar, J. Phys. Chem. Lett. **1**, 3323 (2010)
20. W. Vogel, Phys. Rev. Lett. **67**, 2450 (1991)
21. W. Vogel, Phys. Rev. A **51**, 4160 (1995)
22. C.H. Schulte, J. Hansom, A.E. Jones, C. Matthiesen, C.L. Gall, M. Atature, Nature **525**, 222 (2015)
23. U.L. Andersen, T. Gehring, C. Marquardt, G. Leuchs (2015). arXiv:1511.03250
24. R. Slusher, L. Hollberg, B. Yurke, J. Mertz, J. Valley, Phys. Rev. Lett. **55**, 2409 (1985)
25. C. McCormick, A.M. Marino, V. Boyer, P.D. Lett, Phys. Rev. A **78**, 043816 (2008)
26. L.A. Wu, H. Kimble, J. Hall, H. Wu, Phys. Rev. Lett. **57**, 2520 (1986)

27. T. Eberle, S. Steinlechner, J. Bauchrowitz, V. Händchen, H. Vahlbruch, M. Mehmet, H. Müller-Ebhardt, R. Schnabel, Phys. Rev. Lett. **104**, 251102 (2010)
28. P. Grünwald, W. Vogel, Phys. Rev. Lett. **109**, 013601 (2012)
29. L. Knöll, S. Scheel, D.G. Welsch, *Coherence and Statistics of Photons and Atoms*, chap. QED in dispersing and absorbing dielectric media, (Wiley, New York, 2001), pp. 1–60
30. D. Dzsotjan, J. Kästel, M. Fleischhauer, Phys. Rev. B **84**, 075419 (2011)
31. H.T. Dung, L. Knöll, D.G. Welsch, Phys. Rev. A **65**, 043813 (2002)
32. L.W. Li, P.S. Kooi, M.S. Leong, T.S. Yee, I.E.E.E. Trans, Microw. Theory Tech. **42**, 2302 (1994)
33. H.T. Dung, L. Knöll, D.G. Welsch, Phys. Rev. A **64**, 013804 (2001)
34. N.M. Mojarad, V. Sandoghdar, M. Agio, J. Opt. Soc. Am. B **25**, 651 (2008)
35. X.W. Chen, V. Sandoghdar, M. Agio, Phys. Rev. Lett. **110**, 153605 (2013)
36. A. Batalov, C. Zierl, T. Gaebel, P. Neumann, I.Y. Chan, G. Balasubramanian, P.R. Hemmer, F. Jelezko, J. Wrachtrup, Phys. Rev. Lett. **100**, 077401 (2008)
37. L. Novotny, N. van Hulst, Nat. Photon. **5**, 83 (2011)
38. D.R. Lide (ed.), *CRC Handbook of Chemistry and Physics*, 87th edn. (CRC Press, Boca Raton, FL, 2006)
39. F. Kaminski, V. Sandoghdar, M. Agio, J. Comput. Theor. Nanosci. **4**, 635 (2007)
40. C.F. Bohren, Am. J. Phys. **51**, 323 (1983)
41. A.G. Curto, G. Volpe, T.H. Taminiau, M.P. Kreuzer, R. Quidant, N.F. van Hulst, Science **329**, 930 (2010)
42. M. Collett, D. Walls, P. Zoller, Opt. Commun. **52**, 145 (1984)
43. Z. Ou, C. Hong, L. Mandel, J. Opt. Soc. Am. B **4**, 1574 (1987)
44. L. Knöll, W. Vogel, D.G. Welsch, Phys. Rev. A **42**, 503 (1990)
45. L. Rogobete, F. Kaminski, M. Agio, V. Sandoghdar, Opt. Lett. **32**, 1623 (2007)
46. A. Huck, S. Smolka, P. Lodahl, A.S. Sorensen, A. Boltasseva, J. Janousek, U.L. Andersen, Phys. Rev. Lett. **102**, 246802 (2009)
47. Z. Ficek, R. Tanas, Phys. Rep. **372**, 369 (2002)
48. H.T. Dung, L. Knöll, D.G. Welsch, Phys. Rev. A **66**, 063810 (2002)
49. R. Tanas, Z. Ficek, in ed. by H. Moya-Cessa, R. Jauregui, S. Hacyan, O. Castanos, *Proceedings of the 8th International Conference on Squeezed States and Uncertainty Relations*, (Rinton Press, Princeton, 2003), pp. 416–423
50. R. Tanaś, Z. Ficek, Fortschr. Phys. **51**, 230 (2003)
51. H. Saito, M. Ueda, Phys. Rev. Lett. **79**, 3869 (1997)
52. G.V. Varada, G.S. Agarwal, Phys. Rev. A **45**, 6721 (1992)
53. C. Hettich, C. Schmitt, J. Zitzmann, S. Kühn, I. Gerhardt, V. Sandoghdar, Science **298**, 385 (2002)
54. W.K. Wootters, Phys. Rev. Lett. **80**, 2245 (1998)
55. L. Pezzé, A. Smerzi, Phys. Rev. Lett. **102**, 100401 (2009)
56. K. Hammerer, A.S. Sørensen, E.S. Polzik, Rev. Mod. Phys. **82**, 1041 (2010)
57. A. González-Tudela, D. Martin-Cano, E. Moreno, L. Martin-Moreno, C. Tejedor, F.J. Garcia-Vidal, Phys. Rev. Lett. **106**, 020501 (2011)
58. J. Hou, K. Słowik, F. Lederer, C. Rockstuhl, Phys. Rev. B **89**, 235413 (2014)
59. Z. Ficek, R. Tanas, Quantum optics. J. Eur. Opt. Soc. B **6**, 95 (1994)
60. M. Gullans, T.G. Tiecke, D.E. Chang, J. Feist, J.D. Thompson, J.I. Cirac, P. Zoller, M.D. Lukin, Phys. Rev. Lett. **109**, 235309 (2012)
61. A. González-Tudela, D. Porras, Phys. Rev. Lett. **110**, 080502 (2013)
62. A. Delga, J. Feist, J. Bravo-Abad, F. Garcia-Vidal, Phys. Rev. Lett. **112**, 253601 (2014)

Chapter 3
Coupling of Quantum Emitters to Plasmonic Nanoguides

Shailesh Kumar and Sergey I. Bozhevolnyi

Abstract Surface plasmon-polaritons represent electromagnetic waves that propagate along metal dielectric interfaces and can be guided by metallic nanostructures, plasmonic waveguides, beyond the diffraction limit. This remarkable capability has unique prospects for the design of highly integrated photonic signal-processing systems. Due to strong mode confinement, quantum emitters located in plasmonic waveguides feature enhanced emission rates, with the emission being channeled into the waveguide modes. In this chapter, we consider different plasmonic waveguides and emitters that have been investigated for the realization of their efficient coupling. First, we discuss mode properties of relevant plasmonic waveguides, followed by an overview of coupling theory that reveals important system parameters affecting the coupling characteristics. We then describe quantum emitters that have been used in coupling experiments and found promising for quantum plasmonics. Finally, several experimentally realized configurations with quantum emitters efficiently coupled to plasmonic waveguides are reviewed.

3.1 Introduction

Nanostructured metal-dielectric configurations can support propagating electromagnetic modes representing electromagnetic excitations (in dielectrics) coupled to free electron oscillations (in metals), i.e., surface plasmon-polariton modes. There exist various configurations of plasmonic waveguides that feature a unique property of supporting strongly confined modes, far below the diffraction limit [1]. This remarkable capability opens exciting prospects for the design of highly integrated photonic

S. Kumar (✉) · S.I. Bozhevolnyi
Centre for Nano Optics,
University of Southern Denmark, 5230 Odense M, Denmark
e-mail: shku@iti.sdu.dk

S.I. Bozhevolnyi
e-mail: seib@iti.sdu.dk

S.I. Bozhevolnyi et al. (eds.), *Quantum Plasmonics*,
Springer Series in Solid-State Sciences 185,
DOI 10.1007/978-3-319-45820-5_3

signal-processing systems. It also enables the enhancement and channeling of emission from a quantum emitter into the waveguide [2]. Realization of configurations with quantum emitters being efficiently coupled to waveguides is very important from the perspective of quantum computing. Many of the protocols for quantum information processing rely on high photon flux and emission of radiated photons into single spatial optical modes [3, 4]. A single emitter efficiently coupled to a single optical mode can yield single photons on demand [2], or allow for non-linear gate operations at the level of a few photons [3]. Single photons are also essential for the realization of various protocols in quantum cryptography. Bright single photon sources will increase the communication rate and will make quantum cryptography more practical [5].

Single photon emitters can be coupled to different kinds of dielectric or plasmonic structures to increase their decay rates in order to obtain high rate of single photons. Dielectric structures such as photonic crystal cavity [6, 7], photonic crystal waveguide [8] or waveguides with high refractive index dielectric material have been utilized for increasing the decay rate and channeling the emission from quantum emitters [9, 10]. The confinement that can be achieved in a dielectric material is ultimately limited by the diffraction, whereas plasmonic structures allow for confinement beyond the diffraction limit [1]. Deep subwavelength mode confinement available with plasmonic waveguides allows the decay rate of a quantum emitter coupled to the correspondingly designed waveguides to become extremely high. In passing, we note that spatially confined plasmonic nanostructures can support localized surface plasmons, which also enhance the decay rate of quantum emitters and can act as an antenna as well [11–15]. Plasmonic waveguides, on the other hand, not only enhance the decay rate of the quantum emitter coupled to the waveguide but also channel its emission into the propagating waveguide modes [16–20].

In this chapter, we focus on coupling of quantum emitters to plasmonic waveguides. In Sect. 3.2, we discuss different plasmonic waveguides and the theory of coupling between those waveguides and quantum emitters. In Sect. 3.3, we first present different quantum emitters that are found useful for coupling to plasmonic waveguides, and then we review the coupled emitter-waveguide systems that have been realized experimentally.

3.2 Theory of Coupling an Emitter to a Plasmonic Waveguide

As briefly mentioned in the introduction, the coupling between a quantum emitter and a waveguide depends on the properties of the waveguide and that of the emitter. The total decay rate depends on the dipole moment of the emitter, the position and orientation of the emitter in the waveguide cross-section and the confinement of the waveguide mode. Properties of different quantum emitters will be discussed in Sect. 3.3. In this section, we first discuss modes of different plasmonic waveguide structures and then the coupling of a quantum emitter to those plasmonic waveguides.

3.2.1 Modes in Plasmonic Waveguides

Plasmonic waveguides support waveguide modes beyond diffraction limit. This is due to the presence of free electrons in metals which makes the real part of permittivity in metals negative, that is, $\Re(\varepsilon) < 0$. Modes of a waveguide can be obtained by solving the wave equation with appropriate boundary conditions. The wave equation for electric wave is given by,

$$\nabla \times \frac{1}{\mu_{\mathbf{r}}} (\nabla \times \mathbf{E}(\mathbf{r})) - k_0^2 \varepsilon(\mathbf{r}) \mathbf{E}(\mathbf{r}) = 0, \tag{3.1}$$

where, $\mathbf{r} \equiv (x, y, z)$, $k_0 = \omega\sqrt{\varepsilon_0 \mu_0}$ is the vacuum wavenumber, ω is the angular wave frequency, and ε_0 and μ_0 are permittivity and permeability of vacuum, respectively. $\varepsilon(\mathbf{r})$ is dielectric function relative to vacuum, and $\mu_{\mathbf{r}}$ denotes the relative permeability constant. $\mu_{\mathbf{r}}$ will be assumed 1 for all calculations involved in this chapter. For infinitely long waveguides with propagation direction along z-axis, the solution to the wave equation must have the following form:

$$\mathbf{E}(x, y, z) = \mathbf{E}_\alpha(x, y) e^{-i(\omega t - kz)} \tag{3.2}$$

For any guided mode in a waveguide (plasmonic or dielectric), at a specific angular frequency ω, α and k define a complete set of orthogonal modes.

For simple structures such as a silver wire with circular cross-section, analytical methods can be used to solve the Maxwell's equations [21]. However, for more complicated structures, one needs to use numerical methods. Guided plasmonic modes can be calculated using a finite element method (FEM). We use commercial software (COMSOL multiphysics) for calculating the propagating plasmonic modes with FEM. A vacuum wavelength of 700 nm is chosen for our calculations, because many of the quantum emitters emit around this wavelength. For example, *NV* center has a peak around 700 nm at room temperature, *SiV* center emits around 738 nm, quantum dots can also be found to be emitting around that wavelength. The electric permittivity of silver at 700 nm, i.e. $\varepsilon_{Ag} = -20.437 - 1.284i$, is used [22]. For fused silica substrate and poly-(methamethacrylate) (PMMA), a refractive index of 1.46 and 1.5, respectively, are used. In Fig. 3.1, we present fundamental modes for different plasmonic waveguides. Figure 3.1a shows the mode supported by a silver nanowire of radius 50 nm. The mode is symmetric, and the field decays exponentially with the distance from the surface of the silver wire. For two parallel wires the

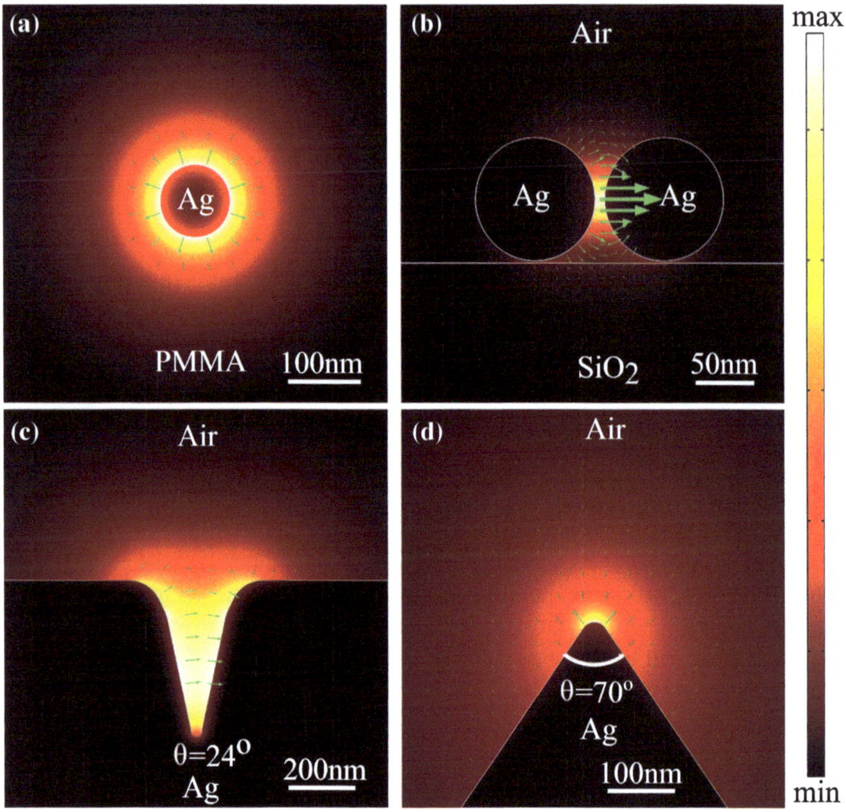

Fig. 3.1 Electric field distribution for the fundamental mode of **a** a silver wire of radius 50 nm, **b** two parallel wires, each of radius 55 nm and a gap of 9 nm, **c** a V-groove of angle 24° and **d** a wedge of angle 70°. The tip radius for V-groove and the wedge are 15 nm and 20 nm, respectively. The *arrows* denote the direction of the electric field

field is confined in the gap between the two nanowires (Fig. 3.1b). V-groove supports the mode that is confined within it (Fig. 3.1c). The wedge structure supports a mode that is confined near the tip of the wedge (Fig. 3.1d).

Figure 3.2a shows the effective mode index for the fundamental modes supported by the structures presented in Fig. 3.1, as a function of wavelength. In Fig. 3.2b, we show propagation lengths, defined as the distance for the mode intensity to decrease by a factor of 1/e, for the same structures. Propagation lengths for the same structures where silver is replaced by gold are smaller in this wavelength range. However, silver tends to corrode, which has been the reason why some of the experimentally demonstrated structures are fabricated of gold [19].

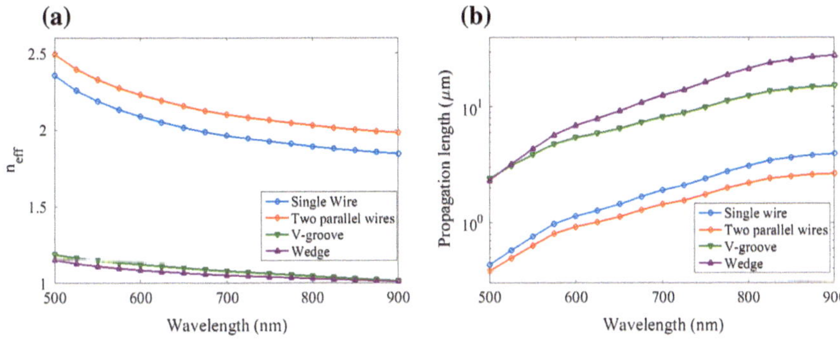

Fig. 3.2 **a** Effective mode index and **b** propagation lengths for fundamental modes supported by silver structures presented in Fig. 3.1, as indicated in the graphs

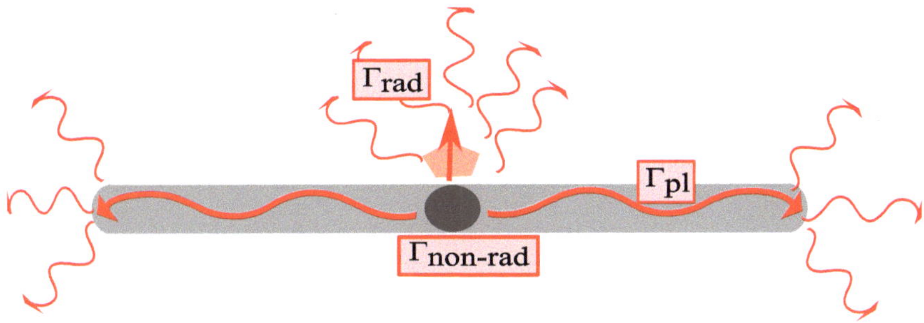

Fig. 3.3 Schematics showing three channels of decay when a quantum emitter is placed near a plasmonic waveguide: radiative decay (Γ_{rad}), decay into plasmonic mode (Γ_{pl}) and non-radiative decay ($\Gamma_{non-rad}$)

3.2.2 Theory of Coupling

For the emitter near a plasmonic waveguide, there are three decay channels. First channel is the emission to the radiative mode, second channel is emission into the waveguide mode and third channel is non-radiative decay. This is shown schematically in Fig. 3.3.

When an emitter is coupled to a waveguide, the useful channel out of the three channels is decay into the plasmonic waveguide. So, a parameter of coupling, called β-factor, is defined as the ratio of decay rate into the plasmonic mode to the total decay rate of the emitter. In the following, we first present a way to calculate the decay rate into the plasmonic mode, and subsequently, we discuss total decay rate and β-factor.

3.2.2.1 Decay Rate of an Emitter into a Plasmonic Mode

The electric field dyadic Green's function for a specific guided plasmonic mode can be constructed from the numerical calculation of the plasmonic mode's electric and magnetic fields, which can subsequently be used for calculation of projected local density of states (LDOS) for the plasmonic mode [23, 24]. The electric dyadic Green's function $\bar{\bar{\mathbf{G}}}(\mathbf{r}, \mathbf{r}', \omega)$ is defined as follows,

$$\left[\nabla \times \nabla \times -k_0^2 \varepsilon (\mathbf{r})\right] \bar{\bar{\mathbf{G}}}(\mathbf{r}, \mathbf{r}', \omega) = \bar{\bar{\mathbf{I}}}\delta(\mathbf{r} - \mathbf{r}'), \tag{3.3}$$

where $\bar{\bar{\mathbf{I}}}$ is a unit dyad, and δ is the dirac delta function. Projected LDOS for the plasmonic mode is given by,

$$\rho_{pl}(\mathbf{r}, \omega) = \frac{6\omega \left[\bar{\mathbf{n}}_D \cdot \Im\{\bar{\bar{\mathbf{G}}}\left(\mathbf{r}, \mathbf{r}', \omega\right)\} \cdot \bar{\mathbf{n}}_D\right]}{\pi c^2}, \tag{3.4}$$

where $\bar{\mathbf{n}}_D$ is a unit dipole at position $\mathbf{r}$, and c is the velocity of light in vacuum. The LDOS for the plasmonic mode in terms of electric and magnetic fields is given by,

$$\rho_{pl}(\mathbf{r}, \omega) = \frac{6\left|\mathbf{E}\left(x, y\right) \cdot \bar{\mathbf{n}}_D\right|^2}{2\pi \Re\{\int_{A_\infty} (\mathbf{E} \times \mathbf{H}^*) \cdot \bar{\mathbf{z}} dA\}} = \frac{6\left|\mathbf{E}\left(x, y\right) \cdot \bar{\mathbf{n}}_D\right|^2}{N v_g}, \tag{3.5}$$

where,

$$v_g = \frac{\Re\{\int_{A_\infty} (\mathbf{E} \times \mathbf{H}^*) \cdot \bar{\mathbf{z}} dA\}}{\int_{A_\infty} \varepsilon_0 \varepsilon(x, y) \left|\mathbf{E}\left(x, y\right)\right|^2 dA} \tag{3.6}$$

and

$$N = 2\pi \int_{A_\infty} \varepsilon_0 \varepsilon(x, y) \left|\mathbf{E}\left(x, y\right)\right|^2 dA \tag{3.7}$$

are group velocity and normalization factor, respectively. $\bar{\mathbf{z}}$ is a unit vector along +z direction. The integration is over the entire transverse plane to +z direction. All other symbols have the same meaning as in Sect. 3.2.1.

From 3.5, it is clear that to maximize LDOS, v_g should be small, the mode should be confined (for higher E) and the dipole orientation of the emitter should be parallel to the electric field. The decay rate of an emitter in terms of LDOS is given by,

$$\Gamma_{pl}\left(\mathbf{r}, \omega\right) = \frac{\pi \omega_0}{3\hbar \varepsilon_0} \left|\mu_D\right|^2 \rho_{pl}(\mathbf{r}, \omega) \tag{3.8}$$

where μ_D is the dipole moment of the emitter and $\hbar$ is Planck's constant divided by 2π. Thus,

$$\frac{\Gamma_{pl}(\mathbf{r},\omega)}{\Gamma_0} = \frac{6\pi^2 c^3 \left|\mathbf{E}(x,y)\cdot\bar{\mathbf{n}}_D\right|^2}{\omega^2 N v_g} \tag{3.9}$$

where,

$$\Gamma_0 = \frac{\omega^3 \left|\mu_D\right|^2}{3\pi\hbar\varepsilon_0 c^3} \tag{3.10}$$

is the decay rate of the emitter in vacuum. In terms of electric and magnetic fields, the decay rate of an emitter, normalized to its emission in vacuum, is given by,

$$\frac{\Gamma_{pl}}{\Gamma_0} = \frac{3\pi c\varepsilon_0 \left|\mathbf{E}(x,y)\cdot\bar{\mathbf{n}}_D\right|^2}{\Re\{k_0^2 \int_{A_\infty} (\mathbf{E}\times\mathbf{H}^*)\cdot\bar{\mathbf{z}}dA\}} \tag{3.11}$$

In Fig. 3.4, we present the decay rate to the plasmonic mode normalized to the decay rate in vacuum according to the formula above for the waveguide structures

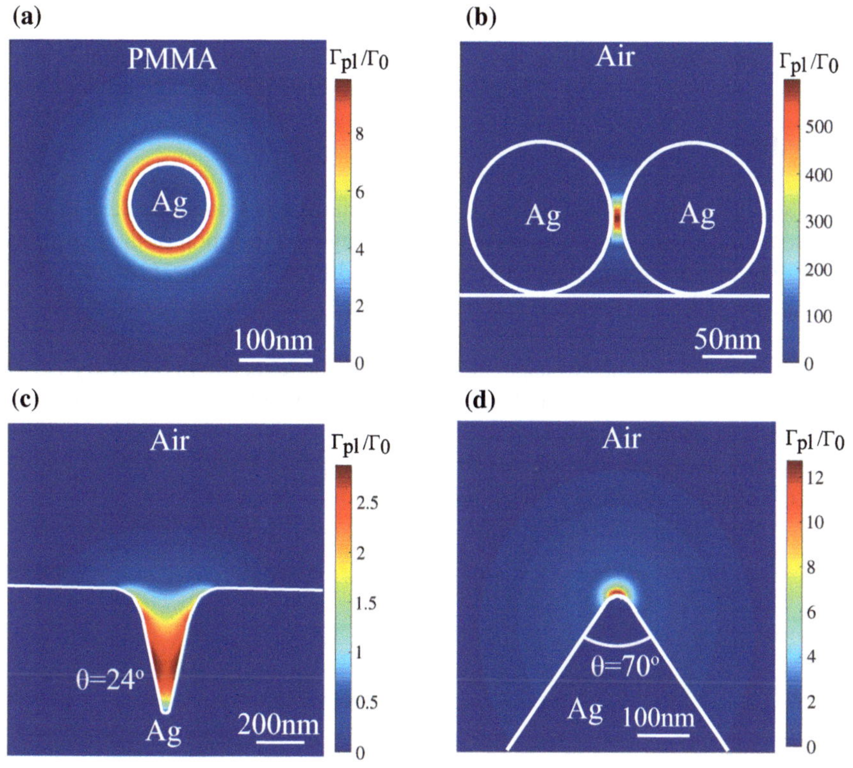

Fig. 3.4 Rate of decay into the plasmonic mode normalized to the decay rate in vacuum for **a** a silver wire in PMMA, **b** two parallel wires, **c** a V-groove and **d** a wedge structure. The dimensions of the structures are the same as in Fig. 3.1

presented in Fig. 3.1. The decay rates are optimized with respect to the emitter orientation, that is, the dipole is assumed to be parallel to the electric field at each point in the cross-section. It is clear from Fig. 3.4 that the more confined the mode is, the higher the decay rate into the plasmonic mode. Also, the distributions of the decay rates clearly suggest that the point with higher field results in higher decay rate into the plasmonic mode.

3.2.2.2 Total Decay Rate of an Emitter

We have briefly described the method followed for calculating the decay rate into the plasmonic mode by calculating the Green's dyadic function. This is possible due to confinement of the plasmonic modes. A similar approach to calculate the total decay rate due to the non-confined radiation modes cannot be followed. To calculate the total decay rate, one needs to take into account the non-radiative and radiative channels of decay, in addition to the plasmonic channel of decay. A 3D model is needed to include the radiation modes as well as the non-radiative contributions. The following equation with a harmonic source (time dependent current source) term is solved to calculate the total decay rate:

$$\left[\nabla \times \nabla \times -k_0^2 \varepsilon(\mathbf{r})\right] \mathbf{E}(\mathbf{r}, \omega) - i\omega\mu_0 \mathbf{J}(\omega) = 0 \tag{3.12}$$

where $\mathbf{J}(\omega)$ is the current density.

Above equation is solved by formulating an appropriate functional, which can be solved with FEM using COMSOL. Proper truncation of computational domain is crucial. In the direction transverse to the propagation direction of the plasmonic mode, the computational domain is truncated with perfectly matched layers with a thickness of half a wavelength in vacuum. Along the propagation direction, the computational domain is terminated by a mode matching boundary condition, which behaves as a sink for electromagnetic waves. The mode matching boundary condition is obtained from the mode field calculation for the plasmonic mode in 2D, as described before in Sect. 3.2.1.

The total decay rate is extracted from the total power dissipation of the current source coupled to the nearby metallic waveguide as follows,

$$\frac{\Gamma_{total}}{\Gamma_0} = \frac{P_{total}}{P_0}, \tag{3.13}$$

where

$$P_{total} = \frac{1}{2} \int\int\int \Re\left(\mathbf{J}^* \cdot \mathbf{E}\right) dV \tag{3.14}$$

and

$$P_0 = \frac{1}{2} \int\int\int \Re\left(\mathbf{J}^* \cdot \mathbf{E}_0\right) dV \tag{3.15}$$

are power dissipated by the current source near the waveguide and in the vacuum, respectively. Above method can also be used to compare the decay rates of an emitter in two different environments.

3.2.2.3 β-Factor

β-factor, as defined before in Sect. 3.2.1, can be calculated from the decay rate into the plasmonic mode and total decay rate, and is given as,

$$\beta = \frac{\Gamma_{pl}}{\Gamma_{total}}. \tag{3.16}$$

In Fig. 3.5, we present β-factor for two parallel wire structure presented in Fig. 3.1b, which supports a very confined mode. It is clear that such a structure can be used to obtain β-factor around 0.9. It should also be noted that β-factor is not maximum where Γ_{pl} is highest. Γ_{pl} is highest near the metal-dielectric interface, but the nonradiative decay also increases near to the metal surface. Therefore, the emitter should be placed at an optimum position in order to maximize the β-factor. This is an experimental challenge towards which many research groups have been working and continue to work. In the next section, we give an overview of experimental demonstrations of coupling a single photon emitter to different plasmonic waveguides.

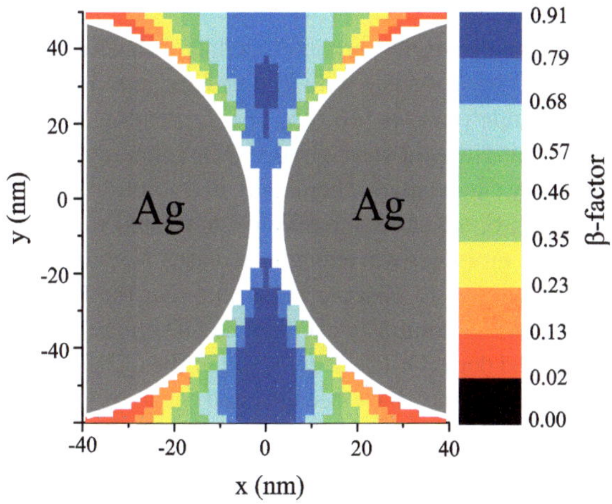

Fig. 3.5 β-factor for the structure presented in Fig. 3.1b

3.3 Experimental Demonstrations of Coupling a Quantum Emitter to Plasmonic Nanoguides

From the previous section, we know that if a quantum emitter is placed at an optimum position, the emission can be channeled into the waveguides efficiently. In this section, we present the experimental results obtained by coupling different kinds of emitters to some of the structures presented in Sect. 3.2. First, we describe different quantum emitters in Sect. 3.3.1. In Sect. 3.3.2, we present an overview of various coupled systems.

3.3.1 Quantum Emitters

There are many sources of single photons, for example, single atoms, ions, molecules, quantum dots and defect centers in diamonds. Atoms and ions require elaborate experimental set-up to trap them. Molecules do not need to be trapped, but they tend to bleach fast. Quantum dots (QDs) and defect centers in diamonds do not need to be trapped, and they are also, relatively, long lived. We briefly describe QDs and two kinds of defect centers in diamonds, namely, nitrogen vacancy (*NV*) center and silicon-vacancy (*SiV*) center, which can be useful for coupling to propagating modes in a plasmonic waveguide.

3.3.1.1 Quantum Dots

Quantum dots (QDs) are semiconductor structures confined in three dimensions surrounded by materials with higher bandgap. They act as "artificial atoms" and can emit single photons [16, 25, 26]. Depending on the size and structure of the QD, they emit different wavelengths. QDs can be synthesized in different ways, for example, they can be made chemically in a solution, which are called colloidal QDs [26]. In Fig. 3.6, we show measurements related to CdSe/CdS QDs. It can be observed that the emission wavelength increases as the size of QDs increases (Fig. 3.6b–e). Another point to observe is the fact that the QD shows blinking behavior, that is, when it is continuously excited, the QD emission switches randomly between bright (ON) and dark (OFF) states. Some of the QDs also bleach out, that is, they stop fluorescing after they have been excited for some time duration. Blinking and bleaching are two undesirable properties of these QDs and research groups continue to work to solve these shortcomings [26]. One other method of fabricating QDs is molecular beam epitaxy where they are formed as islands due to the lattice mismatch between the substrate and the QD materials, these are called Stranski-Krastanov QDs [25]. Lithography in combination with molecular beam epitaxy is also used for fabrication of QDs. In this chapter, we present the coupling of colloidal QDs to plasmonic waveguides as they can be placed, relatively easily, within the mode area of plasmonic waveguides, compared to other QDs [16, 20].

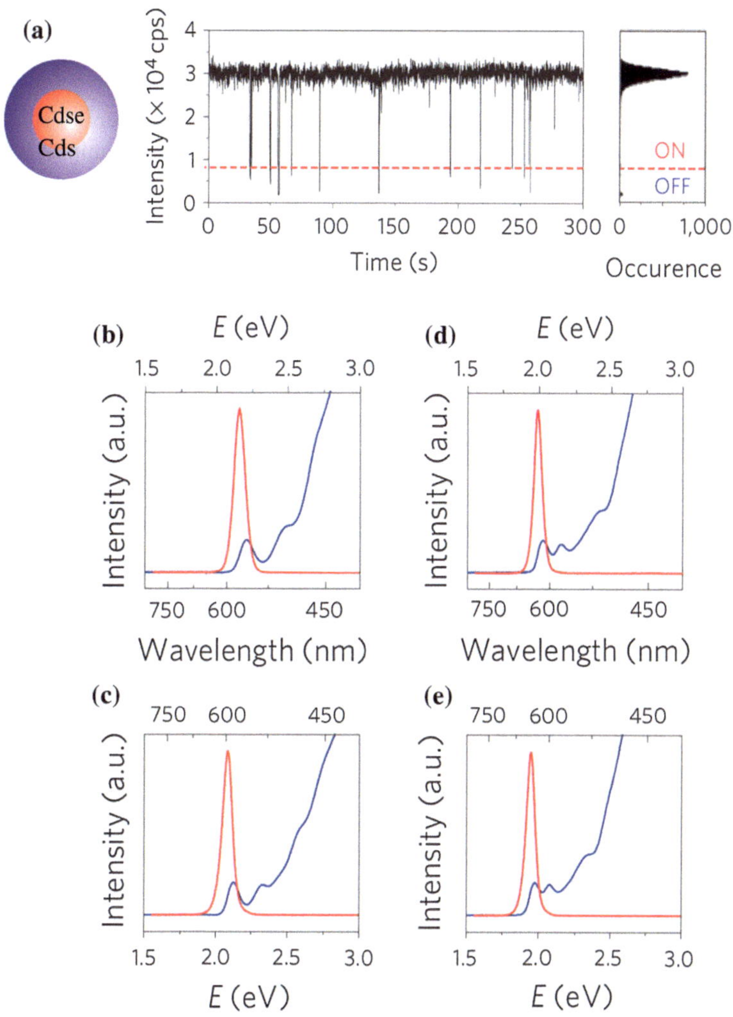

Fig. 3.6 **a** A schematic of a quantum dot showing CdSe core and CdS shell. Representative photoluminescence blinking for a quantum dot of core diameter of 2.2 nm and a shell thickness of 2.4 nm is also shown. **b–e** Absorption (*blue curves*) and photoluminescence (*red curves*) spectra for four different core diameters of **b** 2.7 nm, **c** 3.4 nm, **d** 4.4 nm and **e** 5.4 nm, where shell thickness is 2.4 nm (Figure is adapted by permission from Macmillan Publishers Ltd: [Nat. Mater.] [26], © 2013)

3.3.1.2 Nitrogen Vacancy Center

There are many kinds of defect centers in diamond, one of them is nitrogen vacancy (*NV*) center. As the name suggests, it consists of a substitutional nitrogen atom and a vacancy adjacent to the nitrogen atom. In Fig. 3.7a, we present the atomic structure of an *NV* center. *NV* center can be in two charged states, it can be neutral (called NV^0

center) or it can have a single negative charge (called NV^- center). A simplified electronic structure for NV^0 and NV^- centers is presented in Fig. 3.7b and c, respectively. NV^0 and NV^- centers have their zero phonon lines (ZPLs) at 575 nm (2.156 eV) and 637 nm (1.945 eV), respectively. The ground state of the NV^- center is a spin triplet state and the coherence time of this state is long so that many manipulations and readout can be performed before it decoheres. The electronic spin state can be initialized as well as read out optically. All of this is possible at room temperature for the NV^- center [27]. Due to the stability, and the properties of its electronic structure, it has been used the most, among defect centers in diamonds, in quantum information related experiments so far [28–33]. In addition, it has been used for ultra-sensitive magnetic field, electric field and temperature measurements [34–37].

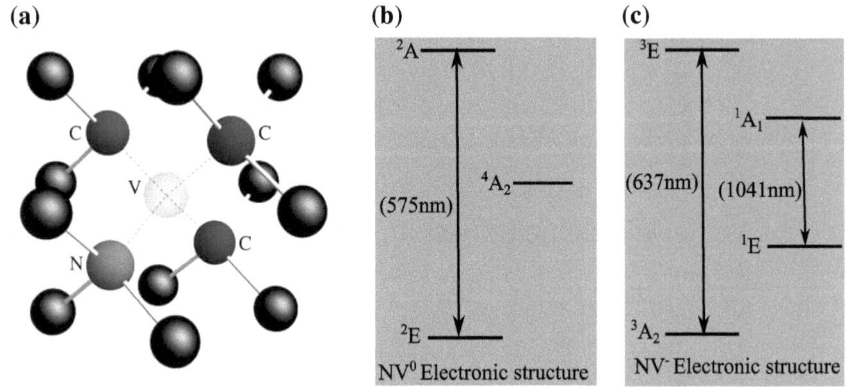

Fig. 3.7 **a** Atomic structure of an NV-center in diamond. N denotes nitrogen atom, V denotes vacancy and the 3 Cs denote the carbon atoms nearest to the vacancy. **b** and **c** electronic structure of NV^0 and NV^- centers, respectively. The relative energies of the doublet (^{2}A, ^{2}E) and quartet states (^{4}A$_2$) of NV^0 and the triplet (^{3}E, ^{3}A$_2$) and singlet states (^{1}A$_1$, ^{1}E) of NV^- are currently unknown (Figure is adapted with permission from Elsevier, [27] © 2013)

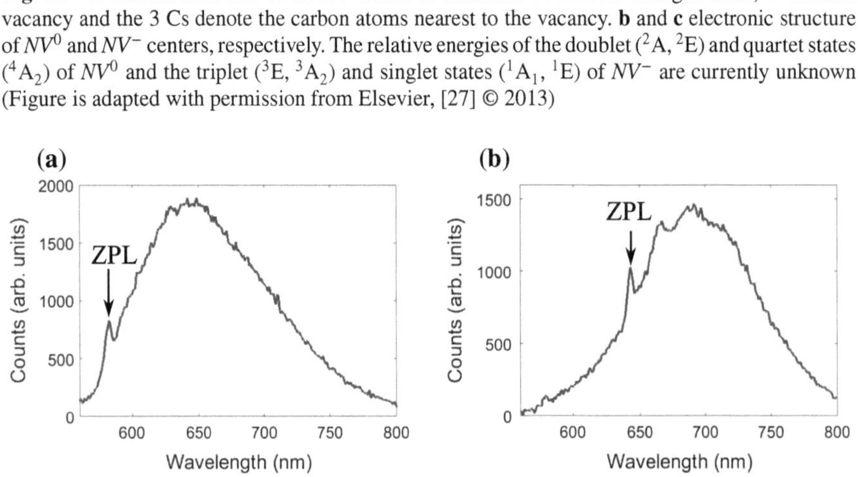

Fig. 3.8 Spectra of emission from **a** a single NV^0 and **b** a single NV^- center. Zero phonon lines (ZPL) are indicated with *arrows*

NV centers in nanodiamonds are also a stable single photon emitter at room temperature. However, NV^0 and NV^- centers have a drawback that the emission from *NV* centers is spread over more than 100 nm in wavelength due to interaction of the optical transition with phonons. In Fig. 3.8, we present the spectrum of the NV^0 and NV^- centers at room temperature when excited by a 532 nm laser. Lifetime of the excited state of NV^- center in bulk diamond is ~11.6 ns. In nanodiamonds, the lifetime varies a lot. It depends on the optical density of states at the site of the emitter. The quantum efficiency of an *NV* center in a nanodiamond can also be lower than that of an *NV* center inside a bulk diamond [38].

3.3.1.3 Silicon Vacancy Center

Silicon vacancy (*SiV*) center is also a defect center in diamond which forms when two adjacent carbon atoms are missing and a silicon atom is situated in between the two vacancies [39]. The atomic structure of a *SiV* center is presented in Fig. 3.9a. *SiV* centers can also be in two charged states, similar to *NV* centers, singly negatively charged (SiV^-) and neutral (SiV^0). The zero phonon line (ZPL) of SiV^- and SiV^0 centers are at 737 nm (1.68 eV) and 946 nm (1.31 eV), respectively [41]. SiV^0 center has not been demonstrated as a single photon source yet. In the following, we only discuss the SiV^- center. Figure 3.9b shows the electronic structure of an SiV^- center. In Fig. 3.10, we present spectra of a single SiV^- center, when it is excited by a 671 nm laser, taken at different temperatures. From the spectra, it is clear that as the temperature is decreased more and more emission from the SiV^- center goes into the zero phonon line [40]. SiV^- centers have the advantage over NV^- centers that more than 70 % of the emission from an SiV^- center can be into the zero phonon line. SiV^- emits single photons, and it has been shown that emission from different

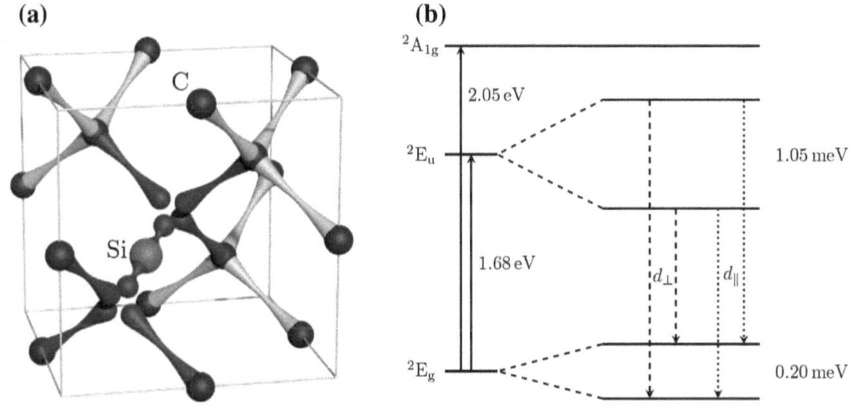

Fig. 3.9 **a** Atomic structure of a *SiV* center. **b** Electronic structure of a negatively charged *SiV* center (Figure is adapted from [39])

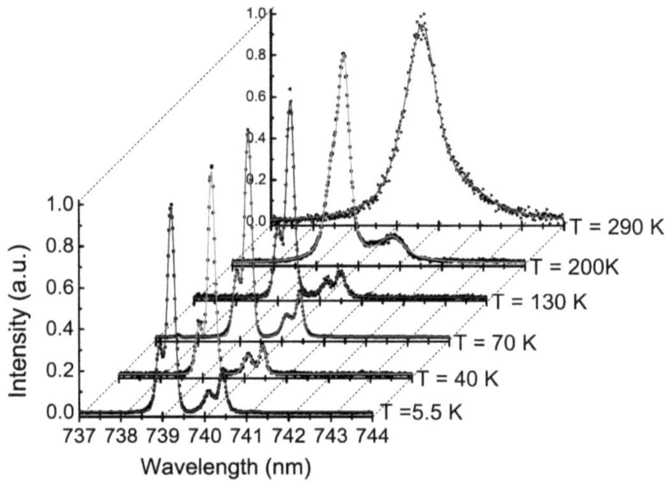

Fig. 3.10 Temperature dependent photoluminescence spectra of a single SiV-center under a laser excitation at 671 nm (Figure is adapted from [40] © IOP Publishing & Deutsche Physikalische Gesellschaft. CC BY-NC-SA)

SiV^- centers can be indistinguishable [42, 43]. This property of indistinguishability will eventually be useful in making a quantum network. One of the disadvantage for SiV center could be its lower quantum efficiency of less than 10 % [44].

3.3.2 Coupling of Quantum Emitters to Plasmonic Waveguides

In this section, we present various coupled systems that have been realized by different groups within the last decade. The coupling of a single quantum emitter, a QD, to a silver nanowire was first demonstrated, in 2007, by Akimov et. al. [16]. In 2009, coupling of an *NV* center to silver nanowires was demonstrated [17]. Later, coupling of a single *NV* center to propagating gap plasmons in two parallel wires was achieved [18]. In recent years, coupling of a quantum dot to wedge waveguide and coupling of an *NV* center to a V-groove have been realized experimentally [19, 20]. In this section, we present experiments of coupling a quantum emitter to a single nanowire and to a wedge waveguide. Coupling of a single emitter to gap mode of two parallel silver nanowires and V-groove is presented in Chap. 4. We, first, present a simplified schematic of the experimental set-up that is commonly used for characterization of the coupled systems.

3.3.2.1 Experimental Set-Up for Characterization of Coupled Systems

The coupled systems are, in general, characterized using a confocal fluorescence microscope. The coupling needs to be observed and the emitters need to be characterized for their lifetimes, single photon fluorescence and the emission spectrum. To achieve these characterization goals, different groups may have slightly different set-ups, but in essence, they are similar.

In Fig. 3.11, we present a schematic of experimental set-up that has been used for characterization of single quantum emitters coupled to different kinds of plasmonic waveguides. An excitation laser is used which has a shorter wavelength, in general, than the emission wavelengths of the quantum emitter. The laser is focused using an objective of high numerical aperture onto the sample. The sample can be raster scanned in a plane (xy in Fig. 3.11). The emission is then collected from the same objective that is used for focusing the excitation laser. A dichroic mirror is used to separate the excitation laser from the emission of quantum emitter. Some more filters might be used depending on the quantum emitter to minimize the background. A 50/50 beam splitter is used to split the emission into two paths. In one path, the emission is spatially filtered with a pin-hole and the pin-hole is imaged onto an avalanche photo-diode (APD). In another path, a galvanometric mirror is used to image the

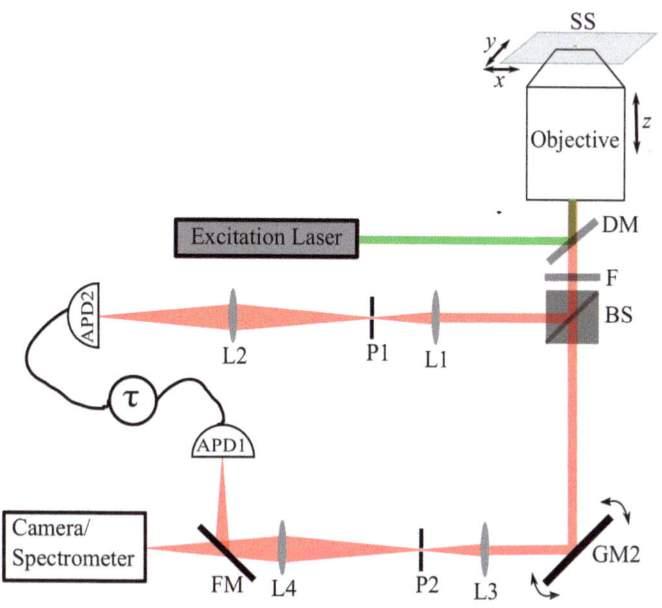

Fig. 3.11 Schematic of an optical set-up that is generally used for characterization of the coupled system. *SS* sample stage, *DM* dichroic mirror, *F* filters, *BS* beam-splitter, *L1*, *L2*, *L3*, *L4* lenses, *P1*, *P2* pin-holes, *GM* galvanometric mirror, *APD1* and *APD2* avalanche photo diodes. APD1 and APD2 are connected to counting electronics which relates the timing of detection of photons in the two channels

sample plane while exciting a quantum emitter. This feature helps in observation of coupling of quantum emitter emission and also helps in the measurement of correlation between the excitation point and other emission points that might appear in the sample plane due to the coupling to a waveguide. A spectrometer is used to obtain the emission spectrum from the quantum emitter as well as from the waveguide ends.

3.3.2.2 Coupling of a Single Quantum Dot to a Silver Nanowire

Coupling of a single quantum emitter to a plasmonic waveguide was first demonstrated with QDs and silver nanowires [16]. In this experiment, chemically synthesized silver wires and colloidal QDs were used. The sample was created by, first, spin coating the QD solution onto the glass substrate. A poly-(methamethacrylate) (PMMA) layer of ~30 nm was, subsequently, spin-coated. The silver wires were stamped onto the PMMA layer. After stamping the silver wires, a thick layer of PMMA was spin-coated on top. The 30 nm layer between the QDs and the silver wires acted as a spacer between the wires and the QDs. The QDs were of radius 5 nm. Therefore, the minimum distance between the silver wire surface and the center of QDs in this experiment was 35 nm. The diameter of the silver wires used in this experiment was 102 ± 24 nm.

The sample was characterized using a set-up similar to the one described in Sect. 3.3.2.1. First, the silver wire was imaged which can be seen in Fig. 3.12a Ch I. Then, a fluorescence image of the area was obtained which shows many spots. By matching the silver wire image and the fluorescence scan image, one can observe that a QD is situated close to the silver nanowire. That is indicated by the red circle in both the images (Ch I and Ch II). To find out whether the QD actually coupled to the silver nanowire, a fluorescence image in the sample plane was obtained by continuously exciting the QD. The image obtained is shown in Fig. 3.12a Ch III and shows two extra spots, other than that corresponding to the QD, matching with the position of the silver wire ends. This shows that the emission from QD is coupled to the silver nanowire mode, which propagates and gets scattered at the ends of the nanowire. The time trace of the emission directly from the QD and that from the end of the silver nanowire, presented in Fig. 3.12b, demonstrates that the fluctuation of the emission from the QD and that from the end of the nanowire are correlated. Thus, the source for these emissions are the same QD(s).

To demonstarte that the coupled QD is indeed a single photon source, second order correlation measurement was performed. To measure the photon coincidences, the free space radiation from the QD was incident on a balanced beam splitter and the coincidences between the detections from two APDs is recorded as a function of τ, that is, the time delay between photons detected by the two APDs. For a single photon source, the photon can be detected only by one of the APDs. So, the correlation at zero time delay should be zero. In Fig. 3.12c, one can see that the correlation for $\tau = 0$, is close to zero. The slight increase of the correlation is due to stray light, dark count of detectors and finite resolution of the counting electronics. Similarly, the photon correlation between the emission directly from the quantum dot and the

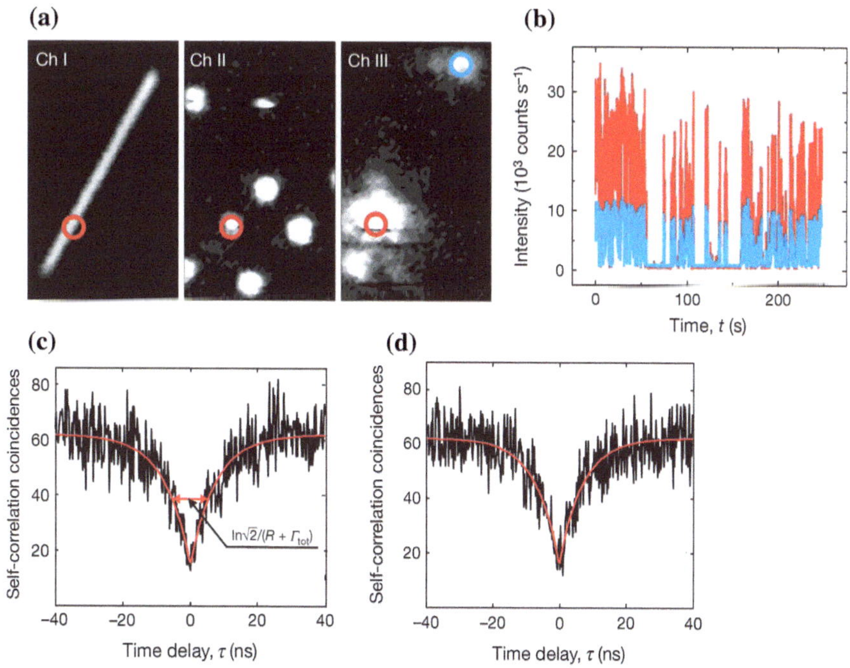

Fig. 3.12 **a** Ch I shows microscope image of a silver wire, Ch II shows a fluorescence image obtained with confocal microscope and Ch III shows the fluorescence image of the plane while the QD indicated by *red circle* in the figure is excited. The image in Ch III clearly shows the emission from the ends of the nanowire as well (one of the wire ends is indicated with a *blue circle*). **b** Time trace of the emission from the QD (*red*) and that from one of the ends of the nanowire which is indicated in (**a**) Ch III with a *blue circle*. **c** Second order correlation of QD fluorescence. **d** Second order correlation between fluorescence from QD and that from the wire-end. The *black* and *red* traces in **c** and **d** show the measured data and best fits, respectively (Figure is adapted by permission from Macmillan Publishers Ltd: [Nature] [16], © 2007)

emission from the silver wire end, in Fig. 3.12c, also shows a dip that is close to zero and further proves that the emission source for the middle and end of the wire are the same, and is a single QD.

To measure the change in decay rates of the QDs due to coupling, the decay rates for coupled QDs and uncoupled QDs were obtained. Photon correlation was measured at different powers and fitted to a model to estimate the decay rate of a QD. The model gives the width of the anti bunching dip as $\Delta\tau = ln\left(\sqrt{2}/\left(R + \Gamma_{tot}\right)\right)$, where R is the rate of excitation proportional to the excitation power and Γ_{tot} is the total decay rate. By extrapolating the width of the anti-bunching dip to $R = 0$, one obtains Γ_{tot}. In Fig. 3.13, we present a histogram showing the distribution of lifetimes for the coupled QDs and uncoupled QDs. From these measurements, it was found that average change in decay rate was 1.7.

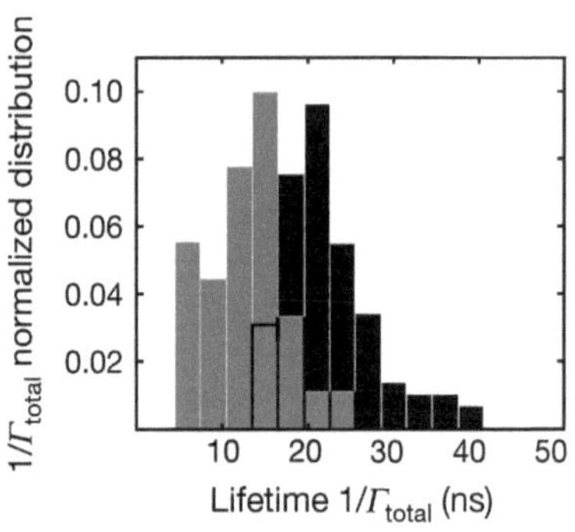

Fig. 3.13 Normalized histograms of quantum dot lifetimes. The *black* (*grey*) *bars* denote the distribution of uncoupled (coupled) quantum dots. Overlapping parts of the histograms are indicated by *outlined* and *vertically stacked bars* (Figure is adapted by permission from Macmillan Publishers Ltd: [Nature] [16], © 2007)

This was the first demonstration of coupling a single colloidal QD to a plasmonic waveguide. But, colloidal QDs tend to blink and bleach. *NV* centers are a stable source of single photons. In the next section, we present coupling of a single *NV* center to a silver nanowire [17].

3.3.2.3 Coupling of an *NV* Center to a Silver Nanowire

In this experiment, *NV* centers contained in nanodiamonds of average size 50 nm and silver nanowires with average diameter of 70 nm were used. The sizes were chosen in order to have an optimum combination of coupling, propagation length and out-coupling from the nanowires. In this experiment, the nanodiamond solution was mixed with the nanowire solution and the resulting solution was spin coated on a glass substrate. The self assembly process resulted in single nanodiamonds sticking to the silver nanowires. An atomic force microscope (AFM) image is shown in Fig. 3.14a where nanodiamonds can be seen on the surface of the nanowire. The sample was, first, chacterized in a manner similar to what is presented for the single quantum dot coupled to a silver nanowire. The lifetimes of the *NV* centers coupled to silver wires and that for the *NV* centers in nanodiamonds away from the silver wire were measured. By comparing the lifetimes, an average decay rate enhancement by a factor of 2.5 was found. When compared to the coupling of a QD to a silver nanowire, nanowires with smaller diameter was used and the distance between the emitter and the silver nanowire surface on average was 25 nm. So, higher decay rate enhancement, in this case, could be expected.

This experiment demonstrated wave-particle duality of single plasmon polaritons also. Figure 3.14b shows schematics of measuring correlation for a photon source and an equivalent set-up for measuring correlation for the emitter coupled to plasmons. In Fig. 3.14d, correlation measured for the *NV* center demonstrates that the

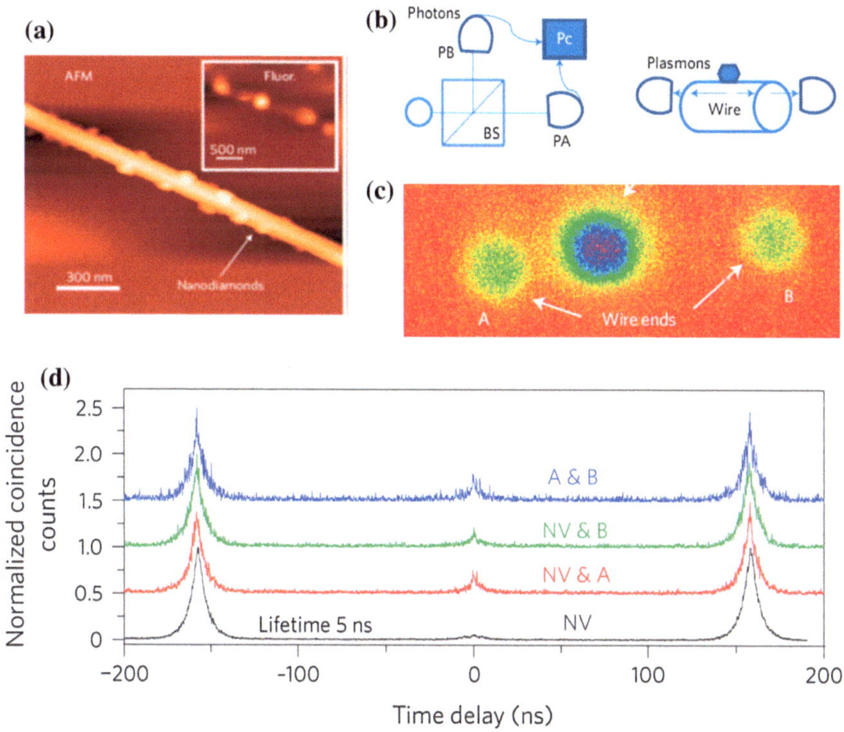

Fig. 3.14 **a** AFM image of a silver nanowire on which single nanodiamonds are stuck and are distributed over the nanowire surface. In the *inset* a fluorescence image of the area can be seen where some of the nanodiamonds fluoresced. **b** The Hanbury BrownTwiss experiment for single photons (*left*) and single surface plasmon polaritons (*right*). *PA*/*PB* photodiode A/B; *Pc* photon correlator; *BS* beam splitter. **c** The fluorescence image of a single quantum emitter (optically excited single *NV* center in diamond) coupled to a silver wire. **d** Second-order intensity correlation function of a single-photon *NV* emitter in the far-field (*black curve*). The *blue line* shows the measurement of cross-correlation between the two ends of the wire. Here, the missing peak at zero delay is proof of the particle-like behavior of single surface plasmon polaritons. *Red*, *green* and *blue curves* are shifted vertically by 0.5, 1 and 1.5, respectively, for clarity (Figure is adapted by permission from Macmillan Publishers Ltd: [Nat. Phys.] [17], © 2009)

coupled *NV* center is a single photon emitter. A & B shows a correlation measured between two ends of the nanowire. The missing coincidence peak at zero delay time indicates that the emission is originating from a single quantum system and demonstrates that plasmons behave as a single quantum particle. When the coincidence is recorded between the *NV* center photon and either end of the wire, the missing peak at zero delay indicates that single surface plasmon polaritons originating from the single-photon source are coupled out at the ends of the wire (red and green lines).

Figure 3.15a shows a sketch of a set-up, that can be used for demonstration of self-interference of single photons, where one of the mirrors can be moved to change the

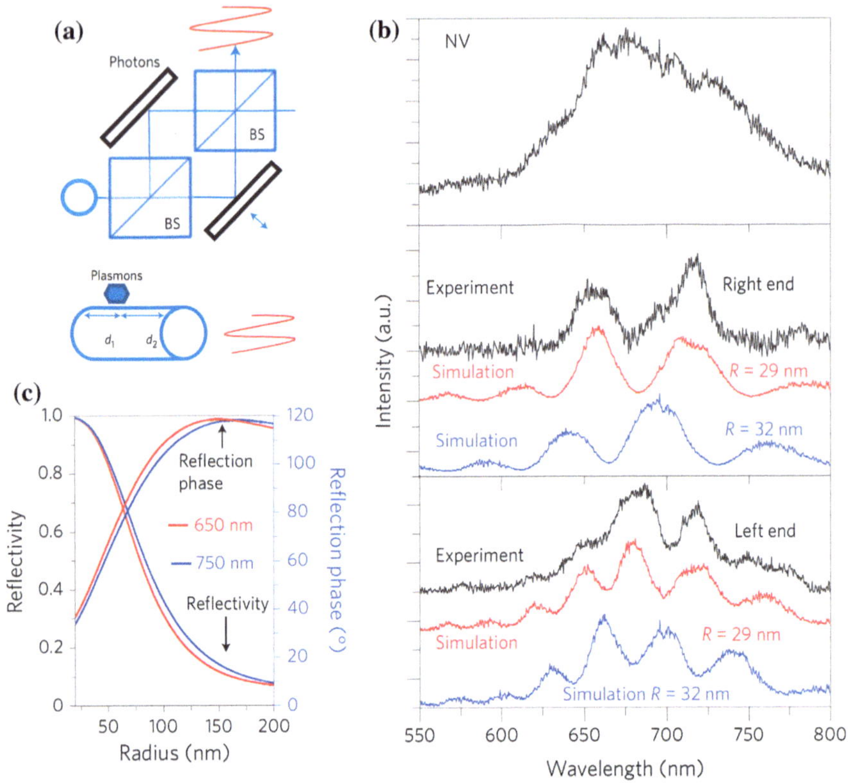

Fig. 3.15 **a** Diagram showing single photon (*top*) and single plasmon (*bottom*) self-interference experiments. **b** Fluorescence emission spectra of a single *NV* centre (*top*) and of single plasmons coupled out from the ends of the wire (two *bottom* graphs, *black curves*). The *red lines* show results of simulations taking into account losses during propagation of plasmons in the wire as well as dispersion for wire radius $R = 29$ nm. The *blue lines* are results of simulations for wire radius $R = 32$ nm. **c** The graph shows numerical simulations of the reflection coefficient and the reflection phase for two different wavelengths covering the range of *NV* center emission (Figure is adapted by permission from Macmillan Publishers Ltd: [Nat. Phys.] [17], © 2009)

path length in one of the arms. The detection probability of a photon is modulated depending on the path length. Similarly, a quantum emitter coupled to a nanowire, where the nanowire ends reflection is significant can be used to demonstrate self-interference of plasmons as illustrated in Fig. 3.15a. The coupling acts as the first beam splitter where the quantum emitter emits symmetrically in two directions. Plasmons propagating towards right end interfere with the plasmons that are emitted towards the left end and are reflected from the left end. The phase acquired by different wavelengths varies due to their wavelength, dispersion of the waveguide as well as the wavelength dependent reflection phase. Therefore, the plasmons which scatter to the far field from the wire ends are modulated as a function of wavelength.

Figure 3.15b shows the spectra that is obtained directly from the NV center and that obtained from the two ends of the silver nanowire.

This experiment demonstrated wave-particle duality for single plasmon polaritons in a system where quantum emitter was coupled to a silver nanowire. In Sect. 3.3.2.4, we present coupling of a quantum dot to a wedge waveguide.

3.3.2.4 Coupling a Quantum Dot to a Wedge Waveguide

Wedge is another promising structure that has been utilized, recently, for the demonstration of coupling a single quantum emitter to propagating modes it supports [20]. To realize the coupled system, the process of fabrication of wedge waveguides was optimized. To fabricate the wedge waveguide, first, triangular trenches in 100-oriented silicon wafer were formed via anisotropic etching. The sides of trenches which are defined by the Si(111) atomic planes are very smooth with a precise relative angle of 70.54°. Then a silver layer of thickness > 350 nm was deposited. For silver deposition, the rate of deposition was kept high at 2.5 nm/s and a low residual gas pressure (3×10^{-8} Torr) was used. This resulted in silver films with optical properties comparable to that of single crystalline silver. The silver film was then peeled off via a process called template stripping. This produced smooth wedges with sharp apexes. To deposit QDs on the apex of a wedge waveguide, electro-hydro-dynamic (EHD) printing was utilized. This technique allows deposition of a countable number of QDs down to individual QDs. Figure 3.16h shows a single QD deposited on the apex of a wedge waveguide.

Figure 3.16a shows a wedge waveguide with fluorescence from 3 QDs near the apex of the waveguide. The image is obtained by avaeraging 1000 1 s frames. The three QDs are labelled QD1, QD2 and QD3. The blinking behaviour of these QDs show that these are, indeed, single QDs. The blinking of QD2 could be correlated to the blinking of the fluorescence extracted from the waveguide end. This proved the coupling of QD2 to the wedge waveguide. The blinking of QD1 and QD3 were not correlated to the emission at the waveguide end, which could result from their position not close enough to the wedge apex.

The decay rate of the QDs was measured to estimate the change in the lifetime of the QDs. By comparing the lifetime for QDs in tetradecane and that on the apex of the wedge waveguide, a decay rate enhancement of 4.64 was observed. This experiment shows that the coupling depends critically on the position of the quantum emitter near to the waveguide. If the quantum emitters are placed deterministically at right positions, coupling can be enhanced and channeled into the waveguide.

In Chap. 4, experiments where quantum emitters are deterministically coupled to silver nanowires, gap mode between two parallel silver nanowires and V-grooves are presented.

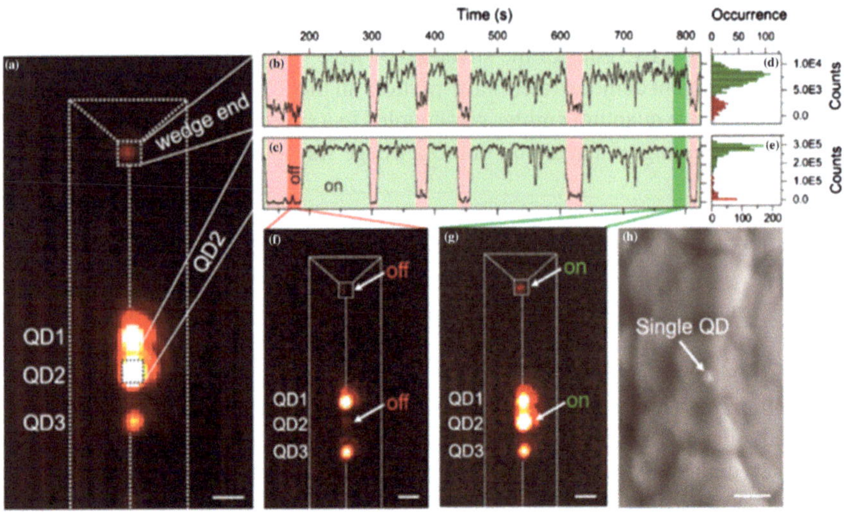

Fig. 3.16 **a** False-color fluorescence micrograph of three individual quantum dots (QD1, QD2 and QD3) onto the apex of a silver wedge waveguide. Plasmons generated by the three QDs scatter at the ends of the wedge waveguide. **b** and **c** Time series for the fluorescence intensities extracted from the wedge end and QD2, respectively. The two signals are strongly correlated and exhibit same on and off periods. **d** and **e** The bimodal distribution in intensities is also seen in histograms for the wedge end and QD2, respectively. **f** and **g** Fluorescence intensities integrated over 20 frames of the off state and the on state, respectively, as indicated. *Scale bars* in **a, f** and **g** are 1 μm. **h** A scanning electron micrograph of an individual QD on the apex of the silver wedge (*scale bar* 40 nm) (Figure is adapted from [20] © 2015 ACS)

3.4 Conclusion and Outlook

We have, in this chapter, presented various waveguide structures that can be useful for coupling to a quantum emitter. There are some theoretical proposals which show that plasmonic waveguide system is suitable for entanglement of two quantum emitters [4]. If this is demonstrated, then it will open up a way towards a quantum network based on quantum emitter plasmonic waveguide system.

To realize a quantum network based on quantum emitters, quantum emitters with narrowband emission and high quantum efficiency along with their emission being indistinguishable from each other is required. Quantum dots at cryogenic temperatures are stable and have high quantum efficiency. However, to find two quantum dots whose emission are indistinguishable is usually challenging. *NV* centers have stable emission, but they are very broadband. *SiV* centers have stable emission, and it has been shown that many *SiV* centers in the same diamond sample have indistinguishable emission. A drawback could be their quantum efficiency, which is below 10 % [44]. To find an emitter with suitable properties is a requirement to build a quantum network.

Many groups have demonstrated coupling of quantum emitters to plasmonic waveguides. But, the problem of the plasmonic waveguides is their intrinsic losses, which are high. The size of the network will ultimately be limited by the losses. Different materials are being explored to reduce the losses in plasmonic waveguides [45, 46]. So far, silver and gold have been the materials of choice for plasmonic waveguides, which are coupled to quantum emitters. Given an inevitable increase in propagation losses for plasmonic waveguides with progressively stronger mode confinement (that is desirable for boosting up the emission rate in coupled emitter-waveguide systems), we would suggest making use of hybrid waveguide configurations with a short section of a plasmonic waveguide being coupled to a dielectric waveguide [47], that can further be coupled to an optical fiber [48]. We believe, this integration would allow one to combine the best of two waveguiding configurations (deep subwavelength confinement in plasmonic waveguides and low loss propagation in dielectric waveguides), and might eventually bring plasmonic waveguides close to a realm of practical applications in quantum information processing.

Acknowledgements The authors gratefully acknowledge the financial support from the European Research Council, grant No. 341054 (PLAQNAP).

References

1. D.K. Gramotnev, S.I. Bozhevolnyi, Plasmonics beyond the diffraction limit. Nat. Photonics **4**, 83–91 (2010)
2. D.E. Chang, A.S. Sørensen, P.R. Hemmer, M.D. Lukin, Quantum optics with surface plasmons. Phys. Rev. Lett. **97**, 053002 (2006)
3. D.E. Chang, A.S. Sørensen, E.A. Demler, M.D. Lukin, A single-photon transistor using nanoscale surface plasmons. Nat. Phys. **3**, 807–812 (2007)
4. A. Gonzalez-Tudela, D. Martin-Cano, E. Moreno, L. Martin-Moreno, C. Tejedor, F.J. Garcia-Vidal, Entanglement of two qubits mediated by one-dimensional plasmonic waveguides. Phys. Rev. Lett. **106**, 020501 (2011)
5. M. Leifgen, T. Schröder, F. Gdeke, R. Riemann, V. Métillon, E. Neu, C. Hepp, C. Arend, C. Becher, K. Lauritsen, O. Benson, Evaluation of nitrogen- and silicon-vacancy defect centres as single photon sources in quantum key distribution. New J. Phys. **16**(2), 023021 (2014)
6. J. Riedrich-Möller, L. Kipfstuhl, C. Hepp, E. Neu, C. Pauly, F. Mücklich, A. Baur, M. Wandt, S. Wolff, M. Fischer, S. Gsell, M. Schreck, C. Becher, One-and two-dimensional photonic crystal microcavities in single crystal diamond. Nat. Nanotechnol. **7**(1), 69–74 (2012)
7. A. Faraon, C. Santori, Z. Huang, V.M. Acosta, R.G. Beausoleil, Coupling of nitrogen-vacancy centers to photonic crystal cavities in monocrystalline diamond. Phys. Rev. Lett. **109**, 033604 (2012)
8. M. Arcari, I. Söllner, A. Javadi, S. Lindskov, Hansen, S. Mahmoodian, J. Liu, H. Thyrrestrup, E.H. Lee, J. D. Song, S. Stobbe, P. Lodahl. Near-unity coupling efficiency of a quantum emitter to a photonic crystal waveguide. Phys. Rev. Lett. **113**, 093603 (2014)
9. J. Claudon, J. Bleuse, N.S. Malik, M. Bazin, N. Gregersen, C. Sauvan, P. Lalanne, J.-M. Gérard, P. Jaffrennou, A highly efficient single-photon source based on a quantum dot in a photonic nanowire. Nat. Photonics **4**(3), 174–177 (2010)
10. T.M. Babinec, B.J.M. Hausmann, M. Khan, Y. Zhang, J.R. Maze, P.R. Hemmer, M. Lončar, A diamond nanowire single-photon source. Nat. Nanotechnol. **5**(3), 195–199 (2010)

11. S. Kühn, U. Håkanson, L. Rogobete, V. Sandoghdar, Enhancement of single-molecule fluorescence using a gold nanoparticle as an optical nanoantenna. Phys. Rev. Lett. **97**, 017402 (2006)
12. P. Anger, P. Bharadwaj, L. Novotny, Enhancement and quenching of single-molecule fluorescence. Phys. Rev. Lett. **96**, 113002 (2006)
13. S. Schietinger, M. Barth, T. Aichele, O. Benson, Plasmon-enhanced single photon emission from a nanoassembled metal-diamond hybrid structure at room temperature. Nano Lett. **9**(4), 1694–1698 (2009)
14. A. Kinkhabwala, Z. Yu, S. Fan, Y. Avlasevich, K. Müllen, W.E. Moerner, Large single-molecule fluorescence enhancements produced by a bowtie nanoantenna. Nat. Photonics **3**(11), 654–657 (2009)
15. G.M. Akselrod, C. Argyropoulos, T.B. Hoang, C. Cirací, C. Fang, J. Huang, D.R. Smith, M.H. Mikkelsen, Probing the mechanisms of large purcell enhancement in plasmonic nanoantennas. Nat. Photonics **8**(11), 835–840 (2014)
16. A.V. Akimov, A. Mukherjee, C.L. Yu, D.E. Chang, A.S. Zibrov, P.R. Hemmer, H. Park, M.D. Lukin, Generation of single optical plasmons in metallic nanowires coupled to quantum dots. Nature **450**(06230), 402–406 (2007)
17. R. Kolesov, B. Grotz, G. Balasubramanian, R.J. Stöhr, A.A.L. Nicolet, P.R. Hemmer, F. Jelezko, J. Wrachtrup, Wave-particle duality of single surface plasmon polaritons. Nat. Phys. **5**, 470–474 (2009)
18. S. Kumar, A. Huck, U.L. Andersen, Efficient coupling of a single diamond color center to propagating plasmonic gap modes. Nano Lett. **13**(3), 1221–1225 (2013)
19. E. Bermúdez-Ureña, C. Gonzalez-Ballestero, M. Geiselmann, R. Marty, I.P. Radko, T. Holmgaard, Y. Alaverdyan, E. Moreno, F.J. Garcia-Vidal, S.I. Bozhevolnyi, R. Quidant, Coupling of individual quantum emitters to channel plasmons. Nat. Commun. **6**, 7883 (2015)
20. S.J.P. Kress, F.V. Antolinez, P. Richner, S.V. Jayanti, D.K. Kim, F. Prins, A. Riedinger, M.P.C. Fischer, S. Meyer, K.M. McPeak, D. Poulikakos, D.J. Norris, Wedge waveguides and resonators for quantum plasmonics. Nano Lett. **15**(9), 6267–6275 (2015)
21. Z. Han, S.I. Bozhevolnyi, Radiation guiding with surface plasmon polaritons. Rep. Prog. Phys. **76**(1), 016402 (2013)
22. E.D. Palik, *Handbook of Optical Constants* (Academic Press, 1985)
23. Y. Chen, T.R. Nielsen, N. Gregersen, P. Lodahl, J. Mørk, Finite-element modeling of spontaneous emission of a quantum emitter at nanoscale proximity to plasmonic waveguides. Phys. Rev. B **81**, 125431 (2010)
24. D. Martín-Cano, L. Martín-Moreno, F.J. García-Vidal, E. Moreno, Resonance energy transfer and superradiance mediated by plasmonic nanowaveguides. Nano Lett. **10**(8), 3129–3134 (2010)
25. P. Lodahl, S. Mahmoodian, S. Stobbe, Interfacing single photons and single quantum dots with photonic nanostructures. Rev. Mod. Phys. **87**, 347–400 (2015)
26. O. Chen, J. Zhao, V.P. Chauhan, J. Cui, C. Wong, D.K. Harris, H. Wei, H. Han, D. Fukumura, R.K. Jain, M.G. Bawendi, Compact high-quality cdse-cds core-shell nanocrystals with narrow emission linewidths and suppressed blinking. Nat. Mater. **12**(5), 445–451 (2013)
27. M.W. Doherty, N.B. Manson, P. Delaney, F. Jelezko, J. Wrachtrup, L.C.L. Hollenberg, The nitrogen-vacancy colour centre in diamond. Phys. Rep. **528**(1), 1–45 (2013)
28. H. Bernien, L. Childress, L. Robledo, M. Markham, D. Twitchen, R. Hanson, Two-photon quantum interference from separate nitrogen vacancy centers in diamond. Phys. Rev. Lett. **108**, 043604 (2012)
29. A. Sipahigil, M.L. Goldman, E. Togan, Y. Chu, M. Markham, D.J. Twitchen, A.S. Zibrov, A. Kubanek, M.D. Lukin, Quantum interference of single photons from remote nitrogen-vacancy centers in diamond. Phys. Rev. Lett. **108**, 143601 (2012)
30. E. Togan, Y. Chu, A.S. Trifonov, L. Jiang, J. Maze, L. Childress, M.V.G. Dutt, A.S. Sørensen, P.R. Hemmer, A.S. Zibrov, M.D. Lukin, Quantum entanglement between an optical photon and a solid-state spin qubit. Nature **466**(7307), 730–734 (2010)
31. H. Bernien, B. Hensen, W. Pfaff, G. Koolstra, M.S. Blok, L. Robledo, T.H. Taminiau, M. Markham, D.J. Twitchen, L. Childress, R. Hanson, Heralded entanglement between solid-state qubits separated by three metres. Nature **497**(7447), 86–90 (2013)

32. W. Pfaff, B.J. Hensen, H. Bernien, S.B. van Dam, M.S. Blok, T.H. Taminiau, M.J. Tiggelman, R.N. Schouten, M. Markham, D.J. Twitchen, R. Hanson, Unconditional quantum teleportation between distant solid-state quantum bits. Science **345**(6196), 532–535 (2014)
33. B. Hensen, H. Bernien, A.E. Dreaú, A. Reiserer, N. Kalb, M.S. Blok, J. Ruitenberg, R.F.L. Vermeulen, R.N. Schouten, C. Abellán, W. Amaya, V. Pruneri, M.W. Mitchell, M. Markham, D.J. Twitchen, D. Elkouss, S. Wehner, T.H. Taminiau, R. Hanson, Loophole-free bell inequality violation using electron spins separated by 1.3 kilometres. Nature **526**(7575), 682–686 (2015)
34. L. Rondin, J.-P. Tetienne, T. Hingant, J.-F. Roch, P. Maletinsky, V. Jacques, Magnetometry with nitrogen-vacancy defects in diamond. Rep. Prog. Phys. **77**(5), 056503 (2014)
35. F. Dolde, H. Fedder, M.W. Doherty, T. Nöbauer, F. Rempp, G. Balasubramanian, T. Wolf, F. Reinhard, L.C.L. Hollenberg, F. Jelezko, J. Wrachtrup, Electric-field sensing using single diamond spins. Nat. Phys. **7**(6), 459–463 (2011)
36. P. Neumann, I. Jakobi, F. Dolde, C. Burk, R. Reuter, G. Waldherr, J. Honert, T. Wolf, A. Brunner, J.H. Shim, D. Suter, H. Sumiya, J. Isoya, J. Wrachtrup, High-precision nanoscale temperature sensing using single defects in diamond. Nano Lett. **13**(6), 2738–2742 (2013)
37. G. Kucsko, P.C. Maurer, N.Y. Yao, M. Kubo, H.J. Noh, P.K. Lo, H. Park, M.D. Lukin, Nanometre-scale thermometry in a living cell. Nature **500**(7460), 54–58 (2013)
38. A. Mohtashami, A.F. Koenderink, Suitability of nanodiamond nitrogenvacancy centers for spontaneous emission control experiments. New J. Phys. **15**(4), 043017 (2013)
39. L.J. Rogers, K.D. Jahnke, M.W. Doherty, A. Dietrich, L.P. McGuinness, C. Müller, T. Teraji, H. Sumiya, J. Isoya, N.B. Manson, F. Jelezko, Electronic structure of the negatively charged silicon-vacancy center in diamond. Phys. Rev. B **89**, 235101 (2014)
40. E. Neu, C. Hepp, M. Hauschild, S. Gsell, M. Fischer, H. Sternschulte, D. Steinmüller-Nethl, M. Schreck, C. Becher, Low-temperature investigations of single silicon vacancy colour centres in diamond. New J. Phys. **15**(4), 043005 (2013)
41. U.F.S. D'Haenens-Johansson, A.M. Edmonds, M.E. Newton, J.P. Goss, P.R. Briddon, J.M. Baker, P.M. Martineau, R.U.A. Khan, D.J. Twitchen, S.D. Williams, Epr of a defect in cvd diamond involving both silicon and hydrogen that shows preferential alignment. Phys. Rev. B **82**, 155205 (2010)
42. A. Sipahigil, K.D. Jahnke, L.J. Rogers, T. Teraji, J. Isoya, A.S. Zibrov, F. Jelezko, M.D. Lukin, Indistinguishable photons from separated silicon-vacancy centers in diamond. Phys. Rev. Lett. **113**, 113602 (2014)
43. L.J. Rogers, K.D. Jahnke, T. Teraji, L. Marseglia, C. Mller, B. Naydenov, H. Schauffert, C. Kranz, J. Isoya, L.P. McGuinness, F. Jelezko, Multiple intrinsically identical single-photon emitters in the solid state. Nat. Commun. **5**, 4739 (2014)
44. E. Neu, M. Agio, C. Becher, Photophysics of single silicon vacancy centers in diamond: implications for single photon emission. Opt. Express **20**(18), 19956–19971 (2012)
45. G.V. Naik, V.M. Shalaev, A. Boltasseva, Alternative plasmonic materials: beyond gold and silver. Adv. Mater. **25**(24), 3264–3294 (2013)
46. U. Guler, A. Boltasseva, V.M. Shalaev, Refractory plasmonics. Science **344**(6181), 263–264 (2014)
47. A. Melikyan, L. Alloatti, A. Muslija, D. Hillerkuss, P.C. Schindler, J. Li, R. Palmer, D. Korn, S. Muehlbrandt, D. Van Thourhout, B. Chen, R. Dinu, M. Sommer, C. Koos, M. Kohl, W. Freude, J. Leuthold, High-speed plasmonic phase modulators. Nat. Photonics **8**(3), 229–233 (2014)
48. T.G. Tiecke, K.P. Nayak, J.D. Thompson, T. Peyronel, N.P. de Leon, V. Vuletić, M.D. Lukin, Efficient fiber-optical interface for nanophotonic devices. Optica **2**(2), 70–75 (2015)

Chapter 4
Controlled Interaction of Single Nitrogen Vacancy Centers with Surface Plasmons

Esteban Bermúdez-Ureña, Michael Geiselmann and Romain Quidant

Abstract Efficient light-matter interaction lies at the heart of many emerging technologies that seek on–chip integration of solid-state photonic systems. In the areas of quantum computation and single photon sources it is fundamental to achieve precise control of the emission dynamics of a single quantum emitter, and exploiting the unique properties of surface plasmons is one way of achieving such control. In this chapter we review several works in the field where single nitrogen vacancy centres in nanodiamond particles have been deterministically assembled together with plasmonic antennas and plasmonic waveguides using state of the art nano-positioning techniques.

4.1 Introduction

The nitrogen vacancy (NV) center in diamond has emerged as one of the most promising quantum emitters in solid state systems [1]. As introduced in Chap. 3, several characteristics provide these defect centers with such praise, enabling their use in a wide range of fields. Their bright and stable emission at room temperature allows their use as bio-markers in fluorescence microscopy, while their ground state spin triplet enables them as the main building block in future quantum computation platforms. For many applications, it is desirable to have certain degree of control of

E. Bermúdez-Ureña (✉) · R. Quidant (✉)
ICFO-Institut de Ciencies Fotoniques, Barcelona Institute of Science and Technology, 08860 Castelldefels, Barcelona, Spain
e-mail: esteban.bermudez@icfo.es

R. Quidant
e-mail: romain.quidant@icfo.es

M. Geiselmann
École Polytechnique Fédérale de Lausanne (EPFL), 1015 Lausanne, Switzerland
e-mail: michael.geiselmann@epfl.ch

R. Quidant
ICREA-Instució Catalana de Recerca i Estudis Avançats, 08010 Barcelona, Spain

S.I. Bozhevolnyi et al. (eds.), *Quantum Plasmonics*,
Springer Series in Solid-State Sciences 185,
DOI 10.1007/978-3-319-45820-5_4

the emission dynamics and directionality, for example, to achieve faster and more efficient photon extraction that in turn yields higher data transfer and computational power. The emission properties of quantum emitters depend not only on their intrinsic structure but also on the local electromagnetic environment. Surface plasmons, which are excited oscillations of the electron gas density in metals, provide the means to modify the electromagnetic landscape at very small volumes [2]. The excitation of plasmon nanostructures leads to highly localized photon fields and therefore to an enhancement of the excitation as well as of the radiative and nonradiative decay rates of nearby emitters.

Fully exploiting the capability of plasmon-NV coupling requires controlling, with high accuracy, the relative position between the emitter and the electromagnetic modes supported by the plasmonic structures. Such assembly can enable precise control of the emission dynamics of the NV centers (excitation/emission rate enhancement), directivity tuning in free-space, on-chip plasmon waveguiding, and quantum bit entanglement among others. These features make this task highly relevant for applications in integrated solid state quantum devices such as single photon sources or quantum networks, both for free-space and on-chip platforms [3].

The ability to position the quantum emitter at an optimum position within the landscape of the local density of states (LDOS) supported by the nanostructure is at the moment one of the challenging aspects in current plasmonic research. The main challenge relies in placing the emitter at the appropriate positions for ideal coupling due to the competing decay channels available to the emitter.

In this chapter we will focus on the state of the art of the controlled coupling of single NV centers to different plasmonic structures, limiting to the assembly of hybrid devices involving diamond nanocrystals. It is worth mentioning that the methods presented here have also been used to study coupling of other quantum emitters such as colloidal quantum dots to plasmonic nanostructures, but it is beyond the scope of this chapter to cover those works as well.

4.2 Scanning Probe Assembly

The technique of manipulating nanoscale particles with an atomic force microscope (AFM) cantilever was introduced in 1995, [4] nevertheless it took more than 10 years for it to see an application in the coupling of a quantum emitter to a plasmonic structure [5]. The technique exploits the ability of an AFM system to perform movements both in tapping mode for imaging, and in contact mode to move the particles across a substrate. In the early years, it required custom made modifications of the AFM systems in order to achieve such a dual operation; however, nowadays commercial nano-positioning add-on software is readily available.

This technique must be implemented together with a confocal fluorescence microscope in order to identify and track the movements and changes in the emission dynamics of the emitters. For reference on the essential components of a typical optical characterization setup used with this approach, the reader can refer to Fig. 3.11 in Chap. 3. Working with transparent substrates has the advantage that both the AFM and microscope systems can be integrated within the same platform, offering real time tracking of the coupling dynamics upon the manipulation [5–8]. On the other hand, when working with metallic or non-transparent substrates, the optical characterization and nano-manipulation are preferably performed at separate stages [9]. The key is being able to acquire fluorescent images from the substrates to locate the emitters, and to assess the emission dynamics and single emitter behavior by means of fluorescent lifetime decay and a Hanbury-Brown and Twiss (HBT) experiment respectively.

The general assembling procedure is to first acquire a fluorescence image to identify the single quantum emitters and their relative position to the plasmonic structures. Afterwards, an AFM image in tapping mode is performed in the same area as the fluorescence image, and the selected particle is moved towards the desired position by doing movements with the AFM in contact mode. A second fluorescence image can be acquired in order to confirm the new position and assess the emission dynamics of the coupled system. The procedure is repeated until the hybrid system has been assembled. In the following, we present representative experimental results on the controlled interaction of single NV centers with plasmonic nanoantennas and plasmonic waveguides assembled with this technique.

4.2.1 Control of Emission Dynamics Through Plasmon Coupling

Coupling of NV Centers to Gold Nanoparticles

Metallic nanoparticles can function as nanoscale plasmonic antennas in the visible and near infrared regions of the electromagnetic spectrum [10]. The field confinement at nanoscale volumes provides a rich enhancement of the LDOS available to a quantum emitter. Plasmonic antennas can influence the emission properties of a quantum emitter depending on their design and composition. For example, the resonance of an antenna can be matched to the excitation laser wavelength to achieve excitation enhancement of a quantum emitter, [11] or matching the emission energy to enhance the radiative decay rates and tune the directionality of the emission [12].

In 2009, the group of Oliver Benson demonstrated the controlled coupling between nanodiamonds (NDs) containing single NV centers and spherical gold nanoparticles by positioning them with an AFM cantilever [5]. The samples were prepared by spin-coating aqueous solutions of NDs and gold nanospheres onto a glass coverslip. Only a few percent of the particles in commercial ND solutions contain single NV centers, making it crucial to screen the substrate for fluorescence and single photon emission. The use of a transparent substrate enabled them to use a setup with an AFM system integrated with an inverted confocal microscope, to simultaneously retrieve the optical properties upon the nano-manipulation steps (as shown in Fig. 3.11).

Once a ND with a single NV center is identified, the gold nanospheres are brought in contact with the ND. In this case, the group studied the emission dynamics of a single NV center coupled to one and two gold nanospheres in subsequent steps, as can be seen by configurations A and B respectively in Fig. 4.1. Figure 4.1b shows the corresponding electromagnetic simulations of the enhancement of the excitation light expected for the three configurations. At least a ten-fold enhancement is expected at the center of the nanocrystal, and even stronger enhancement in close proximity to the metallic surfaces. A similar enhancement at the emission wavelength is what leads eventually to an increase in the decay rates of an emitter positioned inside a diamond nanocrystal.

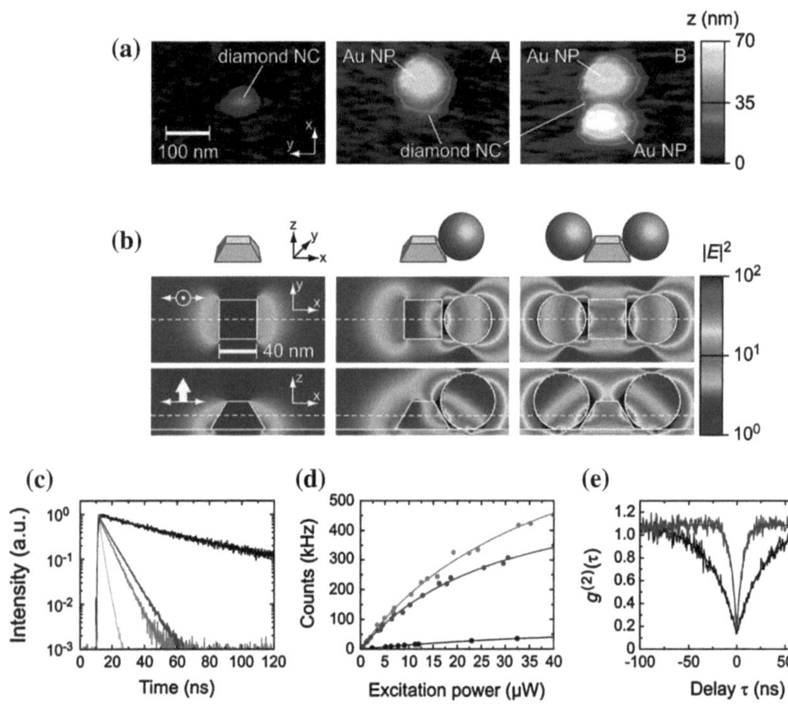

Fig. 4.1 **a** AFM images of a single ND (*left*), to which one (*middle*) and two (*right*) gold nanospheres have been positioned in close proximity. **b** Corresponding numerical simulations of the intensity enhancement of the electric field of the excitation light, with linear polarization along the x axis. The field intensity is normalized to the value at the center of the bare ND and displayed in a logarithmic *color scale*. **c** Fluorescence lifetime traces of the bare ND (*black*), configuration A (*blue*), and configuration B (*red*). **d** Emission intensity (same *color coding*) as a function of the excitation power, corrected for the background emission from the gold. **e** Normalized autocorrelation measurements on the bare diamond (*black*) and for configuration A (*blue*). Adapted from [5]

The optical characterization of the emission dynamics is summarized in the bottom panels of Fig. 4.1. The time-resolved measurements of configurations A (blue) and B (red) in Fig. 4.1c correspond to an increase in the excited-state decay rate by a factor of 7.5 and 9.5 respectively, when compared to the decay rate of the isolated ND (black).

To verify if the enhancement is due to radiative or nonradiative processes, power-dependent measurements were performed (Fig. 4.1d). The data were fitted with a saturation model of the form

$$P = \xi\sigma\Phi P_{exc}/(1 + \sigma\Phi\tau_{rad}P_{exc}) \tag{4.1}$$

where σ is the absorption cross section, τ_{rad} is the radiative lifetime, Φ is the internal quantum efficiency, and ξ is the total collection efficiency of the setup.

In the case of strong excitation, the maximal number of emitted photons is only restricted by the radiative lifetime, reducing Eq. (4.1) to

$$P = \xi/\tau_{rad} \tag{4.2}$$

The authors obtained an increase of the radiative decay rate γ_{rad} by a factor of 5.8 and 8.9 for configuration A and B, respectively, which corresponds to quantum efficiencies $\Phi = \tau_{dec}/\tau_{rad}$ of 0.78 and 0.93.

Under weak excitation, P reduces to

$$P = \xi\sigma\Phi P_{exc} \tag{4.3}$$

and it is possible to calculate the enhancement of the excitation rate $\gamma_{exc} \propto \sigma P_{exc}$ from the slope of the power-dependent measurements at weak excitation intensities, where the fluorescence is linearly dependent on the excitation intensity. A 12- and 14-fold enhancement rate can be deduced for configurations A and B, respectively.

Despite that both, γ_{exc} and γ_{rad} are strongly enhanced, the single photon character of the emission is still preserved (blue trace in Fig. 4.1e).

Mapping Decay Rates Around a Bowtie Nanoantenna

The Benson group also used this AFM technique to map the local electromagnetic environment of a plasmonic bowtie antenna with the NV center as the local probe [6]. They placed and characterized a ND containing a single NV center at various positions around the nanoantenna and in doing so, derived a map of decay rates (inverse of the lifetime) of the excited state of the NV center close to the plasmonic nanoantenna, as can be seen in Fig. 4.2 for 60 nm (a) and 15 nm (b) NDs. The underlay shows the AFM topography image of the nanoantenna, while the points indicate the position of the particle, and their color the measured decay rate. The maximum observed decay rate occurred inside the gap (lifetime τ = 3.6 ns), where it was increased by a factor of 10.7 compared to its decay rate several micrometers away from the antenna.

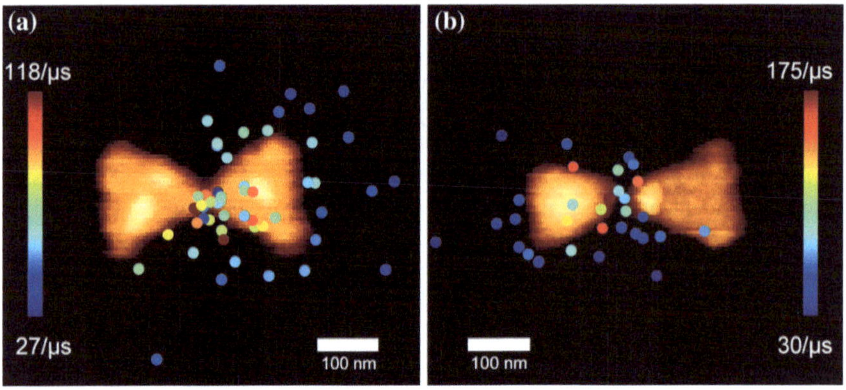

Fig. 4.2 Decay rate maps showing of the excited state of the NV center in the ND for different positions with an underlay of the corresponding AFM image of the bowtie nanoantennas. In **a** a ND 60 nm in height was used while in **b** one of 15 nm. The *color code* represents the decay rate at each position. Adapted from [6]

4.2.2 Coupling of NV Centers to Propagating Surface Plasmons

While the coupling of NV centers to plasmonic nanoantennas allows tuning of the emission dynamics, the emission directivity typically restricts the experiments to free-space processing of the optical signals. In future photonic circuitry applications involving single photon sources or quantum bit networks, it is desirable to achieve an on-chip control of the emission from the quantum emitters.

Plasmonic waveguides support confined surface plasmon modes similarly to the nanoantennas, with the added functionality that the mode can propagate along the waveguide. The most common two-dimensional surface plasmon waveguides studied in the field of plasmonics are illustrated in Fig. 4.3. These include single metallic nanowire waveguides, gap plasmon waveguides composed of joint nanowires, V-groove (VG) plasmonic waveguides supporting channel plasmon polaritons (CPPs), and finally, the wedge plasmonic waveguides (sharp triangular like structures that support wedge plasmons at the tip of the structure). For more details on the supported modes and general coupling of quantum emitters to plasmonic nanoguides the reader should refer to Chap. 3. In this chapter we focus on the controlled interaction of NV centers coupled to single nanowires, gap nanowires and V-groove waveguides.

NV Center Deterministically Coupled to a Single Silver Nanowire
As presented in Chap. 3, the first realization of a hybrid system involving a single NV center and a plasmonic waveguide dates back to the work by the groups of Fedor Jelezko and Jörg Wrachtrup in 2009, where they demonstrated the wave-particle duality of single surface plasmons [13]. Despite the achievement, the

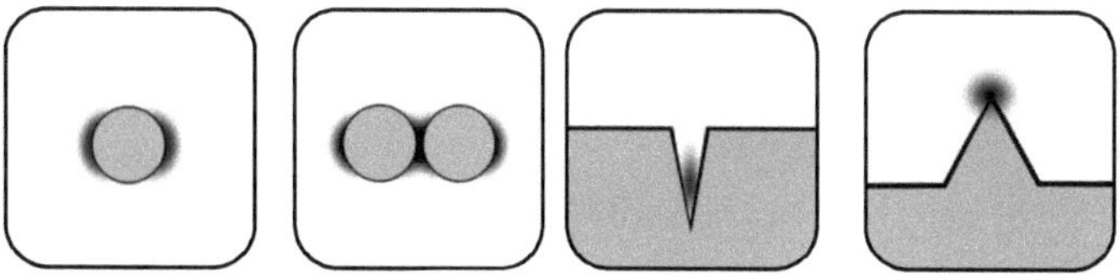

Fig. 4.3 Cross-section and mode location for the most common 2-D plasmonic waveguides. From *left* to *right* single nanowire, gap nanowire, V-groove and wedge plasmonic waveguides

work relied on random assembly, expecting that pairs of particles would be in close proximity to enable the coupling.

In 2011, both the groups of Oliver Benson in Germany and Ulrik Andersen in Denmark reported the use of the AFM manipulation to assemble together single NDs and colloidal silver nanowires and achieve guiding of the emission [6, 7]. Samples were prepared in a similar fashion as described in the previous section, where NDs and silver nanowires are spin coated onto a transparent substrate. The experimental setups resembled that of Fig. 3.11, with the main difference that there is a scanning mirror on the detection channel of one of the APDs, allowing to scan the fluorescence signal in areas surrounding the excitation spot.

In the experiment by the Andersen group, first a fluorescence image was acquired with one APD to locate a suitable ND containing a single NV center. At this point both lifetime measurements and correlation measurements were performed on the particle. The lifetime for the uncoupled particle was found to be 17.3 ns (black trace in Fig. 4.4b).

An AFM-topography image of the same area was then acquired (Fig. 4.4a), and in a second step, the ND was pushed into close proximity to the NW by using the AFM in contact mode. The presence of the nanowire changes the local density of states perceived by the emitter, in this case adding a plasmon decay channel. The emission properties were measured again and it was found that the lifetime was reduced to 4.8 ns, corresponding to reduction factor of 3.6.

At this position, the NV center cannot only decay by emitting a photon into the far field but it can also couple its emission to the supported propagating plasmon mode of the nanowire. To visualize this coupling, a fluorescence image is acquired with a second APD by means of the scanning mirror while fixing the excitation to the NV center position. The resulting image is shown in Fig. 4.4c where two emission spots can be appreciated, one corresponding to a superposition of the free-space emission and that from the lower NW end (A), and a second spot corresponding to the far end of the NW (B). The latter arises from the re emission of the surface plasmon that reaches the end of the NW and scatters into free-space.

NV Center Coupled to Gap Nanowire Waveguides

Building up from the work with NV centers and silver nanowires, the Andersen group went a step further in assembling a single ND between two silver nanowires,

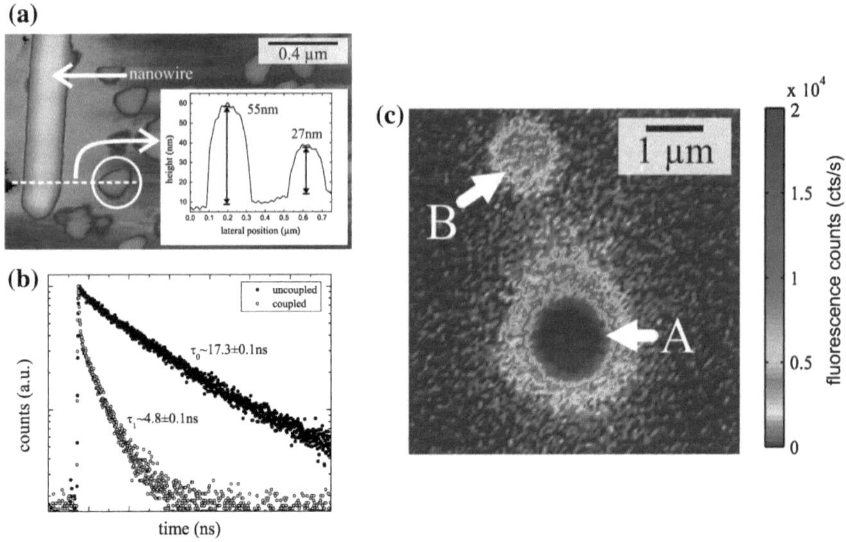

Fig. 4.4 **a** AFM image of the selected ND close-by the metallic nanowire. The inset shows the size of both particles. **b** Fluorescence lifetime measurements of the uncoupled NV center (*black*) and the same NV center when brought in close proximity to the NW (*grey*). **c** Photoluminescence image of the coupled NV center-NW system, taken with the scanning mirror while continuously exciting the emitter. Adapted from [7]

forming a so called gap nanowire plasmon waveguide due to the field confinement generated at the gap between the two nanowires.

In the experiment, a silver nanowire with a length of ~ 10 µm and a diameter of 110 nm was first identified and cut in two halves using the tip of the AFM cantilever. The measured lifetime of the selected ND was of 45.2 ns (black trace in Fig. 4.5e), and a measurement of the second order correlation function $g^{(2)}(\tau)$ confirmed the single emitter characteristics. The nanoparticle was then brought into proximity with one of the silver nanowires. An AFM topography image of the resulting configuration can be seen in Fig. 4.5a. By acquiring a fluorescence image with the scanning mirror, already at this stage coupling into the NW plasmonic mode could be observed, similarly to the previous section (Fig. 4.5b). In addition, the lifetime decreased to a value of 11.9 ns (red trace in Fig. 4.5e), yielding a rate enhancement of 3.8 relative to the position away from the NWs. Finally, the second segment of the NW was re-positioned such that the ND was located at the gap formed between the two nanowires, as can be appreciated in the AFM image in Fig. 4.5c. At this final position the NV center exhibited a lifetime of 5.4 ns for a total enhancement of 8.4 with respect to the isolated ND.

The coupling to the propagating surface plasmon modes was remarkably improved, as can be observed by the enhancement in the intensity at the two outcoupling spots at the ends of the gap nanowire structure in Fig. 4.5d. The

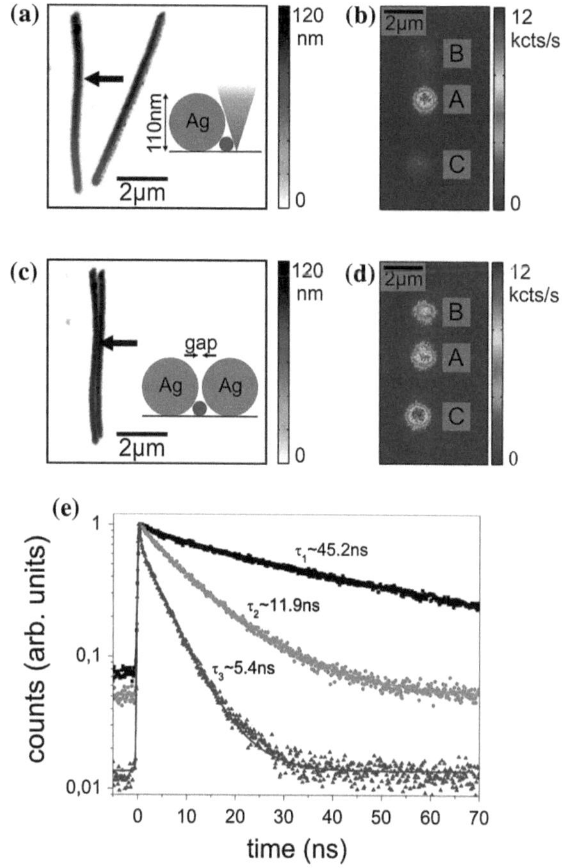

Fig. 4.5 AFM topography and photoluminescence images for the NV center coupled to a single nanowire (**a**, **b**) and to a gap nanowire (**c**, **d**). In both **b** and **d**, the emission spot A corresponds to the position of the ND while B and C correspond to the outcoupling spots at the ends of the nanowires. **e** Lifetime measurements for the case of the bare ND (*black*), coupled to one nanowire (*red*) and to the gap nanowire (*blue*). Adapted from [8]

emission from the wire ends was more than 4 times brighter than for the single nanowire case, while the emission from the position of the ND remained the same.

Coupling NV Emission to Channel Plasmons in V-grooves

The V-groove (VG) plasmonic waveguides are among the most promising candidates for developing a planar plasmonic circuitry platform [3]. These hollow V-shaped channels carved in a metal surface support the propagation of channel plasmon polaritons (CPPs). The CPPs combine unique properties of subwavelength confinement of the electromagnetic fields near the VG bottom, reasonably long propagation and low losses at sharp bends making plasmonic circuitry design flexible and realistic. The integration of nanomirror tapers to the VG terminations also provides an excellent route for the CPP in- and out coupling to free-space optics, opening an easy on-chip access to plasmonic based circuitry [14].

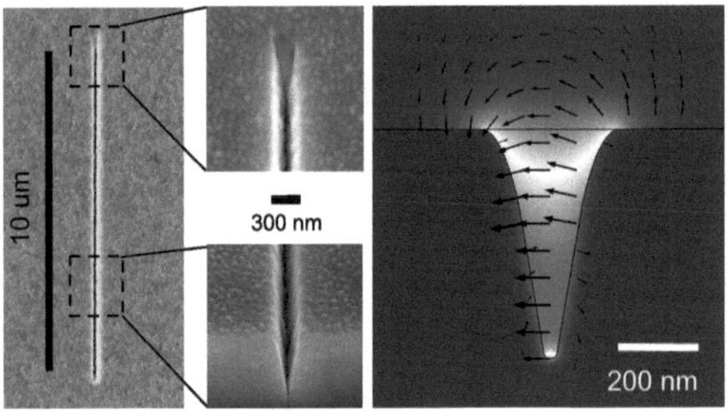

Fig. 4.6 **a** Scanning electron microscope images of a VG fabricated by milling a gold film with a focused ion beam. The *left* image shows a top view of the 10 μm long VG, while the *top right* shows a *top view* of one nanomirror and the *bottom right* a side view of a transversal cut of one VG. **b** Total electric field profile of the VG-supported CPP mode for a wavelength of 650 nm. Adapted from [9]

Recently, together with the group of Sergey Bozhevolnyi and theory work from the group of Francisco García-Vidal, we demonstrated the coupling of a single NV center to the CPPs supported by a VG waveguide [9]. The VGs were prepared by milling a 1.2 μm-thick gold layer by means of a focused ion beam (FIB). They have a width of ~315 nm and a depth of ~510 nm, have an opening angle of ~24°, a length of 10 μm, and are terminated with ~650 nm long width-constant tapers. In Fig. 4.6 we show SEM images of such a VG (Fig. 4.6a), along with the supported propagating mode at a wavelength of 650 nm (Fig. 4.6b).

Simulations on Purcell and Beta Factors

The orientation and position of the dipole moment associated with a quantum emitter are two of the key parameters determining the emission dynamics and the coupling efficiency to the propagating modes. The Purcell and β-factors (coupling efficiency), were calculated using a standard procedure for a bare dipole emitter [15].

To obtain the Purcell factor, the general expression for a radiating point dipole gives:

$$P_F = \frac{6\pi c}{\omega_0} \mathbf{u}_\mu Im\left[\overleftrightarrow{G}(\mathrm{r}_\mu, \mathrm{r}_\mu, \omega)\right] \mathbf{u}_\mu \tag{4.4}$$

where r_μ and $\mathbf{u}_\mu$ are the position of the dipole and the unit vector along the direction of oscillation, respectively. The contraction of the Green's dyadic $\overleftrightarrow{G}(\mathrm{r}_\mu, \mathrm{r}_\mu, \omega)$ is obtained from the real part of the electric field. The coupling efficiency (β-factor) is calculated from the identity

$$\beta = \gamma_{CPP}/\gamma = (\gamma_{CPP}/\gamma_0)/P_F \tag{4.5}$$

where γ and γ_0 are the total decay rates of the emitter inside the VG and in vacuum, respectively, whereas γ_{CPP} is the decay rate into the guided plasmonic modes. In the absence of the diamond host, the ratio γ_{CPP}/γ_0 can be obtained directly from the field profile of the VG eigenmode. The final expression is thus:

$$\beta = \frac{\lambda_0^2}{4\pi c\mu_0 P_F} \frac{|\mathbf{u}_\mu \cdot \mathbf{e}(\mathbf{r}_\mu)|^2}{Re\int dA(e \times \mathbf{h}^*)} \tag{4.6}$$

where **e** and **h** are the electric and magnetic transverse fields of the eigenmode, and μ_0 the vacuum permeability.

When the dipole is oriented transversal to the VG axis, it was found that the β-factor can rise up to 68 % (black open circles in Fig. 4.7a), and that this maximum QE-CPP mode coupling is achieved at distances between 200 and 330 nm from the VG bottom. In this height range, the decay rate increases by a factor of 5

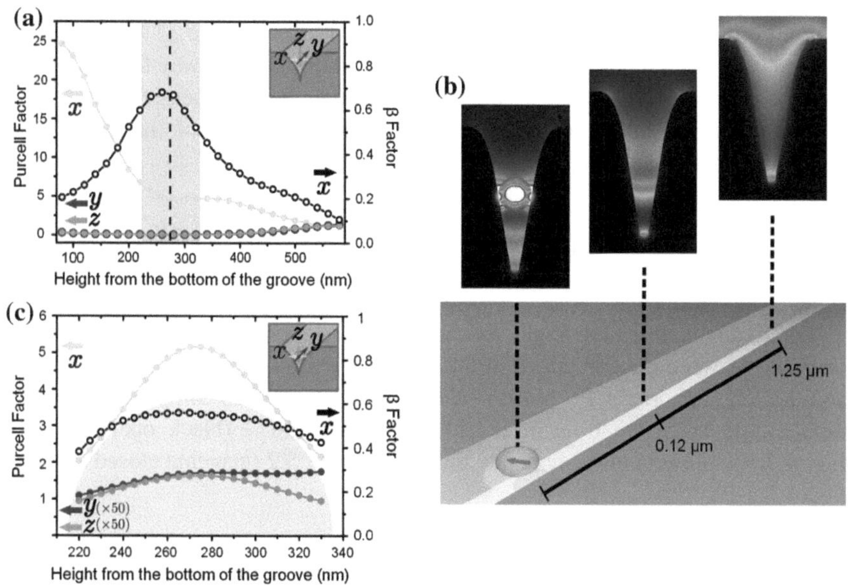

Fig. 4.7 Simulations of a single quantum emitter—channel plasmon coupling. **a** Beta factor (*black open circles*) and Purcell factor (dipole orientation *color codes* as the inset) for a bare dipole placed at different heights from the bottom of the VG. **b** Normalized electric field maps for three different transversal cuts of the VG of the coupled emission from the emitter. **c** Same quantities as in (**a**), but with consideration of a finite diamond shell (60 nm-radius). The β-factor and Purcell factors reach values of 0.56 and 5.2, respectively

as compared to the vacuum decay rate (magenta closed circles in Fig. 4.7a). As the QE approaches the bottom of the VG, the Purcell enhancement reaches higher values, however, the coupling is less efficient, as the decay is dominated by ohmic losses.

These theoretical results shed light on the importance of controlling the position of a single QE inside such a PW (generally, to any plasmonic structure). The NV centers in NDs are ideal candidates to fulfil these conditions required for an efficient coupling to the CPP modes as the diamond host can act as a spacer between the NV center and the metallic surfaces.

We also considered a dipole emitter embedded within a 60 nm-radius ND placed inside the VG and lying within the optimum β-factor region to study the effect of the ND shell on the QE-CPP coupling. When considering the diamond host, however, the system (VG + host) ceases to be translational invariant, and a different expression for the β-factor has to be used. We used the ratio $\beta = W_{CPP}/W$, where W is the total power emitted by the NV inside the VG, and W_{CPP} is the total power emitted into the modes of the VG. The former can be determined from the previously calculated Purcell factor, via $W = P_F W_0$. To calculate the total power coupled to the CPP, the overlap of the fields of the 3D problem, **E**, and the fields of the VG eigenmode, **h** are computed [16].

For convergence reasons it is better to evaluate this overlap at a cross section outside the diamond host, located at a longitudinal distance y_0 away from the dipole (see the panels in Fig. 4.7b). It is necessary to include a compensation factor exp $(2y_0 \mathrm{Im}[\kappa])$ which takes into account the propagation losses of the CPP, which yields the final expression:

$$\beta = \frac{1}{W} e^{2y_0 Im[\kappa]} \langle \mathbf{E}(y_0) | \mathbf{h}(y_0) \rangle = \frac{1}{P_F W_0} e^{2y_0 Im[\kappa]} \frac{\left| \int dA \left(E \times \mathbf{h}^* \right) \right|^2}{Re\left[\int dA \left(e \times \mathbf{h}^* \right) \right]} \tag{4.7}$$

For a dipole oriented along the adequate direction (x-axis in Fig. 4.7), the coupling between the QE placed at the center of the ND (dashed line in Fig. 4.7a) and the VG supported CPP mode is almost as efficient when compared to the bare dipole case, since the β-factor is only reduced to 56 % (black open circles in Fig. 4.7c) while keeping a moderate Purcell factor of 5.2 (magenta closed circles in Fig. 4.7c).

Controlled Assembly of the Hybrid NV-VG System

To position a single ND inside a VG, we combined the use of electron beam lithography (EBL) based assembly of quantum emitters and the scanning probe assembly with the AFM cantilever. First, we coated the substrate containing the VGs with PMMA and exposed an array of 200 nm holes in the vicinity of the structures while protecting the VG (see Fig. 4.8a). Next a solution containing a positively charged polyelectrolyte (poly) diallyldimethylammonium (PDDA) is drop casted onto the structures to perform an electrostatic assisted assembly [17]. After rinsing, we drop casted a solution of NDs (particle radii in the range of

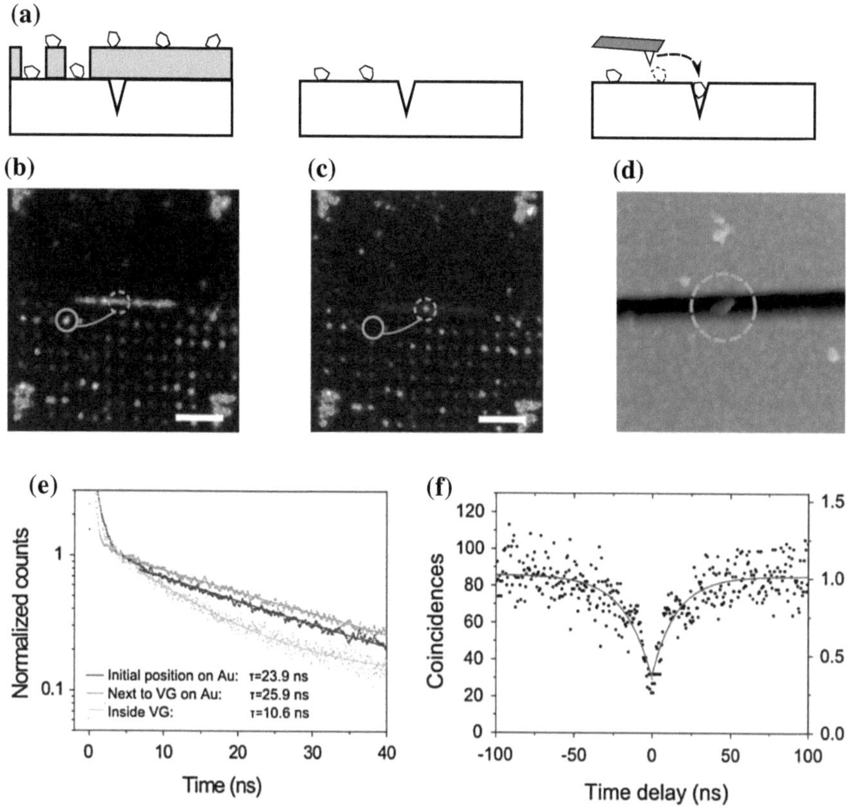

Fig. 4.8 **a** Steps for the positioning of a single ND inside a VG waveguide, from EBL assisted positioning (*left* and *middle*) to AFM precise positioning (*right*). **b–c** Confocal fluorescence microscopy images of the VG and surrounding array of NDs before and after AFM manipulations of the ND hosting a single NV center. **d** AFM image of the ND inside the VG channel. **e** Lifetime measurements of the selected NV at two positions outside the VG (*blue* and *green*) and inside the VG (*orange*). The emitter exhibited a lifetime reduction of ~2.44. (f) Second order autocorrelation of the selected ND. The dip at zero time delay (t = 0) shows the quantum nature of the single NV center. Adapted from [9]

40–80 nm) and let incubate. In a final step we perform a lift-off of the resist to obtain an array of NDs in the vicinity of the VGs (Fig. 4.8a).

The optical setup used for characterization was very similar to the previously described, but independent from the AFM system. Confocal fluorescence microscopy scans under a 532 nm green laser excitation allowed us to locate the optically active NDs (see Fig. 4.8b). Furthermore, with the HBT scheme, we could identify the fluorescent NDs hosting a single NV center ($g^{(2)}(t = 0) < 0.5$) as depicted by the red circle in Fig. 4.8b.

In the second step, we used an AFM in tapping mode for visualization, and in contact mode to move the ND across the Au film and finally into the VG. The red solid circles in Figs. 4.8b–c indicate the location of the chosen ND and the dashed circles show the position at which we intended to relocate this ND inside the VG. We first transferred the ND to a position close to the VG in order to assess the stability of the NV center's emission properties upon the movement. The acquisition of similar lifetime values demonstrates the stability of the emission properties of the chosen NV center under translation of the ND in a homogeneous environment (green and blue traces in Fig. 4.8e).

Finally, the ND was moved inside the VG channel as highlighted with the red dashed circle in the confocal image in Fig. 4.8b–c. A clear visualization of the NV center position within the VG was obtained by setting the excitation polarization parallel to the VG in order to reduce the gold auto-fluorescence from the VG. We observed a lifetime reduction from 25.9 ns to 10.6 ns after positioning the ND inside the VG (Fig. 4.8e), which corresponds to a total decay rate enhancement factor of ~2.44.

To estimate an experimental Purcell factor we did not rely on the measurement performed on the same emitter on the Au film, as the latter can support surface plasmon polaritons that can contribute to the measured lifetime. Instead, we compare the value measured inside the VG to the average of the lifetime distribution measured on single NV centers from the same solution deposited on a glass substrate ($\tau = 24.2 \pm 7.2$ ns), and obtained an experimental Purcell factor of 2.3 ± 0.7.

NV Emission Coupled to Channel Plasmon Modes
To demonstrate the ability of this quantum plasmonic device to couple the NV center emission to the CPP mode supported by the VG, we performed wide-field collection fluorescence imaging around our confocal excitation spot by using an EMCCD camera, allowing us to observe not only the emission from the NV position but also from the two outcoupling spots at the VG ends. The implementation of an EMCCD camera is more versatile than the use of a scanning mirror to image the surrounding area of the emitter as it allows for real-time monitoring of the signals. The polarization dependence of the outcoupling spots was evidenced by the wide-field collection fluorescence images for four combinations of excitation and collection polarizations, i.e. combining polarizations parallel and transversal to VG axis (Figs. 4.9a–e). The observation of outcoupling spots mainly for the collection polarization transversal to the VG axis (Fig. 4.9b–c) is in accordance to the supported CPP mode of the VG (TE polarized as seen in Fig. 4.6b).

Furthermore we could measure the lifetime of the signal collected from the main outcoupling spot and the ones corresponding to the position of the nanomirrors, and fitted the same value regardless of the collection channel (Fig. 4.9g), which is consistent with the fact that the lifetime is a measure of the total decay rate of an emitter and thus not possible to measure independently the contributions from the radiative, non-radiative and plasmon decay channels.

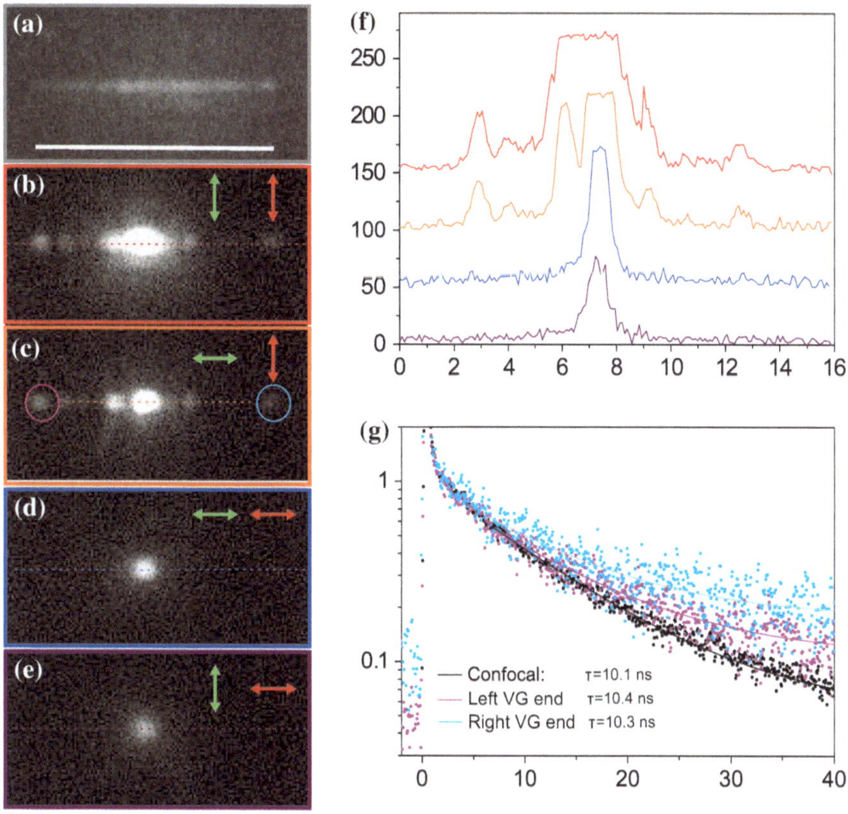

Fig. 4.9 Coupled emission of a single NV center to CPPs in a VG. **a–e** Wide-field collection fluorescence microscopy images of the assembled NV-VG device. **a** Image in wide field illumination and no long-pass filter showing the contour of the VG. **b–c** Collection polarization transversal to the VG axis for two orthogonal excitation polarizations (*green arrows*), the emission spots at the VG ends corresponding to CPP coupling can be observed. **d–e** Collection polarization parallel to the VG axis for two orthogonal excitation polarizations (*green arrows*), only the emission spot at the NV position is appreciated. **f** Fluorescence intensity along the dashed lines in (**b–e**). **g** Lifetime measurements for the polarization configuration (**c**) at three different collection positions, namely at the NV position (*black*) and the VG ends (*cyan* and *magenta*). Adapted from [9]

4.3 Optical Trapping as a Positioning Tool

In 1970, Arthur Ashkin introduced the concept of optical trapping of dielectric particles [18]. This technique relies on the concept that light can exert forces onto dielectric matter by means of momentum exchange. Since then, it has been successfully implemented in a wide variety of fields ranging from biophysics to study the mechanics and interaction of bio-molecules attached to dielectric spheres, to the field of optomechanics to study the fundamentals of motion of these micron and nano-sized objects.

Optical trapping is a suitable tool for achieving dynamical control over the position of a quantum emitter embedded in a dielectric matrix, such as the case of the NV centre in NDs. The group of David Awschalom reported in 2012 optically trapping NDs containing multiple NV centers within an ensemble of particles, and performing electron spin resonance measurements to identify the multiple emitters of the particle [19]. Simultaneously, our group investigated the optical manipulation of individual NDs containing only a single NV center at the focus of the trap, opening the route to control the interaction of such a single emitter with photonic and plasmonic structures [20].

4.3.1 Experimental Platform to Optically Trap a Single NV Center

The schematic in Fig. 4.10 depicts the basic features of our experimental platform. The NDs were dispersed in a 5:1 glycerol/water mix (refractive index n = 1.46) confined in a static fluidic chamber. Due to the small size of the ND, the stability of the trap is enhanced by cancelling out the scattering forces and producing interference along the optical axis. This was achieved by tightly focusing (NA = 1.2) to the same point two equally intense and collinearly polarized counter-propagating beams from an infrared laser (1064 nm). This strategy enabled trapping and manipulating a single ND as small as 50 nm in all three spatial directions. In order to excite the NV centers we superimpose a 532 nm laser with the trapping laser (green beam in Fig. 4.10). The emitted photoluminescence (orange) passes through filters and is sent to different detection channels. One channel has two APDs for the HBT experiments to identify a single emitter (anti-bunching curve as shown in the top left inset of Fig. 4.10). In the top channel we use a highly sensitive EMCCD camera in fluorescence imaging mode to follow the diffusion of the NDs in real time.

To immobilize the fluorescent ND of interest we bring it close to the focus of the dual-beam trap by moving the fluidic chamber with a piezo stage. Additionally a nearby microwave antenna allows us to perform electron spin resonance (ESR) spectroscopy to identify the $m_s = 0$, $m_s = \pm 1$ transition of the spin triplet ground state (lower left inset in Fig. 4.10). Magnetic coils (not shown in the schematic) enable us to apply external magnetic fields which induce a splitting of the $m_s = \pm 1$ states to determine the orientation of the NV axis. We werc able to demonstrate trapping of a single ND containing a single NV center with a particle confinement of less than 70 nm in each direction and allowing movements in the chamber over distances as large as 100 μm, and with velocities up to 10 μ/s.

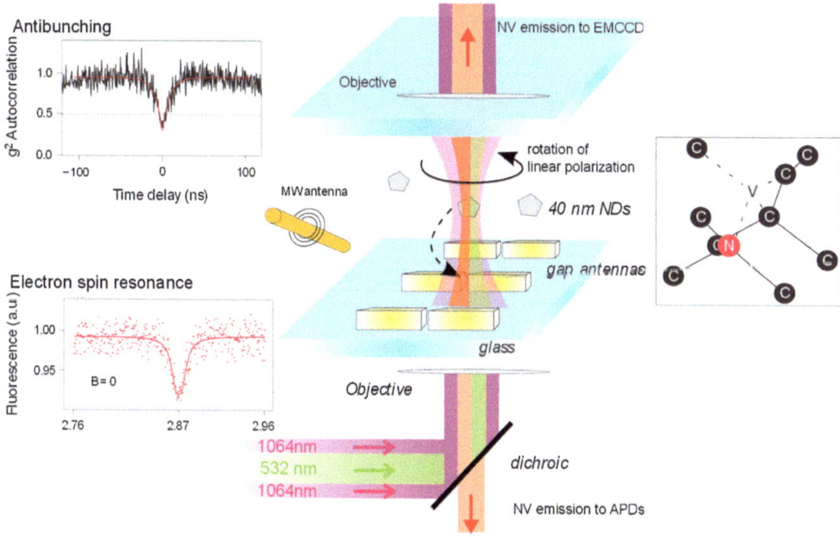

Fig. 4.10 3D Optical trapping and manipulation of a single NV center. A near infrared laser (1064 nm represented in *magenta*) is focused through high NA objectives to trap a ND containing a single NV center. A superimposed 532 nm laser (represented in *green*) excites the NV and its fluorescence (*orange*) is monitored with an EMCCD camera. The *upper left* inset shows the typical auto correlation measurement showing single emitter characteristics, while the *lower left* inset shows an ESR measurement exemplifying the transition frequency of the $m = \pm 1$ ground state. Patterned gold antennas designed to exhibit a resonance at 1064 nm are used for plasmonic based positioning of single NDs.

Control of the NV Axis Orientation

In general, NDs exhibit an asymmetrical shape that give the particles a preferential orientation inside the optical trap with respect to the trapping E-field due to the higher polarizability of the elongated axis of the particle. Consequently, changing the orientation of the polarization axis of the trapping laser leads to a control in the orientation of the NV axis with respect to the optical axis. Figure 4.11 shows the change in the ESR frequency spectrum as a function of the polarization orientation of the trapping beam. This dynamic control over the orientation of the NV center´s dipole moment is unprecedented for any other manipulation technique and highly relevant to the coupling of the NV centers to photonic and plasmonic structures with preferred mode polarization.

4.3.2 Surface Plasmon Based Trapping

Positioning NV centers close to plasmonic structures had been so far achieved in a controlled manner by using EBL approaches, [21] or by mechanically pushing particles with an AFM tip as described in detail in the previous sections. The ability

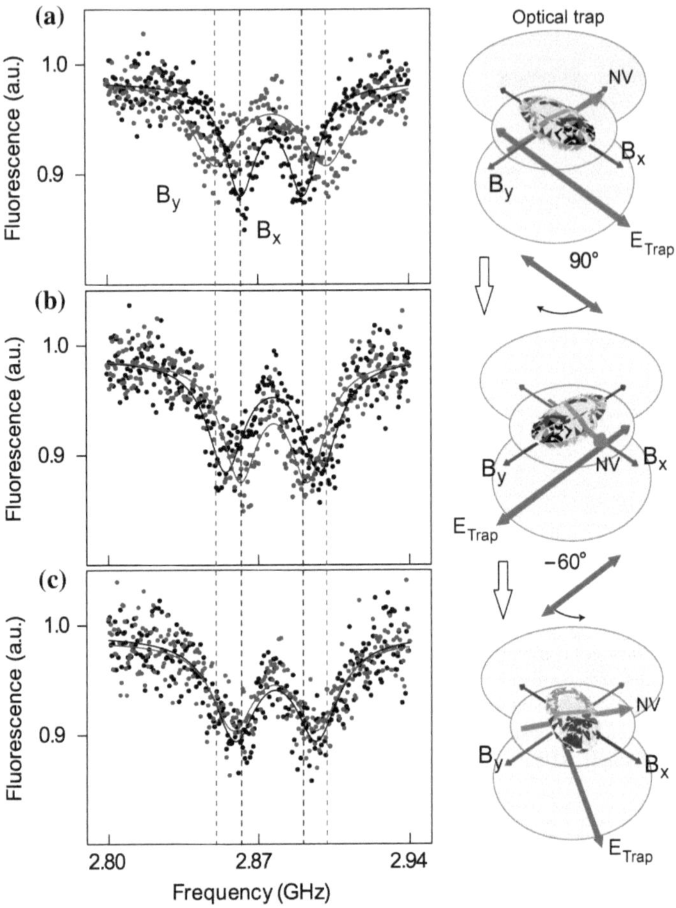

Fig. 4.11 Orientation control of an optically trapped single spin. Control of the NV axis orientation by rotating the polarization of the trapping laser. **a** Two consecutive microwave frequency sweeps were performed for each orientation of the ND with a constant magnetic field directed in X and Y, respectively. **b–c** Same measurements after rotation in the polarization by an angle of 90° and 60° respectively with respect to the previous orientation. The *dashed lines* indicate the initial position of the magnetic field splitting. Due to internal strain, the ground state for this NV was already degenerate for B = 0

to trap single NDs and even control their orientation enables optical trapping as a suitable tool to position single emitters in the vicinity of a plasmonic structure. However, since the trapping strength is proportional to the volume of the particle, small particles require high powers to be trapped and such intensities can damage the plasmonic nanostructures due to heat generation. Localized surface plasmons offer the ability to trap and implant small nanoparticles due to the force field produced by the hotspots in plasmonic nanostructures, enabling to trap particles down to 10 nm upon relatively low incident laser intensities.

Recently, our group demonstrated near-field assisted trapping and positioning of NDs containing NV centers onto plasmonic structures [22]. In the experiment, a glass cover slip was patterned with a periodic array of gold gap antennas, each of them formed by two identical rods separated by a 40 nm gap. The dimensions of the rods (145 nm long, 56 nm wide, and 55 nm high) were chosen such that the gap antenna mode overlaps with the trapping wavelength at 1064 nm.

The approach consisted in focusing the 1064 nm laser on a predefined antenna (3 mW incident power, and linear polarization along the antenna long axis). A ND diffusing near the antenna experiences an optical potential that arises from the combined effect of the trapping beam and the near field hotspots. As a result of that, the ND gets automatically guided towards the regions of maximum intensity and eventually sticks to the antenna. The color coded SEM images in Fig. 4.12a display the three most frequent configurations observed after the trapping process and correspond to the field enhancement positions of such a gap nanorod antenna.

In addition to SEM imaging, we compare the fluorescence intensity map of the antenna array before and after ND trapping and positioning (Fig. 4.12b–c) upon excitation with a 532 nm green laser. As can be seen in Fig. 4.12c, we intentionally positioned the trapping laser on a sequence of gap-antennas forming a cross-like pattern, for which the antennas showed a clear enhancement of the fluorescence intensity after the particle was trapped.

The interaction of the immobilized NV with the nano-antenna is quantified by analyzing the change in its fluorescence lifetime (Fig. 4.12d). Figure 4.12e shows an histogram of the lifetime measured on different hybrid ND gap-antenna structures (red bars) compared with the lifetime measurements of NDs adsorbed on an un-patterned glass coverslip (blue bars). We observed a lifetime distribution centered around 8.5 ns $\pm$ 7 ns, compared to the 17 ns $\pm$ 3 ns for the uncoupled case. The reduction in lifetime and the broadening of the distribution were in accordance to theory simulations, where the former is a consequence of the Purcell effect due to the presence of the hot-spot in the vicinity of the NV center, while the latter results from the broad dispersion of the magnitude of the Purcell factor over the different regions where the particles are most likely trapped (center of the gap and extremities of the nanorods).

Optical Switch Mediated by Plasmon Coupling

One of the ultimate goals for future optical communication technologies is to rely on single quantum emitters that we can switch on and off in a controlled way to construct so called optical transistors or optical switches. Recently, our group demonstrated optical switching in NV centers by illuminating with a non-resonant near infrared (NIR) laser while exciting the NV with a green laser [23]. The 1064 nm laser drives the excited state back to the ground state via a non-radiative decay, thus achieving a modulation of the fluorescence intensity when gating the NIR laser. A fluorescence drop of nearly 80 % was demonstrated with NIR gating laser powers of almost 80 mW for NDs lying on a glass substrate. This laser power

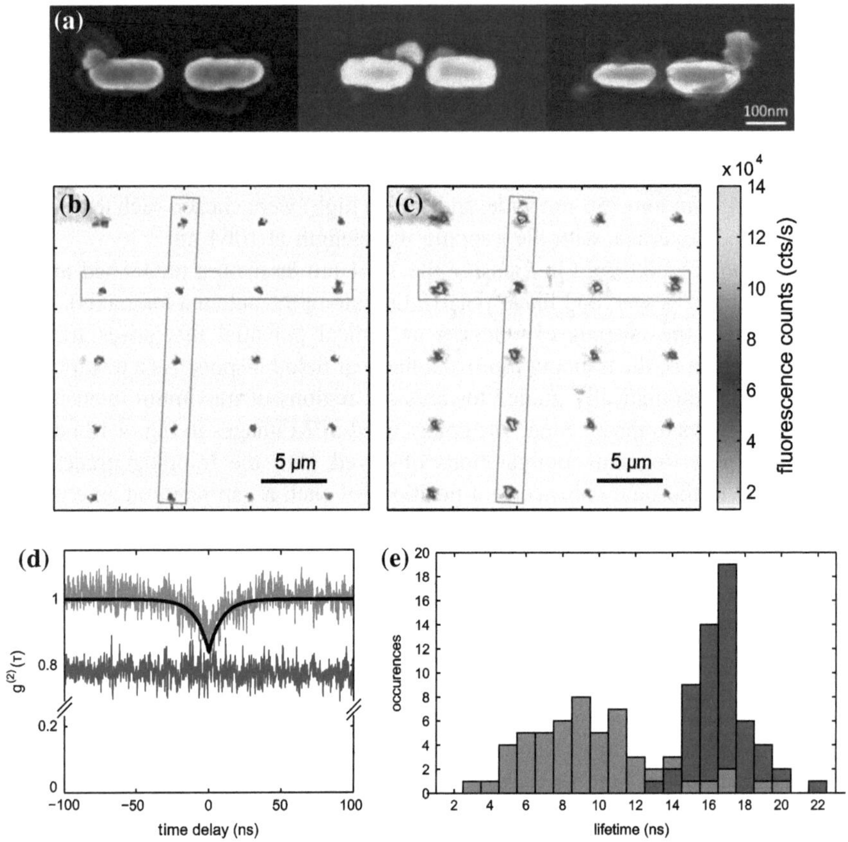

Fig. 4.12 **a** SEM images of three different antennas with positioned NDs (*false colors*). The positions match with the location of the maximum field enhancements for these antennas. **b**–**c** Fluorescence maps before and after the positioning of single NDs containing NV centers, showing a clear increase in the fluorescence intensity for the hybrid NV-antenna devices. **d** Second order correlation function from NV centres positioned in the hot spot of a gold gap antenna. The *blue curve* is the correlation measurement of the background fluorescence from the antenna. **e** Statistical distribution of the lifetime of NVs adsorbed to a glass surface (*blue*) and those positioned at the gap antennas (*red*) via the plasmon assisted trapping and positioning. Adapted from [22]

in practice is still very high for application to an optical modulator. Nevertheless we can circumvent this limitation by exploiting the coupled NV centers to the plasmonic antennas shown earlier. They are conveniently designed to have a resonance at the NIR laser and thus provide an enhancement of the NIR energy at the position of the NV centers. These gap antennas exhibit a plasmon enhancement of the excitation energy preferentially when the polarization is along the long axis of the antenna. In Fig. 4.13 we show the switching properties of the hybrid NV-antenna system for two different polarization directions on the NIR laser, where we can

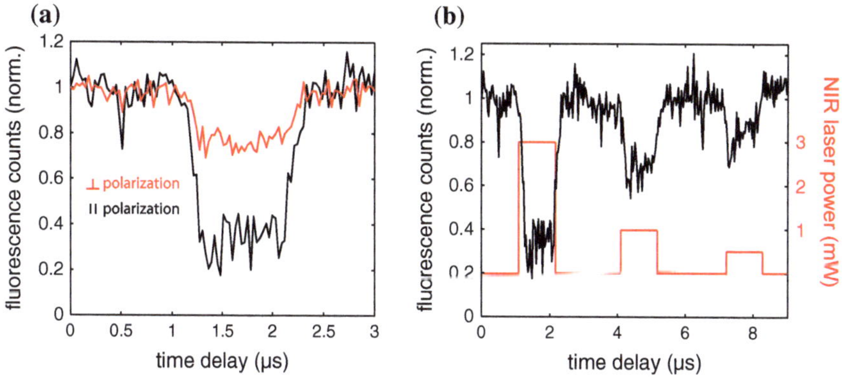

Fig. 4.13 Modulation of the fluorescence from an NV that has been positioned at a gap antenna. **a** With 3.3 mW of NIR laser, while maintaining the *green* laser excitation power constant, a modulation of the NV fluorescence is observed. The modulation is stronger when the polarization of the NIR laser is along the length of the antenna, as opposed to the transversal polarization where the hot-spots are not excited. **b** Using the near-field enhancement associated with the antenna plasmon resonance, we show that a NIR laser with power as low as 0.5 mW is sufficient to modulate the fluorescence

appreciate that for the same laser power an 80 % intensity modulation is observed for the polarization along the antenna while only a 20 % modulation when the polarization is across the antenna. Furthermore we demonstrated that only a 3 mW laser power was needed to modulate the signal to 80 % of its intensity compared to the 80 mW required in the absence of the antenna.

4.4 Conclusions and Outlook

Surface plasmon modes provide confinement of electromagnetic energy at very small volumes and can influence the excitation and emission dynamics of a single quantum emitter placed in the mode volume. The accuracy in assembling a solid state quantum emitter with a plasmonic structure is crucial, and has been one of the focus points in the field of quantum plasmonics. In this chapter we have reviewed the main realizations on the controlled interaction between a single NV center in NDs with a variety of metallic structures supporting surface plasmon polaritons modes. It was shown that it is possible to modify the emission rates of the NV centers by coupling them to metallic nanoantennas, and also to guide their emission along different plasmonic waveguides.

Now that the assembly methods are well established and at reach to research groups with access to lithography, atomic force microscopes or optical tweezing setups, the field is expected to move into concrete proof-of-concept applications such as on-chip routing of single photons or plasmon mediated entanglement between multiple defect centers. In addition, in recent years there has been progress in the development of ND particles with a controlled number of defect centers and with specific geometries, such as diamond nanopilars with implanted defects, [24] and not limited to the nitrogen-vacancy but also with other promising defects such as the silicon-vacancy center [25]. We expect that advances in material science will contribute to improve the performances of hybrid systems based on the coupling of color centers and plasmonic structures.

Acknowledgements The authors acknowledge the European Community's Seventh Framework Programme (grant ERC- Plasmolight; no. 259196), the Severo Ochoa Excellence Grant and Fundació privada CELLEX. E.B.-U. acknowledges financial support from the Spanish MICINN and Universidad de Costa Rica.

References

1. M.W. Doherty et al., The nitrogen-vacancy colour centre in diamond. Phys. Rep. **528**, 1–45 (2013)
2. W.L. Barnes, A. Dereux, T.W. Ebbesen, Surface plasmon subwavelength optics. Nature **424**, 824–830 (2003)
3. T.W. Ebbesen, C. Genet, S.I. Bozhevolnyi, Surface-plasmon circuitry. Phys. Today **61**, 44–50 (2008)
4. T. Junno, K. Deppert, Controlled manipulation of nanoparticles with an atomic force microscope. Appl. Phys. Lett. **66**, 3627–3629 (1995)
5. S. Schietinger, M. Barth, T. Aichele, O. Benson, Plasmon-enhanced single photon emission from a nanoassembled metal—diamond hybrid structure at room temperature. Nano Lett. **9**, 1694–1698 (2009)
6. A.W. Schell et al., Single defect centers in diamond nanocrystals as quantum probes for plasmonic nanostructures. Opt. Express **19**, 7914–7920 (2011)
7. A. Huck, S. Kumar, A. Shakoor, U.L., Andersen, Controlled coupling of a single nitrogen-vacancy center to a silver nanowire. Phys. Rev. Lett. **106**, (2011)
8. S. Kumar, A. Huck, U.L. Andersen, Efficient coupling of a single diamond color center to propagating plasmonic gap modes. Nano Lett. **13**, 1221–1225 (2013)
9. E. Bermúdez-Ureña et al., Coupling of individual quantum emitters to channel plasmons. Nat. Commun. **6**, 20 (2015)
10. L. Novotny, N. van Hulst, Antennas for light. Nat. Photonics **5**, 83–90 (2011)
11. E. Bermúdez-Ureña et al., Excitation enhancement of a quantum dot coupled to a plasmonic antenna. Adv. Mater. **24**, 314–320 (2012)
12. A.G. Curto et al., Unidirectional emission of a quantum dot coupled to a nanoantenna. Science **329**, 930–933 (2010)
13. R. Kolesov et al., Wave-particle duality of single surface plasmon polaritons. Nat. Phys. **5**, 470–474 (2009)
14. Radko, I. P., Holmgaard, T., Han, Z., Pedersen, K. & Bozhevolnyi, S. I. Efficient channel-plasmon excitation by nano-mirrors. *Appl. Phys. Lett.* **99**, - (2011)
15. D. Martín-Cano, L. Martín-Moreno, F.J. García-Vidal, E. Moreno, Resonance energy transfer and superradiance mediated by plasmonic nanowaveguides. Nano Lett. **10**, 3129–3134 (2010)
16. E. Moreno, S.G. Rodrigo, S.I. Bozhevolnyi, L. Martín-Moreno, F.J. García-Vidal, Guiding and focusing of electromagnetic fields with wedge plasmon polaritons. Phys. Rev. Lett. **100**, 23901 (2008)
17. H. Fredriksson et al., Hole-mask colloidal lithography. Adv. Mater. **19**, 4297–4302 (2007)

18. A. Ashkin, Acceleration and trapping of particles by radiation pressure. Phys. Rev. Lett. **24**, 156–159 (1970)
19. V.R. Horowitz, B.J. Alemán, D.J. Christle, A.N. Cleland, D.D. Awschalom, Electron spin resonance of nitrogen-vacancy centers in optically trapped nanodiamonds. Proc. Natl. Acad. Sci. USA **109**, 13493–13497 (2012)
20. M. Geiselmann et al., Three-dimensional optical manipulation of a single electron spin. Nat. Nanotechnol. **8**, 175–179 (2013)
21. W. Pfaff, A. Vos, R. Hanson, Top-down fabrication of plasmonic nanostructures for deterministic coupling to single quantum emitters. J. Appl. Phys. **113**, 024310 (2013)
22. M. Geiselmann, R. Marty, J. Renger, F.J. García De Abajo, R. Quidant, Deterministic optical-near-field-assisted positioning of nitrogen vacancy centers. Nano Lett. **14**, 1520–1525 (2014)
23. M. Geiselmann, R. Marty, F.J. García de Abajo, R. Quidant, Fast optical modulation of the fluorescence from a single nitrogen–vacancy centre. Nat. Phys. **9**, 785–789 (2013)
24. T.M. Babinec et al., A diamond nanowire single-photon source. Nat. Nanotechnol. **5**, 195–199 (2010)
25. E. Neu et al, Single photon emission from silicon-vacancy colour centres in chemical vapour deposition nano-diamonds on iridium. New J. Phys. **13** (2011)

Chapter 5
Hyperbolic Metamaterials for Single-Photon Sources and Nanolasers

M.Y. Shalaginov, R. Chandrasekar, S. Bogdanov, Z. Wang, X. Meng, O.A. Makarova, A. Lagutchev, A.V. Kildishev, A. Boltasseva and V.M. Shalaev

Abstract Hyperbolic metamaterials are anisotropic media that behave as metals or as dielectrics depending on light polarization. These plasmonic materials constitute a versatile platform for promoting both spontaneous and stimulated emission for a broad range of emitter wavelengths. We analyze experimental realizations of a single–photon source and of a plasmonic laser based on two different architectures of hyperbolic metamaterials. At the heart of this material capability lies the high broadband photonic density of states originating from a rich structure of confined plasmonic modes.

M.Y. Shalaginov (✉) · R. Chandrasekar · S. Bogdanov · Z. Wang · X. Meng · O.A. Makarova · A. Lagutchev · A.V. Kildishev · A. Boltasseva · V.M. Shalaev
School of Electrical & Computer Engineering, Birck Nanotechnology Center, and Purdue Quantum Center, Purdue University, West Lafayette, IN 47907, USA
e-mail: shalaginov@purdue.edu

R. Chandrasekar
e-mail: chandrar@purdue.edu

S. Bogdanov
e-mail: sbogdan@purdue.edu

Z. Wang
e-mail: wang1973@purdue.edu

X. Meng
e-mail: meng14@purdue.edu

O.A. Makarova
e-mail: omakarov@purdue.edu

A. Lagutchev
e-mail: lagutch@purdue.edu

A.V. Kildishev
e-mail: kildishev@purdue.edu

A. Boltasseva
e-mail: aeb@purdue.edu

V.M. Shalaev
e-mail: shalaev@purdue.edu

S.I. Bozhevolnyi et al. (eds.), *Quantum Plasmonics*,
Springer Series in Solid-State Sciences 185,
DOI 10.1007/978-3-319-45820-5_5

5.1 Introduction

The advent of nanofabrication has opened entirely new ways for designing optical materials, by virtue of the freedom to generate meta-atoms smaller than the wavelength with tailored electromagnetic responses. This has allowed for improved spatial control of bulk parameters, such as permittivity and permeability, and to engineer the dispersion of optical materials with properties unavailable in nature, termed as optical metamaterials. Metamaterials offer new avenues for manipulation of light, leading to such unusual applications as high-resolution imaging, high-resolution lithography, and emission lifetime engineering. This chapter is focused on a subclass of metamaterials exhibiting hyperbolic dispersion that serves as a flexible platform for engineering optical phenomena.

Hyperbolic metamaterials (HMM), uniaxial nanostructured materials that combine the properties of transparent dielectrics and reflective metals, first attracted the attention of researchers in the middle of last century. These efforts were stimulated by the problem of propagation of radio waves in the Earth's ionosphere [1, 2], and, more generally, by the behavior of the electromagnetic waves in a plasma of electrons and ions upon the applied permanent magnetic field. Along with the anisotropic plasmas, stratified man-made hyperbolic materials were studied for radiofrequency applications of transmission lines [3].

Hyperbolic composites (also known as media with indefinite permittivity and permeability tensors [4]) and some homogeneous materials with hyperbolic dispersion were experimentally realized across the optical spectrum, from UV to visible, and from near-IR to mid-IR frequencies. Subwavelength imaging [5, 6], high-resolution lithography [7], lifetime engineering [8], and new approaches to

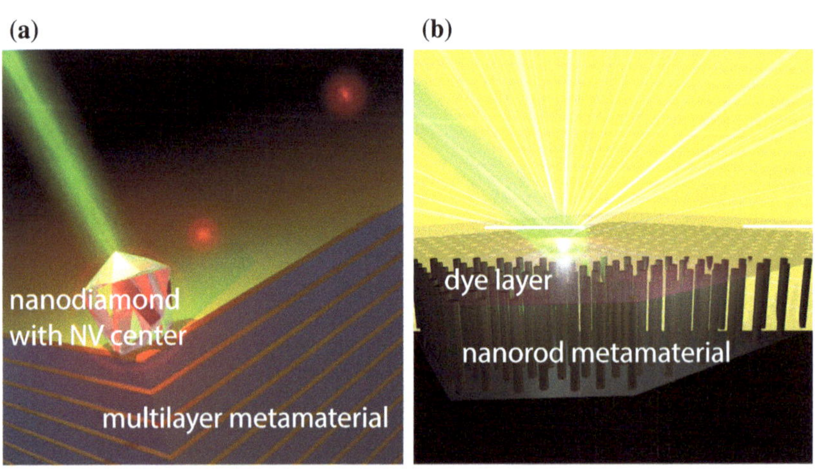

Fig. 5.1 Illustration of **a** multilayer hyperbolic superlattices for enhancing emission properties of single color centers in *diamond* and **b** hyperbolic nanorod arrays for lasing action with dye molecules. Figures reproduced with permission: **a** courtesy of N. Kinsey, **b** [13] from Copyright 2016 ACS

enhance the nonlinear response [9] of optical structures, have all been demonstrated in hyperbolic structures. Hyperbolic media continue to be of great interest to the research community with possible applications emerging in heat transport [10] and acoustics [11]. Although hyperbolic materials also exist naturally, their applications in the visible range are limited by losses and fixed bandwidth [12].

In this chapter, we first discuss the physics of light-matter interaction in hyperbolic metamaterials. Secondly, we consider how multilayer-based metamaterials could enhance the emission from single atom-like defects, such as color centers in diamond (see Fig. 5.1a). Next, we provide experimental observation of lasing using nanorod-based HMMs (see Fig. 5.1b) and corresponding theoretical interpretations.

5.2 Fundamentals of Hyperbolic Metamaterials

Oscillating electric dipoles placed inside or in the vicinity of HMMs see their radiated power enhanced over a broad spectral range. This enhancement is a result of a broadband singularity in the photonic local density of states (LDOS) within the HMM [8]. The photonic LDOS, similar to its electronic counterpart, can be quantified as the volume in k-space between iso-frequency surfaces. For extraordinary waves in a uniaxial anisotropic medium with dielectric tensor $\vec{\varepsilon} = diag[\varepsilon_{\parallel}, \varepsilon_{\parallel}, \varepsilon_{\perp}]$, the iso-frequency surfaces are defined by the following equation:

$$\omega^2/c^2 = k_{\parallel}^2/\varepsilon_{\perp} + k_{\perp}^2/\varepsilon_{\parallel}$$

where subscripts $\perp$ and $\parallel$ indicate the directions perpendicular and parallel to the surface plane, respectively. In the case of dielectric materials with $\varepsilon_{\perp}, \varepsilon_{\parallel} > 0$, LDOS is equivalent to the volume of an infinitesimally thin spheroidal shell in k-space (see Fig. 5.1a) that separates two neighboring iso-frequency surfaces. However, in a medium with extreme optical anisotropy, $\varepsilon_{\perp}$ and $\varepsilon_{\parallel}$ are of opposite signs which produces a hyperboloidal shell, two-sheeted (type I, see Fig. 5.2b) or one-sheeted (type II, see Fig. 5.2c), whose volume is infinitely large (i.e. broadband singularity in LDOS appears). As a result, such a medium allows the propagation of high-k modes with arbitrarily large wavevectors. This hyperbolic regime can be engineered to exist over a broad wavelength range.

The HMM can be practically realized as either a lamellar structure consisting of alternating subwavelength-thick layers of metallic and dielectric materials [14] (see Fig. 5.1a) or as an array of metal nanorods embedded into a dielectric host matrix [15] (see Fig. 5.1b). Both HMM implementations can be well described in the approximation of effective medium theory (EMT). Corresponding expressions for effective permittivities are provided in the Table 5.1. In some cases, the fields in the metamaterials can exhibit strong spatial variation resulting in a nonlocal response. Hence special corrections should be introduced into the EMT formulas [16, 17].

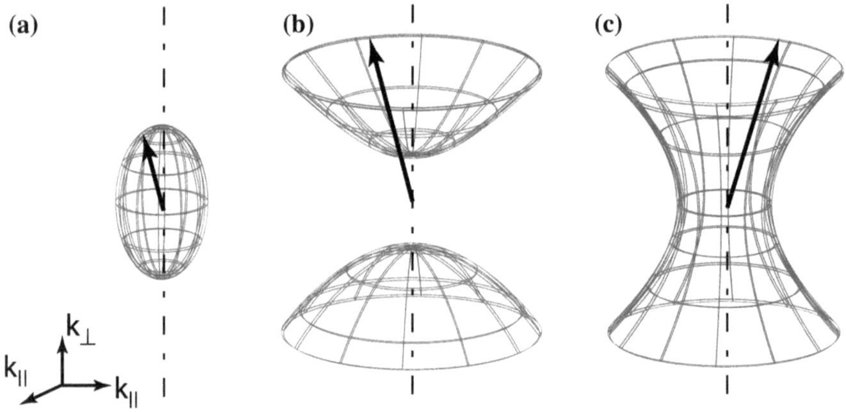

Fig. 5.2 Iso-frequency surfaces in k-space for uniaxial media: **a** spheroid in the case of an anisotropic dielectric ($\varepsilon_\perp, \varepsilon_\parallel > 0$), **b**, **c** hyperboloids (type I and II, respectively) in the cases of extremely anisotropic media ($\varepsilon_\perp < 0$, $\varepsilon_\parallel > 0$ (**b**); $\varepsilon_\perp > 0$, $\varepsilon_\parallel < 0$ (**c**))

Table 5.1 Components of permittivity tensor as defined by effective medium theory (EMT) for metal-dielectric metamaterials implemented as metal nanorod arrays in a dielectric matrix and as a stack of alternating layers of metal and dielectric

Nanorod array		Multilayer stack	
$\varepsilon_\perp = \varepsilon_m f + \varepsilon_d(1-f)$	(1)	$\varepsilon_\perp = \frac{\varepsilon_m \varepsilon_d}{\varepsilon_d f + \varepsilon_m(1-f)}$	(3)
$\varepsilon_\parallel = \varepsilon_d \frac{\varepsilon_m(1+f)+\varepsilon_d(1-f)}{\varepsilon_m(1-f)+\varepsilon_d(1+f)}$	(2)	$\varepsilon_\parallel = \varepsilon_m f + \varepsilon_d(1-f)$	(4)

This broadband optical anisotropy and the resulting high photonic density of states in HMMs can be utilized for engineering integrated optical devices such as deterministic single-photon sources [18] and plasmonic lasers [19].

5.3 Enhancement of Single-Photon Emission from Color Centers in Diamond

The photon is a workhorse of quantum information processing [20], including quantum computation [21], quantum cryptography [22], quantum communication [23] and teleportation [24]. Generation and detection of single photons using quantum emitters plays an important role in applied aspects of quantum photonics.

Over the past 10 years, numerous types of deterministic single-photon sources have been demonstrated [18], including trapped atoms and ions, single molecules, color centers, atomic ensembles, quantum dots, and mesoscopic quantum wells. A nitrogen-vacancy (NV) color center in diamond, formed by a substitutional nitrogen atom and a vacancy at an adjacent lattice site, is of particular interest for photonic quantum technologies [25]. As a single-photon source, it is resistant against photobleaching and operates in the single-photon regime at room temperature [26]. Additionally, an NV center can be well localized, its intrinsic quantum yield is close to unity [25], and its optical transition can be tuned via external electric [27] and/or magnetic fields [28].

The applications of an NV center as a single-photon generator and spin-photon interface could substantially benefit from an enhanced emission rate and collection efficiency. An emission rate enhancement can be achieved by engineering the electromagnetic environment of the emitter utilizing the Purcell effect [29]. So far, this has been accomplished by using resonant photonic structures, such as microspherical resonators [30], photonic crystal microcavities [31], and photonic crystal nanobeams [32], which are all bandwidth limited. Coupling NV centers to HMMs [33, 34] allows for taking advantage of the entire defect radiation spectrum. The HMM provides numerous extra decay channels in a broad spectral range for the coupled emitter [33, 35, 36, 37]. The excitation is transferred into high–k metamaterial modes which result from hybridization of surface plasmon polaritons at the interfaces of the layers constituting the HMM. The high-k waves could be further outcoupled into free space and significantly contribute to the emission signal [35, 37].

For enhancing the NV center emission we epitaxially grew a hyperbolic metamaterial superlattice composed of plasmonic titanium nitride (TiN) and dielectric aluminum scandium nitride ($Al_xSc_{1-x}N$) [14]. TiN is a novel plasmonic material known for its CMOS compatibility, mechanical strength, and thermal stability at high temperatures (melting point >2700°C) [38]. This material can be epitaxially deposited on a variety of different substrates, such as magnesium oxide, aluminum scandium nitride, and c-sapphire, in the form of ultra–thin (<5 nm) films. Successful material deposition with such small thicknesses is of fundamental importance for achieving a significant increase in LDOS, since the LDOS for a multilayer HMM is inversely related to the cube of the layer thickness [39].

Nanodiamonds used in this experiments were dispersed on the HMM surface and had an average size of 50 nm as shown by the SEM scan in Fig. 5.3. NV center emission was collected using high numerical aperture oil immersion objective. A 60-nm–thick layer of polyvinyl alcohol (PVA, 1.5 % w/v) was deposited on top of the sample to immobilize and separate nanodiamonds from the immersion oil. The immersion oil was needed for efficient collection of emitted light by an objective lens with the high numerical aperture.

Nanodiamonds were also deposited by the same method onto standard 150 μm–thick glass coverslips, which were used as control samples. Since the refractive indices of these glass substrates (1.53), immersion oil (1.52) and PVA layer (1.50) were closely matching, the nanodiamonds in the reference samples were effectively immersed into an infinite homogeneous medium with a refractive index of ~1.5.

A $TiN/Al_{0.7}Sc_{0.3}N$ superlattice was epitaxially grown on a 0.5-mm-thick, [001]–oriented magnesium oxide (MgO) substrate using reactive DC magnetron sputtering at 750 °C. The metamaterial was implemented as an epitaxial stack of 10 pairs of layers each consisting of an 8.5-nm-thick film of TiN and a 6.3-nm-thick film of

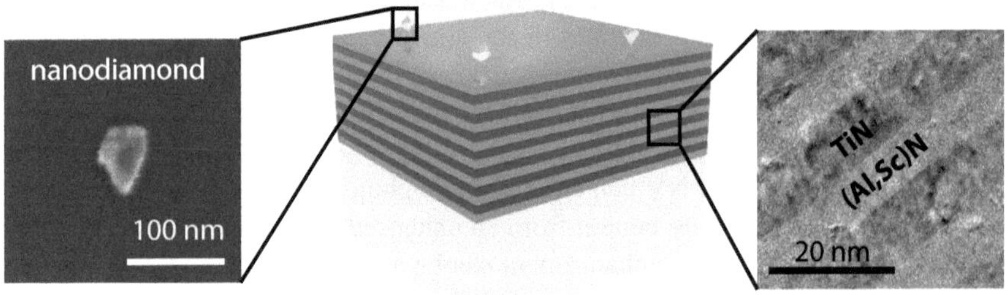

Fig. 5.3 *Center* sample structure composed of TiN/(Al,Sc)N superlattice grown on *top* of MgO and nanodiamonds spin-coated on *top* of the superlattice. *Left* SEM scan of nanodiamonds on *top* of the HMM sample. Average nanodiamond size was 50 nm. *Right* cross-sectional TEM image of TiN-$Al_{0.7}Sc_{0.3}N$ superlattice. Thicknesses of TiN and $Al_{0.7}Sc_{0.3}N$ layers are 8.5 nm and 6.3 nm, respectively

$Al_{0.7}Sc_{0.3}N$. Figure 5.3 demonstrates the cross-sectional TEM image of the superlattice. Since the superlattice layer thicknesses are much smaller than the wavelength of operation (600–800 nm), the HMM can be approximated as a uni-axial anisotropic effective medium with dielectric functions $\varepsilon_{\parallel}$ and $\varepsilon_{\perp}$. The optical constants of the HMM were measured using spectroscopic ellipsometry. The TiN/(Al,Sc)N metamaterial exhibited hyperbolic dispersion with $\varepsilon_{\perp} = 16.4 + i21.1$, $\varepsilon_{\parallel} = -2.3 + i2.1$ at 685 nm, which is the peak emission wavelength of the NV center. Thus, for the electric field, parallel to the interface, the metamaterial at this wavelength behaved as a metal. As an additional benefit, at the excitation

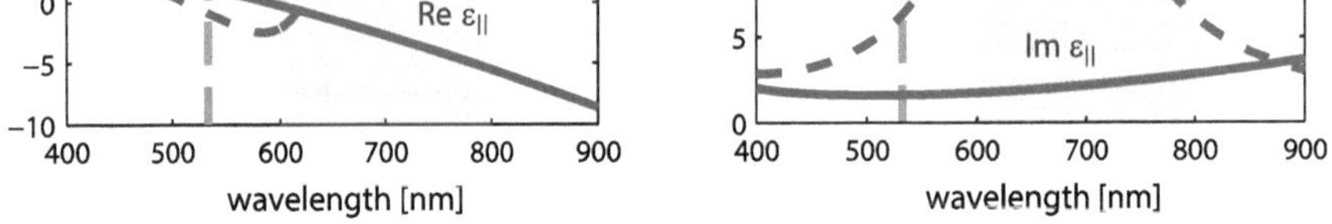

Fig. 5.4 **a** Real and **b** imaginary parts of the dielectric functions of the uniaxial effective medium that approximates the fabricated HMM. The permittivities were obtained by spectroscopic ellipsometry in the spectral range 400–900 nm. Within the range of the NV center emission (600–800 nm-highlighted), the metamaterial shows hyperbolic dispersion ($Re[\varepsilon_{\perp}] > 0$, $Re[\varepsilon_{\parallel}] < 0$). At the excitation wavelength (532 nm, *green line*) the metamaterial behaves as a lossy dielectric. Figures reproduced with permission: [33] from Copyright 2015 Wiley-VCH

wavelength of 532 nm (Fig. 5.4, vertical dashed green line) the TiN/(Al,Sc)N metamaterial behaved as a lossy dielectric allowing the penetration of the tangential component of electric field through the HMM to the emitter.

5.3.1 Calculations of NV Emission Enhancement by HMM

In general, to estimate both the enhancement of the spontaneous emission rate (represented by Purcell factor F_P) and normalized collected emission power f_{rad} one may utilize a semiclassical treatment, where the emitter levels are quantized. Since in our experimental lifetime measurement the quantum emitter is weakly coupled to the metamaterial environment, the results are expected to be in a good agreement with the treatment based on classical electrodynamics [40]. We therefore consider the problem of classical dipole radiation near a planar layered medium, which is described in more detail in [41].

The inset in Fig. 5.5a shows the modelled configuration, which consists of an oscillating dipole elevated at height h above the HMM uppermost layer ($Al_{0.7}Sc_{0.3}N$), the upper half-space (superstrate) with refractive index 1.515 ($\varepsilon_{sup} = 2.295$ simulating PVA/immersion oil), planar multilayer TiN/(Al,Sc)N superlattice ($\varepsilon_m/\varepsilon_d$) and the lower half-space (substrate) made of MgO (ε_{sub}). For this simulation we considered the fabricated HMM composed of 10 pairs of 8.5-nm-thick TiN and 6.3-nm-thick $Al_{0.7}Sc_{0.3}N$ on MgO substrate. The layer thicknesses were chosen to provide optimal performance, which was previously discussed in [14]. The formulas used to calculate the Purcell factor F_P and

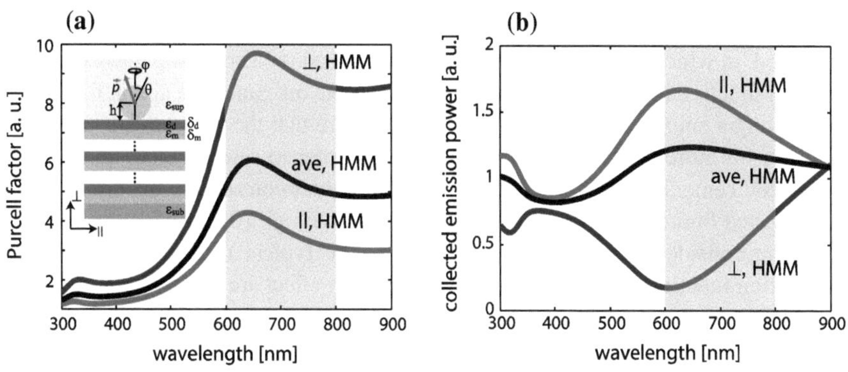

Fig. 5.5 Theoretical estimations of Purcell factor (**a**) and collected emission power (**b**) for a dipole located in immersion oil/PVA ($\varepsilon_{sup} = 2.295$) 25 nm above the HMM surface. *Blue*, *red* and *black curves* correspond to the dipole orientations perpendicular (⊥), parallel (||) to the HMM interface and averaged (ave), respectively. *Highlighted area* indicates the emission spectral range of NV center at room temperature. Collection angle is 79.6°, which corresponds to NA 1.49. Layout of the calculated structure is shown in the inset. Figures reproduced with permission: [33] from Copyright 2015 Wiley-VCH

normalized collected emission power f_{rad} for the dipole orientations: in-plane (||), perpendicular (⊥), and averaged (ave), statistical average over all possible orientations, are given in the Appendix.

The collection angle ($\theta_{max} = 79.6°$) is defined by the numerical aperture of the objective lens (NA = 1.49). The generalized Fresnel's reflection coefficients $\tilde{r}^p$ and $\tilde{r}^s$ for the superlattice were calculated utilizing the recursive imbedding method [42], which is more precise and efficient than the direct transfer matrix approach. Normalization factors included into F_P and f_{rad} are the total radiation power and the power emitted into the collection angle, respectively. Both quantities corresponded to the case of the emitter immersed into a homogeneous medium with dielectric permittivity ε_{sup}, which is a reasonable approximation for the normalization procedure employed in the experiment.

The results of the calculations are demonstrated in Fig. 5.5. Assuming that the NV center is located at the crystal center, the expected Purcell factor (or change in lifetime) for the nanodiamonds with a mean diameter of 50 nm should be on average around 4.5 (we note that placing NV centers closer to the HMM surface can significantly increase the Purcell factor making it on the order of 10^2—see Fig. 5.10). The detected count rates corresponding to the normalized collected emission power for the same type of nanodiamonds are anticipated to increase on average by about 20 percent.

5.3.2 Experimental Demonstration of HMM Enhanced Single-Photon Emission

We have optically examined the samples described earlier consisting of collection of nanodiamond-based NV centers dispersed on HMM surface. A control sample was used, in which the nanodiamonds were dispersed on a coverslip glass substrate, whose refractive index matched that of the objective oil. Since we aim at fabricating and testing a single–photon source, we must ensure that the emitters are single, i.e. cannot emit more than one photon at a time. In order to select nanodiamonds with single NV centers, we first measured the second–order correlation function $g^{(2)}(t)$ of the detected fluorescence spots. Only NV centers with $g^{(2)}(0)$ significantly less than 0.5 were considered for further experiments. The typical measured second-order correlation functions showing photon antibunching effect are shown for a single NV center on the glass coverslip and on the HMM in Fig. 5.6.

We retrieved the total decay lifetimes of the NV centers from the exponential fitting of the fluorescence decays such as the ones shown in Fig. 5.7a. The average values measured were 17.1 and 4.3 ns on glass and HMM, respectively (see Fig. 5.7b). Hence, the NVs on HMM exhibit an average decrease in lifetime by a factor of 4 compared to NVs on coverslip, which is consistent with the above calculations. The lowest recorded lifetime for a single NV center on top of the HMM was 1.5 ns, which corresponds to a Purcell factor of 11.4. The spreads in the

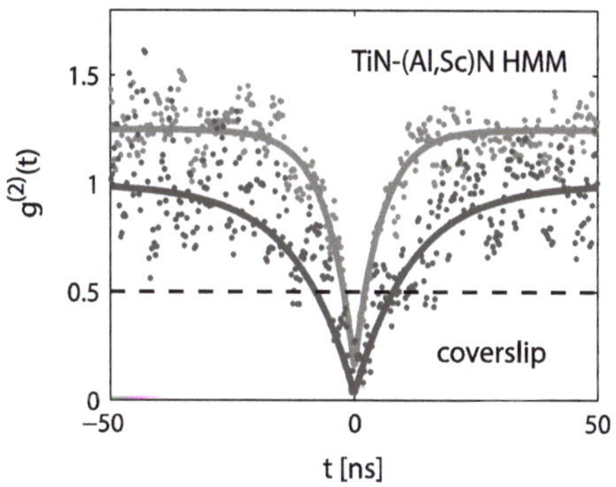

Fig. 5.6 Second-order correlation function $g^{(2)}(t)$ of a representative nanodiamond with single NV center on *top* of HMM. Figure reproduced with permission: [33] from Copyright 2015 Wiley-VCH

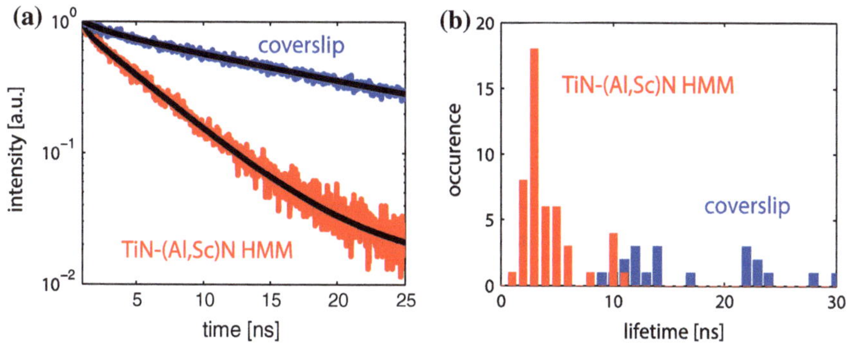

Fig. 5.7 **a** Representative normalized fluorescence decays and **b** histograms of lifetimes for the NV centers located on glass coverslip (reference sample) and on HMM. The average and largest decreases in lifetime are 4 and 11.4, respectively. Figures reproduced with permission: [33] from Copyright 2015 Wiley-VCH

lifetime statistics are likely due to the variation in nanodiamond size, NV center dipole orientation and its distance from the HMM surface.

Finally, we measured the dependence of the single-photon emission rate from the NV centers versus the excitation power. In Fig. 5.8a, we show these dependences for the brightest NV centers on both the HMM and the glass coverslip. Both dependences have been corrected for the background emission. The total count rate was measured at the NV center site and the background count rate was measured at a nanodiamond-free location. Saturation of the emission count rate was observed around 1 mW of the pump power. The experimentally measured saturation curves were fitted using the expression $I(P) = I_0/(1 + P_{sat}/P)$ [43] and yielded the saturated single–photon count rates for each individual NV center. The histogram of the saturated single–photon count rates are shown in Fig. 5.8b for NV centers both on

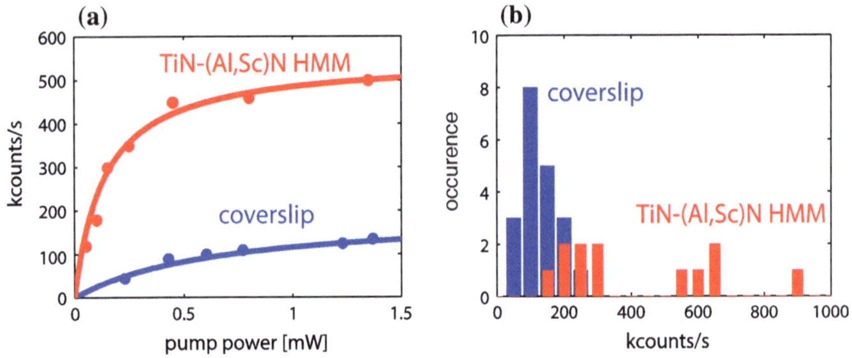

Fig. 5.8 Collected single-photon count rates (corrected for background emission) from NV centers in 50 nm nanodiamonds. **a** Typical saturation curves and **b** histograms of count rates for nanodiamonds on glass coverslip (*blue*) and HMM (*red*). The average enhancements for the first and second statistical maxima are 1.8 and 4.7, respectively. Figures reproduced with permission: [33] from Copyright 2015 Wiley-VCH

glass and the HMM. The histogram displays multiple maxima. For the one around 200 kcounts/s, the average enhancement against the coverslip was 1.8 ± 1.1. This is consistent with predicted value of 1.2, within the error margin. The variation of the enhancement is large because count rates for both HMM and coverslip substrates have significant spreads. The next maximum in the histogram corresponds to an average count rate enhancement of 4.7 ± 2.2. This number is beyond the theory prediction for a planar HMM. Finally, one diamond shows even higher count rate enhancement. The obtained statistics suggests the existence of an additional mechanism responsible for extra enhancement.

5.3.3 *Increasing Collection Efficiency by Outcoupling High-k Waves to Free Space*

We have seen in the simulations of planar HMM presented above that a high Purcell factor did not translate into a higher collected emission power (see Fig. 5.5) because the optical power residing in the high-k metamaterial modes was not outcoupled to free space. However, our experiments (see Fig. 5.8b) indicate that one can in practice collect much more power than what is predicted by theory (see Fig. 5.5).

To get at least a qualitative understanding of this phenomenon we have employed an additional computational effort. For this, we assumed that excessive emission rates from certain nanodiamonds potentially could arise from the influence of either neighboring superlattice defects or adjacent "dark" nanodiamonds lacking NV centers. The proposed mechanism is as follows: HMMs can support surface plasmon-polaritons (SPPs) and bulk electromagnetic modes, both having high propagation constant (high-k modes) [44]. It is known that an excited emitter located in the vicinity of HMM surface has a high probability of transferring energy into such high-k modes [39]. Once excited, these high-k waves propagate through the HMM, never leaving the bulk medium where they are eventually absorbed and do not provide any contribution to the collected photon flux. However, the presence of the surface defects or scattering objects may result in outcoupling of these modes into the far field and therefore provide additional contribution to the emitted signal.

To verify this assumption both substrate defects and "dark" nanodiamonds were modelled as 50-nm-diameter spheres with varied separation from emitting dipole. In essence, the calculation indeed demonstrated that the presence of the dark nanodiamond and TiN particle results in further enhancements of emission rate by factors of 1.3 and 2.0, respectively compared to a perfect HMM surface. This result does support the experimental observations though it clearly requires further rigorous investigation. In particular, instead of relying on random irregularities of the HMM surface, one can place emitting nanodiamonds next to engineered nanostructures with optimized outcoupling efficiency. Our calculations show that the portion of NV center emission coupled into the metamaterial could be quite efficiently recovered prior to dissipation in the HMM. This emission could be directed into the far-field by using even very basic structures, such as gratings, nanodisks, nanoholes, etc.

An example of such artificially created structure is a single circular groove milled in the HMM around the nanodiamond (see Fig. 5.9a). The groove can outcouple the propagating metamaterial modes at its sharp corners. The structure with optimized geometry promises a several times increase in collected power in comparison with a NV center on a coverslip. A simulation using finite-element method (see Fig. 5.9b) showed that for in-plane oriented dipole, the structure should consist of a 100 nm radius cylinder surrounded by a 250 nm wide groove. An additional factor that helps to increase the collected power is the use of a TiN reflecting layer between the MgO substrate and the HMM. The simulation for

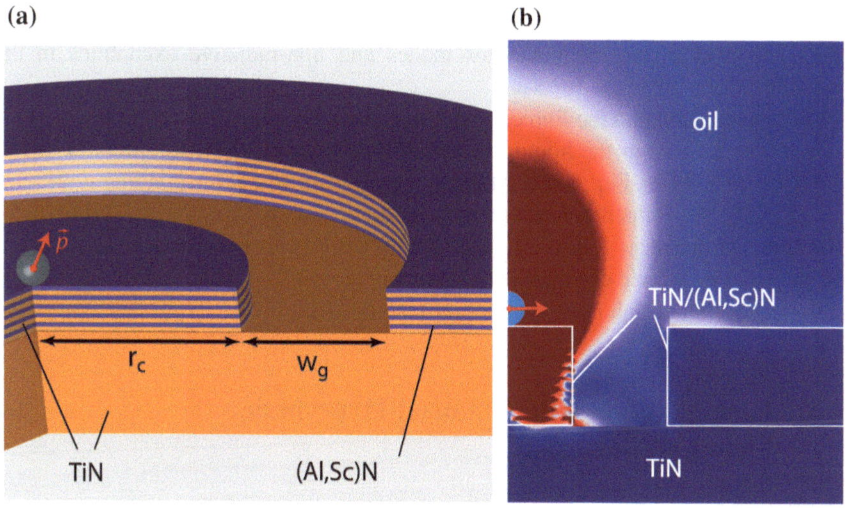

Fig. 5.9 **a** Schematic of a circular groove milled in HMM around a nanodiamond to improve the collected power in the far-field (in 60° aperture angle) by scattering the high-k modes propagating along the surface and inside HMM. **b** Distribution of power density amplitude for the in-plane oriented NV center

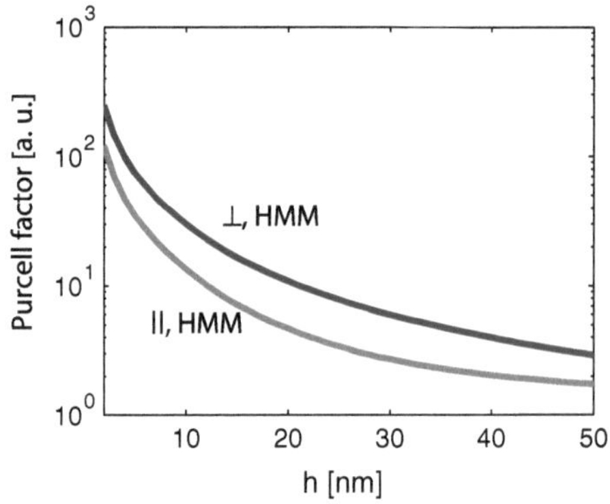

Fig. 5.10 Dependence of the Purcell factor F_P on a dipole position h above the HMM surface. Values of F_P are obtained by averaging the characteristics (shown in Fig. 5.5a) over the wavelength range 600–800 nm. Figure reproduced with permission: [33] from Copyright 2015 Wiley-VCH

optimal structure showed 2.8 and 3.3 increase in collected power compared to coverslip for in-plane and perpendicular dipole orientations, respectively.

In order to understand the theoretical potential of HMMs for enhancing dipole emission, we estimated the Purcell factor F_P increase as function of the separation distance h between the emitter and the surface. This increase is a common feature for dipoles located near metallic surfaces [41] and is expected to be observed as well in the case of an NV center on HMM. In our case, the smallest separation distance is dictated by the ND size. The Purcell factor dependence on h is explained by the fact that at short distances the evanescent fields created by the emitter are better coupled to the metamaterials modes and non-radiative excitations in the metallic layers [39].

We have calculated the Purcell factor as a function of h for different orientations of a dipole located in the homogeneous medium above the HMM structure previously described (see Fig. 5.10). At the distance of 25 nm, corresponding to the size of our NDs, the Purcell enhancement is on the order of 10, as shown in Fig. 5.5a. However, at distances h of a few nanometers the Purcell factor reaches two orders of magnitude, which corresponds to a spontaneous emission lifetime of 0.1 ns, sufficient for single-photon operation at tens of GHz.

5.4 Lasing Action with Nanorod Hyperbolic Metamaterials

Above, we have discussed the role of HMMs in promoting spontaneous emission from single quantum emitters. Strong spontaneous emission is usually seen as a drawback to observe stimulated emission because it diminishes the inversion without contributing photons to the lasing mode. Interestingly, due to their rich mode structure, HMMs can be used to promote lasing as well. High Purcell effect for certain metamaterial modes yields a redistribution of spontaneous emission over wavevector space (*k*-space), with light preferentially coupled to the lasing mode and inhibited in other modes [45].

Several plasmonic lasers have already been demonstrated using various geometries [19], such as metal-insulator-metal waveguides [46], whispering gallery cavities [47], core-shell particles [48, 49], nanohole/particle arrays [50–52], semiconductor-dielectric-metal hybrid cavities [53]. These methods generally used geometries with strong cavity resonances to gain Purcell enhancement and hence lasing as described above. The Purcell enhancement arising from the resonance of metallic structures usually exhibits a relatively narrow bandwidth, thus restraining the frequency of lasing to be achieved. An alternative approach to gain Purcell enhancement is based on non-resonant structures, which can be designed by engineering the dispersion of metamaterials. In this work, we report on the use of nanorod-based HMMs to achieve efficient lasing. It has been shown that metamaterials exhibiting hyperbolic properties can support unique optical waves with very small or large mode indices, allowing for stronger light-matter interaction [16] which could enhance lasing. Our nanorod-based metamaterials are composed of vertically aligned gold nanorods coated with a polyvinyl alcohol (PVA) film. To create a sufficient density of emitters for lasing, rather than using NV centers, we embedded Rhodamine 101 (R101) dye molecules into the PVA film (see the sample schematic in Fig. 5.1b). By adjusting the metal fill ratio, the dispersion of the nanorod metamaterial can be tuned from elliptic to hyperbolic. Our experiments show that HMMs exhibit significant enhancement over elliptic metamaterials (EMM), which is consistent with calculations of Purcell enhancement. The nanorod-based metamaterials are suitable for integration with a broad range of optical gain media to achieve lasing at the desired frequency.

As described earlier, HMMs exhibit a singularity in the photonic LDOS, allowing for spontaneous decay rate enhancement of quantum emitters. On the other hand, it has been shown that the Purcell effect could contribute to efficient optical amplification and lasing action, provided that a portion of the enhanced spontaneous emission feeds into the lasing mode [45]. To our knowledge, so far there is only one report that addressed the possibility to achieve stimulated emission from a HMM comprised of Ag and MgF_2 layer stacks [54]. Although a reduced threshold was observed in a HMM when compared to a reference device based on a bare Ag film, the emission efficiency from the HMM was obviously lower than from the reference. Thus, the full potential of HMM to achieve lasing with low threshold and high efficiency remained to be explored.

In this work, two gold nanorod arrays embedded in anodic alumina templates have been fabricated, exhibiting hyperbolic (labeled HMM) and elliptic dispersion (labeled EMM) at the emission wavelength of R101 ($\lambda = 606$ nm), according to local EMT. The different dispersion characteristics were achieved by altering the metal fill ratio using different nanorod diameters. Arrays of gold nanorods were fabricated by electrodeposition within nanoporous aluminum oxide membranes.

The membranes were prepared by the anodization in H_2SO_4 of Al films deposited on Au-coated glass substrates. Substrate anodization was performed at 30 V for the HMM and at 25 V for the EMM, to yield approximate pore diameters of 40 and 25 nm respectively, surface densities of 35 % and 14 % respectively, and nanopore heights of 250 nm. Gold nanorods were electrodeposited as previously described under galvanostatic conditions [55] using a current density of 0.5 mA/cm^2, with constant electrodeposition up to the maximum height allowed by the nanoporous alumina templates, at which point a distinct drop in voltage (>20 %) was observed. SEM images of the nanorod-based HMM and EMM substrates indicate that the nanorods have uniform diameters and are well dispersed within the Al_2O_3 matrix, which has an approximate pore-to-pore distance of 60 nm for each sample (see insets of Fig. 5.11a,b).

As follows from 1 and 2 in Table 5.1, the EMT model for nanorod HMMs supports two resonances, specifically the epsilon-near-pole (ENP,$\varepsilon_{\parallel} \to \infty$) and

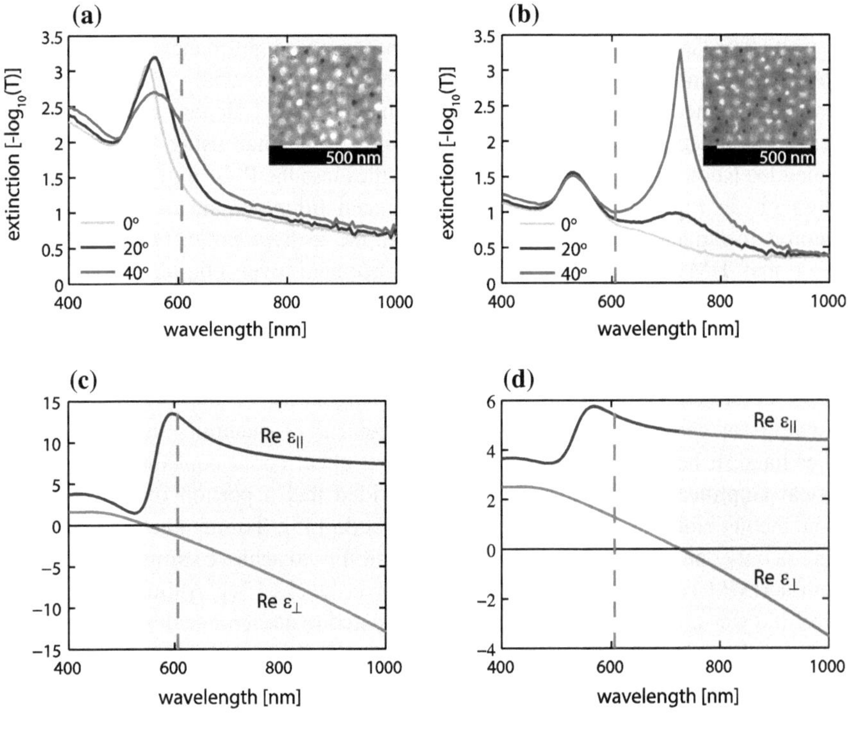

Fig. 5.11 Experimental extinction spectra for **a** HMM and **b** EMM samples. The spectra were obtained using TM–polarized light at 0° (*yellow*), 20° (*green*) and 40° (*pink*) incidence. Effective anisotropic permittivities of **c** HMM and **d** EMM estimated using Maxwell-Garnett theory. *Shaded* regions indicate wavelength range where the metamaterial exhibits hyperbolic Type-I dispersion. *Dotted red line* indicates central emission wavelength of R101 dye (606 nm). SEM scans of the nanorod metamaterials are shown in the *insets*. Figures reproduced with permission: [13] from Copyright 2016 ACS

epsilon-near-zero (ENZ, $\varepsilon_\perp \rightarrow 0$) responses [56]. The ENP resonance is purely dependent on the permittivities of the metal and dielectric, and hence does not vary with the metal fill ratio. However, the ENZ response (and corresponding dispersion crossover wavelength) can be tuned easily by adjusting the fill ratio. In the case of HMM, only one resonance band is observed since the ENP and ENZ responses overlap starting at λ = 530 nm, whereas EMM has distinct ENP (λ = 530 nm) and ENZ (λ = 710 nm) responses, as seen in Fig. 5.11a, b. Effective Maxwell-Garnett permittivities were calculated based on the fill ratios for each sample (see Fig. 5.11c, d) and the permittivity values of bulk Au and amorphous Al_2O_3. Both the extinction and permittivity plots confirm that the HMM exhibits hyperbolic dispersion and the EMM exhibits elliptic dispersion at the same wavelength as the central emission of R101.

5.4.1 Purcell Effect Calculations for Dye Molecules on Nanorod Metamaterials

We estimated the Purcell factors for our HMM and EMM structures by simulating the coupling of dipole emitters with our metamaterials, defined by local EMT. Using the standard Green's function formalism as shown in the previous section, we calculated the Purcell factors for a dipole embedded in PVA (refractive index n = 1.5) placed 20 nm above HMM and EMM deposited on glass substrates (n = 1.5), as shown in Fig. 5.12a, b. In the calculation, we distinguish between in-plane, perpendicular, and "average" dipole orientations. The calculation results show that the HMM provides a Purcell factor of 5.75 for a dipole parallel to the HMM surface, and 12.21 for a dipole perpendicular to the surface at the central wavelength of R101 (λ = 606 nm). On the other hand, the EMM provides a Purcell factor of 1.5 and 2.15 for parallel and perpendicular orientations, respectively. On average the HMM provides an enhancement of 4.6 times over the EMM. The calculation also shows that the Purcell factor strongly depends on the distance of the dipole from the metamaterial surface (see Fig. 5.12c), as it was already discussed for the planar HMM case. For dipoles very close to the metamaterial surface (~5 nm), the Purcell factor for the HMM is extremely large, reaching up to 400, while the EMM only provides an enhancement of 46. As the dipole is moved away from the surface, the Purcell factor for the EMM decays much quicker than that for HMM, reaching ~1 within a 40 nm distance, while it is still slightly larger than 1 for the HMM at a distance of 100 nm.

In order to further study the decay channels providing this Purcell enhancement, we have also calculated the inherent plasmonic modes (named as high-k modes) that can be excited in our HMM and EMM structures. Figure 5.13 shows the *k*-space dissipated power density, $\log_{10}\left[k_0 dF_P^{ave}/dk_{||}\right]$, calculated for a dipole with the averaged orientation, located 20 nm above the surface of HMM and EMM, defined by local EMT. In both HMM and EMM, we can see the modal gap, i.e., wavelength

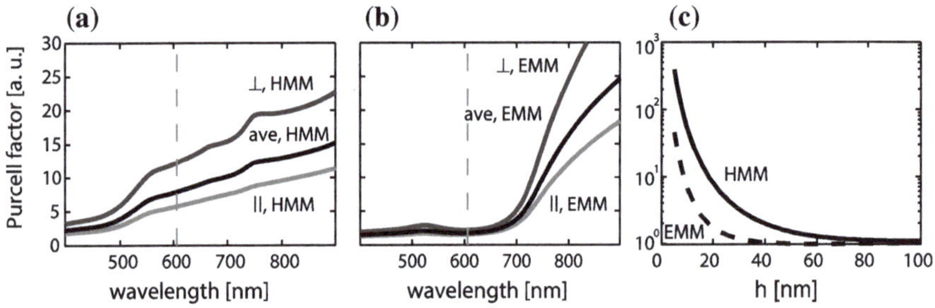

Fig. 5.12 Theoretical estimations of Purcell factor versus emission wavelength for a dipole located 20 nm above HMM (**a**) and EMM (**b**), as well as the Purcell factor dependences on the distance of dipole from metamaterial surface (**c**) at emission wavelength of 606 nm (indicated with a *red dashed line* in (**a**) and (**b**)). *Blue*, *red*, and *black curves* in (**a**) and (**b**) refer to dipoles oriented perpendicular, parallel to the metamaterial surface, and averaged, respectively. The Purcell factor in (**c**) is plotted for the averaged dipole orientation. Figures reproduced with permission: [13] from Copyright 2016 ACS

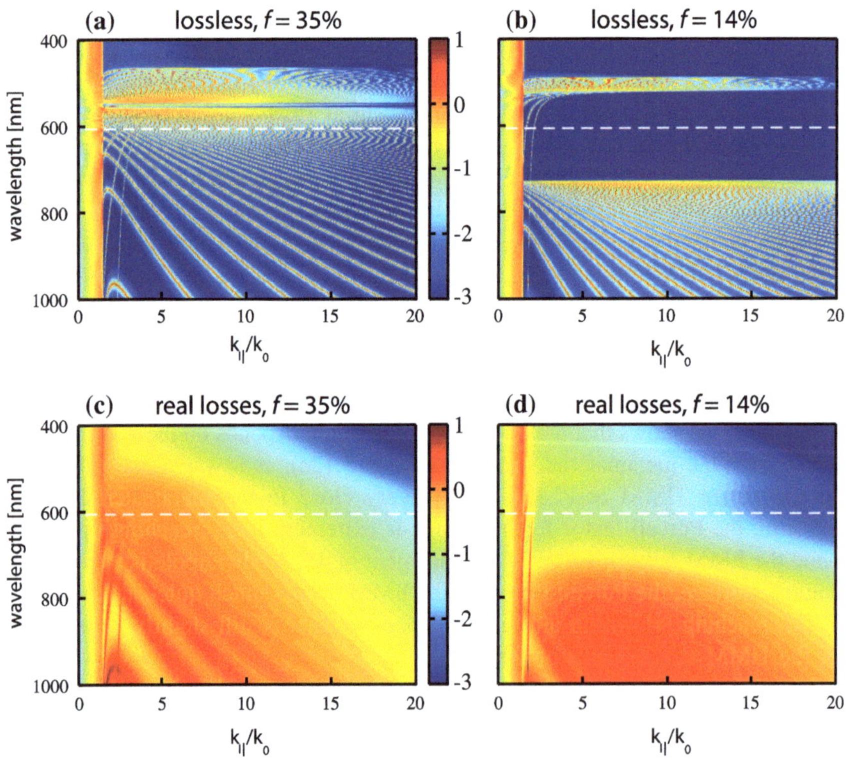

Fig. 5.13 *k*-space dissipated power density for a dipole placed 20 nm above HMM (**a**) and EMM (**b**), without losses. Subfigures (**c**) and (**d**) correspond to the cases of (**a**) and (**b**) with actual losses. At 606 nm (*white dotted line*), HMM provides many more inherent plasmonic modes than EMM. Figures reproduced with permission: [13] from Copyright 2016 ACS

range where very few plasmonic modes are allowed between ENP and ENZ resonances. We see that at 606 nm, EMM falls in this modal gap with almost no propagating modes, while HMM provides much more modes to enhance the emitter's decay rate, for even very large $k_{||}$ (in-plane wavevector). Due to this large number of allowed modes, we clearly see that our HMM provides significant Purcell enhancement over our EMM structure, and is therefore expected to enhance spontaneous emission greatly, which could feed into lasing modes.

5.4.2 Experimental Demonstration of Lasing with Nanorod Metamaterials

The nanorod arrays were spin coated with a 10 mM solution of R101 dye dissolved in PVA at 1000 rpm for 30 s followed by baking at 60 °C for 6 h, which yielded a ~2 μm thick film. A frequency-doubled Nd:YAG picosecond laser (532 nm emission wavelength, 400 ps pulse width and 1 Hz repetition rate) was used to pump the samples at 40°. The pump pulses were focused down to a spot size of ~200 μm in diameter using a 5× objective lens. The emission from the samples was collected with a fiber almost normal to the sample surface, which was fed to a spectrometer equipped with a charge-coupled device.

Figure 5.14a, b show the evolution of the emission spectra with the pump energy for both HMM and EMM. Obvious spectral narrowing was observed for both HMM and EMM when the pump energy was increased up to ~3.2 μJ and 4.0 μJ respectively. At a pump energy of ~5.5 μJ, the emission intensity from HMM was twice as strong as that from EMM. From the plot of pump-dependent peak emission intensity (see Fig. 5.14c, d), we can see a clear threshold for the two systems, with HMM having a lower threshold (~3.2 μJ) than EMM (4 μJ). We attribute the increase in collected emission intensity mainly to the Purcell enhancement, which couples spontaneous emission into lasing modes. Furthermore, it has been reported that the interactions between nanorods may form cavities to provide feedback [52], which could play a role in the occurrence of lasing. In addition, the nonlocal optical waves exhibited in the nanorod medium could enhance spontaneous emission, which could affect lasing action in such metamaterials [16].

As control samples, we studied the emission from a bare glass substrate and a 250-nm thick gold film, both coated with R101 in PVA. Both samples only showed spontaneous emission, without any evident spectral narrowing (see Fig. 5.15a, b). As a comparison, we fabricated a lamellar HMM with ten alternating layers of Au (8.9 nm) and Al_2O_3 (16.1 nm), which has the same metal fill ratio (35 %) and thickness (250 nm) as the nanorod-based HMM. We applied the gain medium and conducted lasing measurements in the same way as described above and enhanced emission was collected from the sample based on the multilayer-based HMM; however, no lasing was observed, as shown in Fig. 5.15c.

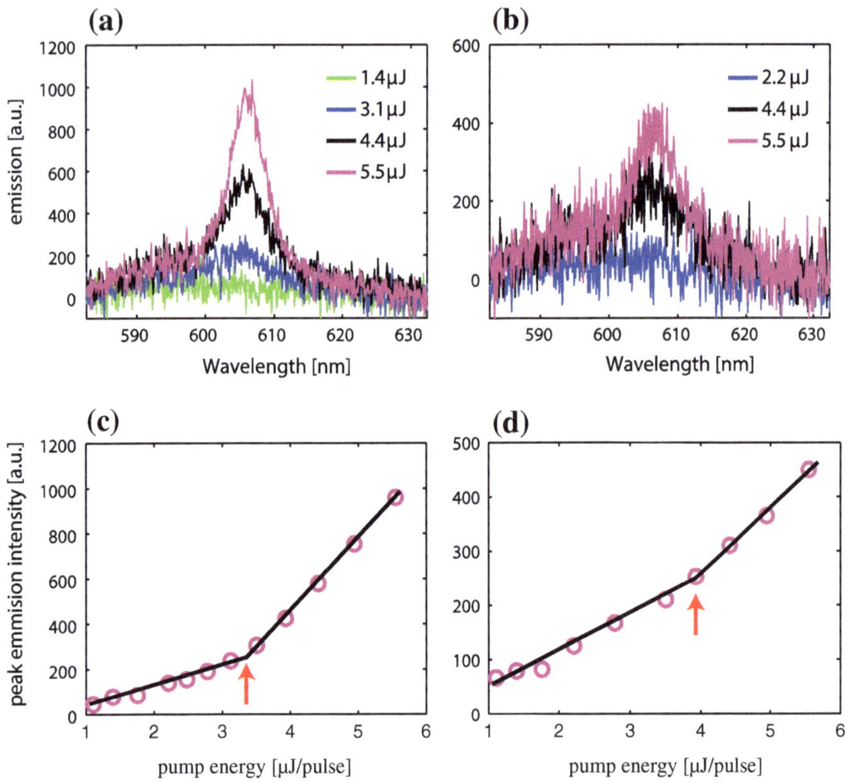

Fig. 5.14 Emission spectra for **a** HMM and **b** EMM at various laser pumping energies. Peak intensity plots for **c** HMM and **d** EMM; the slope kinks correspond to the lasing threshold energy (*red arrows*). *Lines* in **c** and **d** are least-squares fits. Figures reproduced with permission: [13] from Copyright 2016 ACS

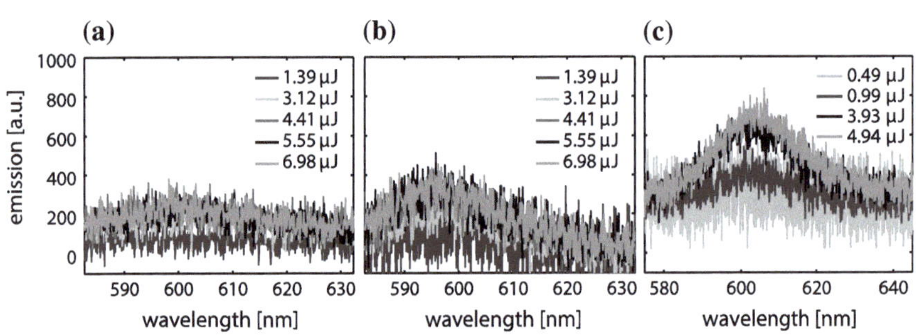

Fig. 5.15 Emission from control **a** glass and **b** gold film samples, as well as **c** lamellar HMM composed of 10 pairs of 8.9 nm gold film and 16.1 nm alumina film, all coated with R101 in PVA. Plots show amplified spontaneous emission, but no threshold behavior for lasing. Figures reproduced with permission: [13] from Copyright 2016 ACS

Therefore, the nanorod-based HMM provides a significant enhancement over its lamellar counterpart, exhibiting low-threshold lasing action. Furthermore, infusion of dyes directly into the matrix of the nanorod-based HMM is far more feasible than embedding them within the lamellar HMM. This should greatly enhance the emitter-field coupling and spontaneous emission, and lead to much lower thresholds for lasing, which is currently under investigation.

5.5 Summary

The chapter explores the opportunities offered by hyperbolic metamaterials to promote both spontaneous and stimulated emission. We have demonstrated implementations of an efficient single-photon source and a plasmonic laser using two different metamaterials designs. We have experimentally demonstrated an improvement in the emission properties of single NV center nanodiamonds placed on top of an epitaxial TiN/(Al,Sc)N HMM compared to a glass substrate. The observed lifetime decreased by a factor of 4, which is in good agreement with theoretical prediction. The collected emission power for NV centers near the HMM increased by a factor of 1.8 on average, although an extremely high enhancement of 4.7 was detected in one particular case. The emission rate could be further enhanced by the presence of an adjacent HMM defect or a deterministically engineered outcoupling feature such as a circular groove.

In a similar fashion, we have demonstrated lasing action supported by metamaterials based on gold nanorod arrays coated with thin films of PVA embedded with R101 dye. The metamaterial could be tuned to exhibit hyperbolic (HMM) or elliptic (EMM) dispersion depending on metal fill ratio. Both regimes supported lasing action, with emission from the HMM being twice as strong as that from the EMM while also exhibiting lower threshold. Embedding lasing dyes directly into the nanorod matrix will likely further enhance the emitter-field coupling, and also lower the HMM lasing threshold. The HMM provides broadband Purcell enhancement, suggesting its application as a source for coherent photon emission in a broadband wavelength range.

Acknowledgements This work was partially supported by AFOSR-MURI grant (FA9550-10-1-0264), NSF-MRSEC grant (DMR-1120923), and ONR grant (N00014-13-1-0649).

Appendix: Semi-analytical Calculations of the Purcell Factor and Normalized Collected Emission Power

High LDOS provided by the HMM can lead to dramatic changes in fluorescence lifetimes of quantum emitters. In practice, one is also interested in the percentage of the dissipated optical power that can be collected by a far field detector. In order to describe effects of a realistic HMM on a single-dipole emitter placed in the vicinity of a planar HMM, we performed calculations of total power flow in the immediate vicinity of the emitter as well as through a far-field collection plane. A single emitter was modeled as an oscillating electric dipole with dipole moment **p** and angular frequency ω. The energy dissipation rate in an inhomogeneous environment is given by [41]

$$P = \frac{\omega}{2}\mathrm{Im}\left[\mathbf{p}^{*}\cdot\left(\mathbf{E}_0(\mathbf{r}_0)+\mathbf{E}_s(\mathbf{r}_0)\right)\right] \tag{A.1}$$

where $\mathbf{E}_0(\mathbf{r}_0)$ and $\mathbf{E}_s(\mathbf{r}_0)$ are the primary dipole field and scattered field at the dipole position (r_0), respectively. These electric fields were calculated using the dyadic Green's function formalism [41]. We analyzed the contribution of each spatial frequency mode using an angular spectrum representation of the Green's functions. The Purcell factors F_P for in-plane (||) and perpendicular (⊥) oriented single-dipole emitters placed at a distance h above a multilayer planar structure were calculated using the following formulae [41]

$$F_P^{\perp} = 1 + \frac{3}{2}\frac{1}{\varepsilon_{\mathrm{sup}}^{3/2}}\int_0^{\infty}\mathrm{Re}\left\{\frac{s^3}{s_{\perp,\,\mathrm{sup}}(s)}\tilde{r}^{\mathrm{p}}(s)e^{2ik_0 s_{\perp,\,\mathrm{sup}}(s)h}\right\}ds \tag{A.2}$$

$$F_P^{\parallel} = 1 + \frac{3}{4}\frac{1}{\varepsilon_{\mathrm{sup}}^{1/2}}\int_0^{\infty}\mathrm{Re}\left\{\frac{s}{s_{\perp,\,\mathrm{sup}}(s)}\left[\tilde{r}^{\mathrm{s}}(s)-\frac{s_{\perp,\,\mathrm{sup}}^2(s)}{\varepsilon_{\mathrm{sup}}}\tilde{r}^{p}(s)\right]e^{2ik_0 s_{\perp,\,\mathrm{sup}}(s)h}\right\}ds \tag{A.3}$$

The value of F_P for the isotropic (ave), statistically averaged dipole orientation is given by

$$F_P^{ave} = \frac{2}{3}F_P^{\parallel} + \frac{1}{3}F_P^{\perp} \tag{A.4}$$

Normalized collected emission powers f_{rad} for the same dipole orientations are shown below

$$f_{rad}^{\perp} = \frac{3}{4}\int_0^{\theta_{\max}}\sin^3\theta\left|e^{-i\varepsilon_{\mathrm{sup}}^{1/2}k_0 h\cos\theta}+\tilde{r}^{\mathrm{p}}(\theta)e^{i\varepsilon_{\mathrm{sup}}^{1/2}k_0 h\cos\theta}\right|^2 d\theta \tag{A.5}$$

$$f_{rad}^{\parallel} = \frac{3}{8}\int_0^{\theta_{\max}}\cos^2\theta\left|e^{-i\varepsilon_{\mathrm{sup}}^{1/2}k_0 h\cos\theta}-\tilde{r}^{\mathrm{p}}(\theta)e^{i\varepsilon_{\mathrm{sup}}^{1/2}k_0 h\cos\theta}\right|^2+\left|e^{-i\varepsilon_{\mathrm{sup}}^{1/2}k_0 h\cos\theta}+\tilde{r}^{\mathrm{s}}(\theta)e^{i\varepsilon_{\mathrm{sup}}^{1/2}k_0 h\cos\theta}\right|^2\sin\theta d\theta \tag{A.6}$$

$$f_{rad}^{ave} = \frac{2}{3}f_{rad}^{\parallel} + \frac{1}{3}f_{rad}^{\perp} \tag{A.7}$$

In equations A.2–A.7, $s = k_{\|}/k_0$, $s_{\perp,\,\mathrm{sup}}(s) = k_{\perp,\,\mathrm{sup}}(s)/k_0 = \left(\varepsilon_{\mathrm{sup}} - s^2\right)^{1/2}$, $k_0 = \omega/c$, θ is a polar angle measured from the $\perp$ direction, the collection angle $\theta_{\max} = 79.6°$. $k_{\|}$ is the in-plane component of the k-vector varying from 0 to infinity. $\tilde{r}^p$ and $\tilde{r}^s$ are generalized superlattice's Fresnel reflection coefficients for p- and s-polarized light. The reflection coefficients were calculated utilizing the recursive imbedding method [42, 57]. The integrals were numerically evaluated by using an adaptive Gauss-Kronrod quadrature method [58]. In the formulae, we assumed that intrinsic quantum yield of NV centers is close to unity [59]. F_P and f_{rad} were normalized by the total radiation power and the power emitted into the collection angle, respectively. Both powers corresponded to the case of the emitter immersed into homogeneous medium with dielectric permittivity $\varepsilon_{\mathrm{sup}}$, which models well the reference sample in the experiment.

References

1. P.C. Clemmow, The theory of electromagnetic waves in a simple anisotropic medium. Proc. Inst. Electr. Eng. **110**(1), 101 (1963)
2. R.H. Ritchie, Plasma losses by fast electrons in thin films. Phys. Rev. **106**(5), 874–881 (1957)
3. S.M. Rytov, Electromagnetic properties of a finely stratified medium. Sov. Phys. JETP **2**(10) (1956)
4. D. Smith, D. Schurig, Electromagnetic wave propagation in media with indefinite permittivity and permeability tensors. Phys. Rev. Lett. **90**(7), 077405 (2003)
5. Z. Jacob, L.V. Alekseyev, E. Narimanov, Optical hyperlens: far-field imaging beyond the diffraction limit. Opt. Express **14**(18), 8247–8256 (2006)
6. Z. Liu, H. Lee, Y. Xiong, C. Sun, X. Zhang, Far-field optical hyperlens magnifying sub-diffraction-limited objects. Science **315**(5819), 1686 (2007)
7. S. Ishii, A.V. Kildishev, E. Narimanov, V.M. Shalaev, V.P. Drachev, Sub-wavelength interference pattern from volume plasmon polaritons in a hyperbolic medium. Laser Photon. Rev. **7**(2), 265–271 (2013)
8. Z. Jacob, J.-Y. Kim, G.V. Naik, A. Boltasseva, E.E. Narimanov, V.M. Shalaev, Engineering photonic density of states using metamaterials. Appl. Phys. B **100**(1), 215–218 (2010)
9. A.D. Neira, N. Olivier, M.E. Nasir, W. Dickson, G.A. Wurtz, A.V Zayats, Eliminating material constraints for nonlinearity with plasmonic metamaterials. Nat. Commun. **6**, 7757 (2015)
10. S.-A. Biehs, M. Tschikin, P. Ben-Abdallah, Hyperbolic metamaterials as an analog of a blackbody in the near field. Phys. Rev. Lett. **109**(10), 104301 (2012)
11. C. Shen, Y. Xie, N. Sui, W. Wang, S.A. Cummer, Y. Jing, Broadband acoustic hyperbolic metamaterial. Phys. Rev. Lett. **115**(25), 1–5 (2015)
12. E.E. Narimanov, A.V. Kildishev, Naturally hyperbolic. Nat. Photonics **9**, 214–216 (2015)
13. R. Chandrasekar, Z. Wang, X. Meng, M.Y. Shalaginov, A. Lagutchev, Y.L. Kim, A. Wei, A. V. Kildishev, A. Boltasseva, V.M. Shalaev, Lasing action with gold nanorod hyperbolic metamaterials. arXiv:1609.05168 (2016)
14. G.V Naik, B. Saha, J. Liu, S.M. Saber, E.A. Stach, J.M.K. Irudayaraj, T.D. Sands, V.M. Shalaev, A. Boltasseva, Epitaxial superlattices with titanium nitride as a plasmonic component for optical hyperbolic metamaterials. Proc. Natl. Acad. Sci. **111**(21), 7546–7551 (2014)
15. J. Elser, R. Wangberg, V.A. Podolskiy, E.E. Narimanov, Nanowire metamaterials with extreme optical anisotropy. Appl. Phys. Lett. **89**(26), 261102 (2006)

16. B.M. Wells, A.V. Zayats, V.A. Podolskiy, Nonlocal optics of plasmonic nanowire metamaterials. Phys. Rev. B **89**(3), 35111 (2014)
17. A.V. Chebykin, A.A. Orlov, A.V. Vozianova, S.I. Maslovski, Y.S. Kivshar, P.A. Belov, Nonlocal effective medium model for multilayered metal-dielectric metamaterials. Phys. Rev. B **84**(11), 115438 (2011)
18. M.D. Eisaman, J. Fan, A. Migdall, S.V. Polyakov, Invited review article: single-photon sources and detectors. Rev. Sci. Instrum. **82**(7), 071101 (2011)
19. R.-M. Ma, R.F. Oulton, V.J. Sorger, X. Zhang, Plasmon lasers: coherent light source at molecular scales. Laser Photonics Rev. **7**(1), 1–21 (2013)
20. C. Monroe, Quantum information processing with atoms and photons. Nature **416**(6877), 238–246 (2002)
21. D. DiVincenzo, Quantum computation. Science **270**(5234), 255–261 (1995)
22. N. Gisin, G. Ribordy, H. Zbinden, Quantum cryptography. arXiv Prepr. quant-ph/0101098 (2001)
23. L.M. Duan, M.D. Lukin, J.I. Cirac, P. Zoller, Long-distance quantum communication with atomic ensembles and linear optics. Nature **414**(6862), 413–418 (2001)
24. D. Bouwmeester, J. Pan, K. Mattle, M. Eibl, H. Weinfurter, A. Zeilinger, Experimental quantum teleportation. 575–579 (1997)
25. A. Gruber, A. Dräbenstedt, C. Tietz, L. Fleury, J. Wrachtrup, C. von Borczyskowski, Scanning confocal optical microscopy and magnetic resonance on single defect centers. Science **276**(5321), 2012–2014 (1997)
26. C. Kurtsiefer, S. Mayer, P. Zarda, H. Weinfurter, Stable solid-state source of single photons. Phys. Rev. Lett. **85**(2), 290–293 (2000)
27. P. Tamarat, T. Gaebel, J.R. Rabeau, M. Khan, A.D. Greentree, H. Wilson, L.C.L. Hollenberg, S. Prawer, P. Hemmer, F. Jelezko, J. Wrachtrup, Stark shift control of single optical centers in diamond **083002**, 1–4 (2006)
28. G. Fuchs, V. Dobrovitski, R. Hanson, A. Batra, C. Weis, T. Schenkel, D. Awschalom, Excited-state spectroscopy using single spin manipulation in diamond. Phys. Rev. Lett. **101** (11), 117601 (2008)
29. E.M. Purcell, Spontaneous emission probabilities at radio frequencies. Phys. Rev. **69** (1946)
30. S. Schietinger, O. Benson, Coupling single NV-centres to high-Q whispering gallery modes of a preselected frequency-matched microresonator. J. Phys. B: At. Mol. Opt. Phys. **42**(11), 114001 (2009)
31. J. Riedrich-Möller, L. Kipfstuhl, C. Hepp, E. Neu, C. Pauly, F. Mücklich, A. Baur, M. Wandt, S. Wolff, M. Fischer, S. Gsell, M. Schreck, C. Becher, One- and two-dimensional photonic crystal microcavities in single crystal diamond. Nat. Nanotechnol. **7**(1), 69–74 (2012)
32. A. Faraon, C. Santori, Z. Huang, V.M. Acosta, R.G. Beausoleil, Coupling of nitrogen-vacancy centers to photonic crystal cavities in monocrystalline diamond. Phys. Rev. Lett. **109**(3), 033604 (2012)
33. M.Y. Shalaginov, V.V. Vorobyov, J. Liu, M. Ferrera, A.V. Akimov, A. Lagutchev, A.N. Smolyaninov, V.V. Klimov, J. Irudayaraj, A.V. Kildishev, A. Boltasseva, V.M. Shalaev, Enhancement of single-photon emission from nitrogen-vacancy centers with TiN/(Al,Sc)N hyperbolic metamaterial. Laser Photon. Rev. **9**(1), 120–127 (2015)
34. M.Y. Shalaginov, S. Ishii, J. Liu, J. Liu, J. Irudayaraj, A. Lagutchev, A.V. Kildishev, V.M. Shalaev, Broadband enhancement of spontaneous emission from nitrogen-vacancy centers in nanodiamonds by hyperbolic metamaterials. Appl. Phys. Lett. **102**(17), 173114 (2013)
35. D. Lu, J.J. Kan, E.E. Fullerton, Z. Liu, Enhancing spontaneous emission rates of molecules using nanopatterned multilayer hyperbolic metamaterials. Nat. Nanotechnol. **9**(1), 48–53 (2014)
36. H.N.S. Krishnamoorthy, Z. Jacob, E. Narimanov, I. Kretzschmar, V.M. Menon, Topological transitions in metamaterials. Science **336**(6078), 205–209 (2012)

37. T. Galfsky, H.N.S. Krishnamoorthy, W. Newman, E.E. Narimanov, Z. Jacob, V.M. Menon, Active hyperbolic metamaterials: enhanced spontaneous emission and light extraction. Optica **2**(1), 62 (2015)
38. G.V. Naik, J.L. Schroeder, X. Ni, V. Alexander, T.D. Sands, A. Boltasseva, Titanium nitride as a plasmonic material for visible and near-infrared wavelengths. Opt. Mater. Express **2**(4), 534–537 (2012)
39. Z. Jacob, I.I. Smolyaninov, E.E. Narimanov, Broadband Purcell effect: radiative decay engineering with metamaterials. Appl. Phys. Lett. **100**(18), 181105 (2012)
40. M.S. Yeung, T.K. Gustafson, Spontaneous emission near an absorbing dielectric surface. Phys. Rev. A **54**(6), 5227–5242 (1996)
41. L. Novotny, B. Hecht, *Principles of nano optics* (Cambridge University Press, Cambridge, 2006)
42. W.C. Chew, S. Chen, Response of a point source embedded in a layered medium. IEEE Antennas Wirel. Propag. Lett. **2**(1), 254–258 (2003)
43. K. Visscher, G.J. Brakenhoff, T.D. Visser, J. Microsc. **175**(2), 162–165 (1994)
44. S.V. Zhukovsky, O. Kidwai, J.E. Sipe, Opt. Express **21**(12), 14982–14987 (2013)
45. M.P. van Exter, G. Nienhuis, J.P. Woerdman, Two simple expressions for the spontaneous emission factor β. Phys. Rev. A **54**(4), 3553–3558 (1996)
46. M.T. Hill, M. Marell, E.S.P. Leong, B. Smalbrugge, Y. Zhu, M. Sun, P.J. van Veldhoven, E. J. Geluk, F. Karouta, Y.-S. Oei, R. Nötzel, C.-Z. Ning, M.K. Smit, Lasing in metal-insulator-metal sub-wavelength plasmonic waveguides. Opt. Express **17**(13), 11107 (2009)
47. S.-H. Kwon, J.-H. Kang, C. Seassal, S.-K. Kim, P. Regreny, Y.-H. Lee, C.M. Lieber, H.-G. Park, Subwavelength plasmonic lasing from a semiconductor nanodisk with silver nanopan cavity. Nano Lett. **10**(9), 3679–3683 (2010)
48. M.A. Noginov, G. Zhu, A.M. Belgrave, R. Bakker, V.M. Shalaev, E.E. Narimanov, S. Stout, E. Herz, T. Suteewong, U. Wiesner, Demonstration of a spaser-based nanolaser. Nature **460** (7259), 1110–1112 (2009)
49. X. Meng, A.V. Kildishev, K. Fujita, K. Tanaka, V.M. Shalaev, Wavelength-tunable spasing in the visible. Nano Lett. **13**(9), 4106–4112 (2013)
50. W. Zhou, M. Dridi, J.Y. Suh, C.H. Kim, D.T. Co, M.R. Wasielewski, G.C. Schatz, T.W. Odom, Lasing action in strongly coupled plasmonic nanocavity arrays. Nat. Nanotechnol. **8**, 506–511 (2013)
51. F. van Beijnum, P.J. van Veldhoven, E.J. Geluk, M.J.A. de Dood, G.W. 't Hooft, M.P. van Exter, Surface plasmon lasing observed in metal hole arrays. Phys. Rev. Lett. **110**(20), 206802 (2013)
52. X. Meng, J. Liu, A.V. Kildishev, V.M. Shalaev, Highly directional spaser array for the red wavelength region. Laser Photonics Rev. **8**(6), 896–903 (2014)
53. R.F. Oulton, V.J. Sorger, T. Zentgraf, R.-M. Ma, C. Gladden, L. Dai, G. Bartal, X. Zhang, Plasmon lasers at deep subwavelength scale. Nature **461**(7264), 629–632 (2009)
54. J.K. Kitur, L. Gu, T. Tumkur, C. Bonner, M.A. Noginov, Stimulated emission of surface plasmons on top of metamaterials with hyperbolic dispersion. ACS Photonics **2**(8), 1019–1024 (2015)
55. J.-M. Moon, A. Wei, Uniform gold nanorod arrays from polyethylenimine-coated alumina templates. J. Phys. Chem. B **109**(49), 23336–23341 (2005)
56. R. Starko-Bowes, J. Atkinson, W. Newman, H. Hu, T. Kallos, G. Palikaras, R. Fedosejevs, S. Pramanik, Z. Jacob, Optical characterization of epsilon-near-zero, epsilon-near-pole, and hyperbolic response in nanowire metamaterials. J. Opt. Soc. Am. B **32**(10), 2074 (2015)

57. M.Y. Shalaginov, S. Bogdanov, V.V Vorobyov, A.S. Lagutchev, A.V. Kildishev, A.V. Akimov, A. Boltasseva, V.M. Shalaev, Chapter 6: Enhancement of single-photon sources with metamaterials. In: From Atomic to Mesoscale: The Role of Quantum Coherence in Systems of Various Complexities, ed. by S.A. Malinovskaya and I. Novikova, World Scientific, 123–148 (2015)
58. L.F. Shampine, Vectorized adaptive quadrature in MATLAB. J. Comput. Appl. Math. **211**(2), 131–140 (2008)
59. M. Frimmer, A. Mohtashami, A.Femius Koenderink, Nanomechanical method to gauge emission quantum yield applied to nitrogen-vacancy centers in nanodiamond. Appl. Phys. Lett. **102**(12), 121105 (2013)

Chapter 6
Strong Coupling Between Organic Molecules and Plasmonic Nanostructures

Robert J. Moerland, Tommi K. Hakala, Jani-Petri Martikainen, Heikki T. Rekola, Aaro I. Väkeväinen and Päivi Törmä

Abstract This chapter introduces the theory behind strong coupling of plasmonic modes, such as surface plasmon polaritons, with electronic transitions that are typical for quantum emitters, such as dye molecules and quantum dots. A brief historical overview of the experimental endeavor of measuring such strong coupling is provided, after which we look more carefully at the dynamics in such systems. We proceed with a more in-depth discussion of strong coupling between emitters and *delocalized* plasmonic modes, called surface lattice resonances, but not before devoting some space to the ideas and theory behind surface lattice resonances for the readers who might not be familiar with the topic. We end the chapter with an outlook on the potential strong coupling has for new and exciting fundamental phenomena and applications of light on the nanoscale.

6.1 Introduction

Strong coupling refers to a phenomenon where the relevant frequencies of the system are significantly modified by couplings between its individual constituents. Two coupled harmonic oscillators are a simple classical physics system which shows normal modes that are different from the original mode frequencies ω_1 and ω_2. For non-damped oscillators with mutual coupling of strength Ω, the normal mode energies are of the form $\omega_+^2 = \omega^2 + 2\Omega^2$ and $\omega_-^2 = \omega^2$ at resonance ($\omega_1 = \omega_2 = \omega$). Thus the modes show a split in energy that depends on the coupling Ω, and the normal modes are linear superpositions of the original modes [1], see Fig. 6.1a. Strong coupling is a somewhat non-rigorous term, but basically it means situations where such a splitting

R.J. Moerland (✉)
Department of Imaging Physics,
Delft University of Technology, NL-2600GA Delft, The Netherlands
e-mail: r.j.moerland@tudelft.nl

T.K. Hakala · J.-P. Martikainen · H.T. Rekola · A.I. Väkeväinen · P. Törmä
COMP Centre of Excellence, Department of Applied Physics,
Aalto University, FI-00076 Aalto, Finland
e-mail: paivi.torma@aalto.fi

S.I. Bozhevolnyi et al. (eds.), *Quantum Plasmonics*,
Springer Series in Solid-State Sciences 185,
DOI 10.1007/978-3-319-45820-5_6

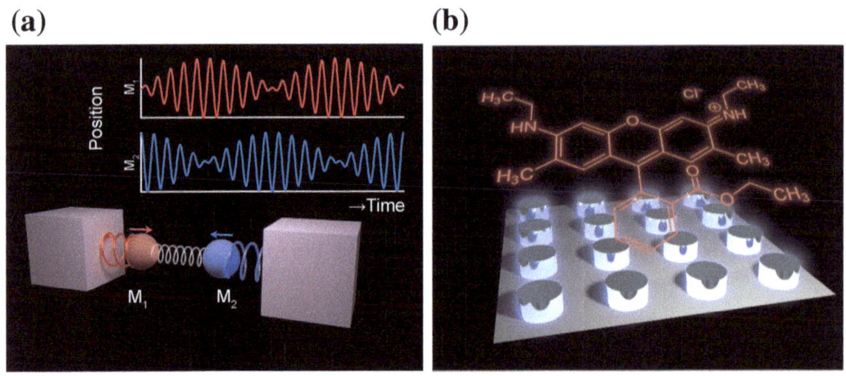

Fig. 6.1 **a** Two coupled harmonic oscillators, M_1 in red and M_2 in *blue*, show normal modes that are linear combinations of the uncoupled modes of the two oscillators. This leads to an exchange of energy between the two oscillators in time, schematically shown in the *inset*. **b** Similarly, electronic transitions in molecules (*red*) may couple strongly with electromagnetic fields, such as the plasmonic fields (*blue*) supported by a silver nanoparticle array, and create light-matter superpositions. The molecular and plasmonic modes are the analogies of the classical oscillators M_1 and M_2. The molecule is not to scale. Figure by Antti Paraoanu

is significant in some sense. Here we resort to the definition that, in the strong coupling regime, the normal mode splitting must be significant compared to the system linewidths. If the two oscillators have similar damping and linewidth properties, then the coupling Ω has to be at least of the same order of magnitude as the linewidths for a normal mode splitting to be experimentally observable. In case one oscillator has a much narrower linewidth than the other, one can still see a splitting at resonance, even when Ω is smaller than the large linewidth, but this would be usually referred to as a Fano-type phenomenon rather than strong coupling.

The concept of normal modes and strong coupling is not limited to harmonic (or anharmonic) oscillators but applies to any kind of system with well-defined modal frequencies and mutual couplings. One can also extend it from the classical to the quantum world: the normal modes now correspond to eigenstates of a coupled system. In the quantum domain, the potential of strong coupling phenomena goes beyond simply finding out what energies the system can have. Namely, two coupled states forming a new eigenstate that is a quantum coherent superposition of the original states basically provides a *qubit*, the basic building block of quantum information processing. If inter-qubit interactions can be realized as well, one can create entanglement, the profoundly quantum mechanical phenomenon that is the basic resource of quantum computing and communication.

The paradigm in studies of strong coupling in quantum mechanics has been the two-level system coupled to a field, for instance a transition between two atomic or molecular electronic states coupled to an electromagnetic field in the visible, infrared or microwave frequencies. Pioneering experiments on strong coupling between ensembles of atoms and microwave or optical high-finesse cavity modes were done in the late 1980s and early 1990s [2–6], eventually reaching the single atom level [7, 8].

Such research expanded into inorganic semiconductor materials [9–16] where excitions were coupled to microcavity modes of various types. Organic semiconductors allowed observing similar phenomena at room temperature [17–22]. The concepts of quantum optics are nowadays beautifully realized in various artificial atom systems such as superconducting circuits, where strong coupling [23, 24] and even ultra-strong coupling [25] have been seen. For a mini-review of strong coupling studies across the fields of modern physics see Sect. 1.3 of the review [26].

This book chapter focuses on discussing the topic of strong coupling between surface plasmon polariton (SPP) modes and emitters, an example is illustrated in Fig. 6.1b. In this field a large amount of literature already exists on experimental observations of *ensembles* of emitters coupled strongly with modes, but experiments performed with single emitters are still a future challenge towards which there has been, however, remarkable progress. The pros and cons of surface plasmon polariton modes with respect to strong coupling basically originate from their near-field character. Strong confinement on a metal surface on one hand provides small mode volumes that benefit strong coupling, but on the other hand it fundamentally imposes losses as discussed in Chap. 13 of this book.

Apart from quantum information processing, strong coupling opens doors to many other important and intriguing phenomena. For instance thresholdless lasing [27] and lasing without inversion [28] may occur in the strong coupling regime. In recent years, coherent superpositions of semiconductor excitons and light, i.e., polaritons, have been shown to form condensates [29]. This opens a rich playground for basic physics and potential for new types of coherent light sources. If strong effective photon-photon interactions can be realized, one may create photon fluids [30] which display highly non-linear and interesting quantum statistical phenomena. Similar concepts may become possible in the field of plasmonics; the high decay rates of plasmonic modes pose a challenge and may lead to new types of condensation phenomena [31].

In this book chapter, we first briefly present the classical, semi-classical and quantum theories of strong coupling in Sect. 6.2. We then provide in Sect. 6.3 a small review of the experiments done on strong coupling between plasmonic modes and emitters. For the topics of Sects. 6.2 and 6.3, a longer and more complete presentation can be found in the review [26]. In contrast, in Sects. 6.4, 6.6.1 and 6.6.2, we review three topics in more depth and detail than in [26]: studies of strong coupling dynamics in plasmonics (Sect. 6.4), strong coupling in periodic arrays of plasmonic nanoparticles—the case depicted in Fig. 6.1b—(Sect. 6.6.1) and spatial coherence of plasmonic strong coupling phenomena (Sect. 6.6.2). Since Sects. 6.6.1 and 6.6.2 deal with periodic nanoparticle arrays, we present in Sect. 6.5 a pedagogical description of the basic theory of the modes in such arrays. While the theory of such nanoparticle arrays has been discussed in a large number of publications [32] we feel that an easy-to-access pedagogical presentation with helpful pictures that guide the intuition has been missing. In Sect. 6.7 we discuss the perspectives of this exciting, expanding research field.

6.2 Theoretical Background

The coupling of emitters and nanostructures can be modeled at various level of detail, ranging from a fully classical approach with damped harmonic oscillators, to a fully quantum-mechanical approach where both emitter and field are quantized. In general, the coupling can range from the weak-coupling regime to the strong-coupling regime. Here, we briefly summarize the theoretical approaches specific to strong coupling between emitters and plasmonic nanostructures.

6.2.1 *Classical Approach*

To see how the possibility of strong coupling between SPPs and emitters might arise from a classical theory, consider emitters as classical Lorentzian emitters on top of a metal surface supporting SPP modes. We can think of a bound electron of the emitter as a harmonically damped oscillator which is driven by a (classical) electromagnetic field. Assuming a harmonic driving field $E(r,t) = E_0 \mathrm{e}^{-\mathrm{i}\omega t}$ we can solve the equation of motion for the position of the electron r, and find the dipole moment $p = -er$ of the electron [26, 33] as

$$p = -er = \frac{e^2}{m}\frac{1}{\omega_0^2 - \omega^2 - \mathrm{i}\gamma\omega}E\,. \tag{6.1}$$

Here ω_0 is the natural frequency of the oscillator and γ is the damping coefficient, and e and m are the elementary charge and electron mass, respectively. For a system containing N such emitters, the average dipole moment per unit volume or macroscopic polarizability P becomes[1]

$$P = \frac{Ne^2}{Vm}\frac{1}{\omega_0^2 - \omega^2 - \mathrm{i}\gamma\omega}E\,. \tag{6.2}$$

On the other hand, the macroscopic polarizability can also be defined in terms of susceptibility χ (and vacuum permittivity ε_0) as $P = \varepsilon_0 \chi E$ from which it follows that

$$\chi(\omega) = \frac{Ne^2}{V\varepsilon_0 m}\frac{1}{\omega_0^2 - \omega^2 - \mathrm{i}\gamma\omega}\,. \tag{6.3}$$

In linear media, as discussed here, the relative permittivity of the material $\varepsilon(\omega)$ then is defined as

$$\varepsilon(\omega) = \varepsilon' + \mathrm{i}\varepsilon'' \equiv 1 + \chi(\omega). \tag{6.4}$$

[1]Here it is implicitly assumed that the density of emitters is sufficiently small so their fields contribute only slightly to the field strength. This amounts to the limit $\alpha N/V\varepsilon_0 \rightarrow 0$, where α is the polarizability.

The connection between the Lorentzian response of the medium containing the emitters and SPPs is via the dispersion relation of SPP mode on a planar metal interface, connecting the wave number k and the frequency ω [34]

$$k = \frac{\omega}{c}\sqrt{\frac{\varepsilon_1\varepsilon_2}{\varepsilon_1 + \varepsilon_2}}\,, \tag{6.5}$$

where c is the speed of light and ε_1 and ε_2 are the permittivities of the metal and the medium containing the emitters, respectively. As a first approximation, we may assume that the permittivity ε_1 of the metal is constant in the frequency range around ω_0 that is of interest here. It is also known that, for metals, ε_1 often has a large negative real part. Then $\varepsilon_1 + \varepsilon_2$ remains negative and the dominant frequency response is from the $\varepsilon_1\varepsilon_2$ term in the numerator. Based on such reasoning we find

$$k^2 = \frac{\omega^2}{c^2}\frac{|\varepsilon_1|}{|\varepsilon_1 + \varepsilon_2|}(1 + \chi(\omega))\,, \tag{6.6}$$

which defines the dispersion relations in the classical case. A somewhat more transparent form of 6.6 can be obtained by scaling the wavenumber to a frequency $\kappa^2 = k^2(|\varepsilon_1 + \varepsilon_2|c^2)/|\varepsilon_1|$:

$$\kappa^2 = \omega^2(1 + \chi(\omega)) = \omega^2\left(1 + \frac{A}{\omega_0^2 - \omega^2 - \mathrm{i}\gamma\omega}\right), \tag{6.7}$$

where $A = Ne^2/(V\varepsilon_0 m)$. We outline the resulting dispersions in Fig. 6.2 from which it can be seen how the presence of emitters can give rise to so-called normal-mode splitting. At resonance, $\kappa = \omega_0$ and the difference in mode frequencies becomes

$$\Omega = \sqrt{\frac{N}{V}}\frac{e}{\sqrt{\varepsilon_0 m}}\,. \tag{6.8}$$

It is one thing for modes to be split and another for this split to be observable since losses cause finite linewidths for the modes. Phenomenologically we can model SPP losses by adding an imaginary part to the SPP frequency $\kappa - \mathrm{i}\gamma_{SPP}/2$. It then turns out that mode splitting is observable if

$$\Omega^2 > \frac{\gamma^2}{2} + \frac{\gamma_{SPP}^2}{2} \Leftrightarrow \frac{Ne^2}{V\varepsilon_0 m} > \frac{\gamma^2}{2} + \frac{\gamma_{SPP}^2}{2}\,, \tag{6.9}$$

which is the condition of strong coupling in the classical theory.

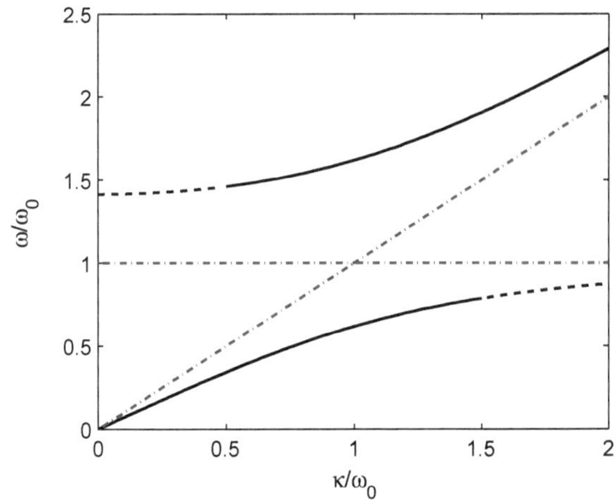

Fig. 6.2 The dispersion of an SPP-emitter system as given by 6.7 is shown by the *black lines*. The *dashed straight lines* are the emitter energy and the SPP dispersion according to 6.7 for $A = 0$. The *black lines* turning from *solid* to *dashed* indicates that, in case of finite damping, the nearly flat modes cease to be well defined further away from the crossing point. We assumed $A = \omega_0$. Figure from [26]. Copyright IOP Publishing. Reproduced with permission. All rights reserved

6.2.2 Semi-Classical Approach

The coupling between emitters and SPPs can also be studied semi-classically by assuming a classical electromagnetic (SPP) field and modeling the emitters as quantum-mechanical two-level systems with an excited state $|e\rangle$ and a ground state $|g\rangle$. The electromagnetic field can then be taken as $\vec{E}\cos(\omega t)e^{i\vec{k}\cdot\vec{r}}$, where $\vec{E}$ tells both the magnitude as well as the direction of the field. As emitters are small, the dipole approximation typically is accurate for the matter-field coupling and the Hamiltonian for a single emitter coupled to the field is given by

$$H = \frac{1}{2}E_e(\vec{I} + \sigma_z) + \frac{1}{2}E_g(\vec{I} - \sigma_z) + \hbar\Omega_0(\sigma_+ + \sigma_-)\cos(\omega t)\,. \tag{6.10}$$

Here E_e and E_g are the energy levels of the two-level system and $\sigma_+ = \begin{pmatrix} 0 & 1 \\ 0 & 0 \end{pmatrix}$ and $\sigma_- = \begin{pmatrix} 0 & 0 \\ 1 & 0 \end{pmatrix}$ are the raising and lowering operators. Furthermore, $\sigma_z = \begin{pmatrix} 1 & 0 \\ 0 & -1 \end{pmatrix}$ and $\Omega_0 = -\vec{d}\cdot\vec{E}/\hbar$ is the semi-classical Rabi frequency with $\vec{d}$ the dipole moment of the two-level system. From the above we can identify $\omega_0 = (E_e - E_g)/\hbar$ as the natural frequency of the emitter. Often Ω is small compared to ω and ω_0 and we can use the rotating wave approximation (RWA) to simplify the Hamiltonian. After applying this approximation, the Hamiltonian becomes

$$H = -\frac{\hbar\delta}{2}\sigma_z + \frac{\hbar\Omega_0}{2}(\sigma_+ + \sigma_-)\,, \tag{6.11}$$

where $\delta = \omega - (E_e - E_g)/\hbar = \omega - \omega_0$ is the detuning between the field and the emitter. This 2×2 system can be promptly diagonalized for eigenvalues $E_1 = \frac{\hbar}{2}\sqrt{\delta^2 + \Omega_0^2}$, $E_2 = -\frac{\hbar}{2}\sqrt{\delta^2 + \Omega_0^2}$, and corresponding eigenstates (also known as dressed states) $|1\rangle$ and $|2\rangle$. If the emitter starts from the ground state, under the presence of the coupling field it will start a periodic Rabi oscillation between ground and excited states. At resonance $\delta = 0$ this is especially apparent with the time-evolution of the two-level system given by

$$|\Psi(t)\rangle = \cos\theta e^{-iE_1t/\hbar}|1\rangle + e^{-iE_2t/\hbar}\sin\theta|2\rangle \equiv \gamma_g(t)|g\rangle + \gamma_e(t)|e\rangle\,. \tag{6.12}$$

Here θ determines the state at time $t = 0$, and $\gamma_g(t)$ and $\gamma_e(t)$ can be obtained from the relation between the original ($|g\rangle, |e\rangle$) and the dressed ($|1\rangle, |2\rangle$) states. Oscillations in occupation probability then happen at the Rabi frequency Ω_0 due to the dependence of E_1 and E_2 on Ω_0 (when $\delta = 0$). A non-zero detuning by δ modifies the oscillation frequency.

The semi-classical solution with a single emitter reveals formal similarity with the earlier classical result with $N = 1$. However, note that the classical derivation was based on assuming a linear response while the semi-classical solution here is exact (when the RWA is accurate). How to recover the classical result from the semi-classical result? To answer this we have to understand how the linear susceptibility for many emitters follows from the semi-classical quantum result for a single emitter. The expectation value of the induced electric dipole moment is

$$\langle\hat{\vec{D}}\rangle = \langle\Psi|\hat{\vec{D}}|\Psi\rangle\,. \tag{6.13}$$

If the wavefunction for the two-level system is given by $|\Psi(t)\rangle = \gamma_g(t)|g\rangle + \gamma_e(t)|e\rangle$, the expectation value can be computed in terms of d_i, which is the dipole matrix element $d = \langle g|\hat{D}_i|e\rangle$ $(i = x, y, z)$:

$$\langle\hat{D}_i\rangle = d_i\gamma_e^*\gamma_g e^{-i\omega_0 t} + c.c. \tag{6.14}$$

As before, the polarizability $\vec{P}$ (now a vector) is the dipole moment per unit volume. In a driven system, the two-level systems are continuously pumped, but they also decay at a rate γ and the dipoles tend to vanish. To get the macroscopic response one should average the single emitter response over many emitters. The result becomes [26]

$$\vec{P} = \frac{N}{V}\frac{d^2}{\hbar}\frac{\omega_0 - \omega + i\gamma}{\gamma^2 + \Omega_0^2 + (\omega - \omega_0)^2}\frac{\vec{E}_0}{2}e^{-i\omega t} + c.c. \tag{6.15}$$

This is almost the same as the earlier result (6.2), based on classical oscillators and a linear response, although Ω_0^2 accounting for saturation effects in a two-level system appears in the denominator of (6.15). However, for weak fields this term is negligible and we find normal mode splitting $\propto d\sqrt{N/V}$ similar the classical case. In this limit the result is independent of the field even though individual emitters oscillate with a Rabi frequency proportional to the field. For more discussion, see Sect. 5 of [26] and references therein.

6.2.3 Fully Quantum-Mechanical Approach

Finally, emitter and field coupling can also be modeled fully quantum mechanically. The Jaynes-Cummings Hamiltonian describes the coupling of a two-level system with an energy difference of $\hbar\omega_0$ between excited and ground state to a single mode of the field with frequency ω

$$H = \frac{1}{2}\hbar\omega_0\sigma_z + \hbar\omega\hat{a}^\dagger\hat{a} + \hbar\left(g\hat{a}\sigma_+ + h.c.\right) . \tag{6.16}$$

The coupling coefficient g is proportional to the dipole moment and the operators $\hat{a}$ and $\hat{a}^\dagger$ are the annihilation and creation operators for the quantized field. $\sigma_z = (|e\rangle\langle e| - |g\rangle\langle g|)/2$ and $\sigma_+ = |e\rangle\langle g|$ and $\sigma_- = |g\rangle\langle e|$ are the raising and lowering operators for the two-level system.

This Hamiltonian only couples states $|e\rangle|n\rangle$ and $|g\rangle|n+1\rangle$ to each other which allows independent diagonalization for each such pair of states in Hilbert space. In this way we find the eigenvalues

$$E_{1n} = \hbar\left(n + \frac{1}{2}\right)\omega - \frac{1}{2}\hbar\sqrt{\delta^2 + 4g^2\,(n+1)} \tag{6.17}$$

and

$$E_{2n} = \hbar\left(n + \frac{1}{2}\right)\omega + \frac{1}{2}\hbar\sqrt{\delta^2 + 4g^2\,(n+1)} . \tag{6.18}$$

The fully quantum mechanical model has some similarities with earlier approaches, but has a dramatic difference for the vacuum where $n = 0$. The modes are split by an energy $\Delta E = 2\hbar g$ even then indicating the presence of vacuum Rabi splitting.

We can generalize the model to many emitters which is more relevant for the experiments that will be discussed in the remainder of the chapter. We can derive a so-called Dicke-Hamiltonian in terms of collective two-level operators (in terms of two-level operators $\sigma_+^{(i)}$, $\sigma_-^{(i)}$, and $\sigma_z^{(i)}$ for the emitter i) $\hat{S}_z = \sum_{i=1}^N \sigma_z^{(i)}$ and $\hat{S}_+ = \sum_{i=1}^N \sigma_+^{(i)}$, with the result

$$H = \frac{1}{2}\hbar\omega_{eg}\hat{S}_z + \hbar\omega\hat{a}^\dagger\hat{a} + \hbar\left(g\hat{a}\hat{S}_+ + h.c.\right) . \tag{6.19}$$

This model can be solved analytically [35], but for our purposes it is enough to consider the limit with many emitters N but only few emitters in the excited state. In this limit the Hollstein-Primakoff transformation can be used to remove two-level system operators and map the Hamiltonian to a bosonic system

$$\hat{H} \simeq \hbar\omega_0 \left(-\frac{N}{2} + \hat{b}^\dagger \hat{b}\right) + \hbar\omega \hat{a}^\dagger \hat{a} + \hbar g \sqrt{N} \left(\hat{a}^\dagger \hat{b} + h.c.\right) , \tag{6.20}$$

where also $\hat{b}$ is now a bosonic operator. Mapping into bosonic operators is justified in this limit since in the superposition state the excitation is shared between a large number of atoms. Since the number of participating atoms is large compared to the number of excitations, adding new excitations is not hindered by the two-level system becoming saturated. The above is a quantum model for two coupled harmonic oscillators which can be diagonalized. The resulting mode splitting turns out to be $\Omega = 2g\sqrt{N}$ and is very similar to the semi-classical result. However, notice that here physical picture is quite different. The oscillators can be thought of as a "superatom" with a large dipole moment. The field then couples to this superatom giving rise to Rabi oscillations between the field and the superatom.

6.3 Coupling of Organic Molecules with Plasmonic Structures

Having introduced the basic theoretical background, we now proceed with a brief overview of experimental work on strong coupling of emitters and plasmon modes. Evidence of strong coupling of emitters and plasmon modes has been reported in the literature numerous times. Because of limited space, however, we will have to restrict ourselves to a few samples and only mention the major scientific advances. As a result, this obviously means that we omit many important works, and the reader is encouraged to see [26] for a more comprehensive list.

Since 6.9 in the previous section implies that a small damping constant helps to see strong coupling phenomena, the class of J-aggregate molecules was among the first to be used in strong coupling experiments with plasmon modes. These molecules, once aggregated, have an exceptionally large effective dipole moment and narrow linewidth [36]. Experimentally, the coupling of the exciton of a J-aggregate and a plasmon mode was first investigated in the seminal work by Pockrand et al. in the early 1980s [37]. The authors used a Kretschmann-type setup [34] to launch surface plasmon polariton (SPP) modes in a thin silver film on top of a glass prism. J-aggregates were placed on top of the silver film, which subsequently coupled to the SPP mode in the silver film. The authors found that the deviation from the 'textbook' SPP dispersion curve (6.5) was significant when the interaction between the exciton states and the SPP mode was the strongest. Only in 2004 was this system investigated again, when Bellessa et al. [38] confirmed that strong coupling between the SPP and the excitons of J-aggregates was taking place in that particular system. The data of

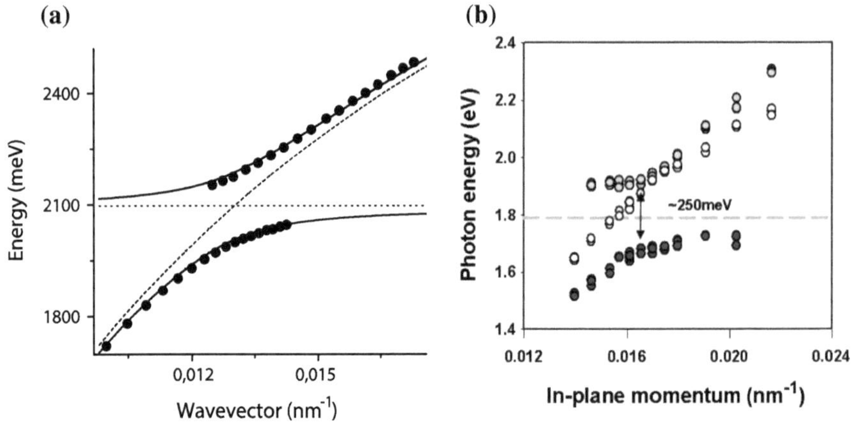

Fig. 6.3 Measured dispersion curves of surface plasmon polariton modes strongly coupled to J-aggregates. **a** The measured dispersion curve of SPPs in a silver film, launched by a Kretschmann configuration. The silver film is loaded with J-aggregates on top. The *dotted lines* are the expected dispersion curves for the uncoupled J-aggregates and SPPs, the *circles* are the measured dispersion of the strongly coupled system, where the in-plane wavevector is tuned by changing the angle of incidence. Reprinted with permission from [38]. Copyright (2004) by the American Physical Society. **b** The measured dispersion curve of a silver film, with periodic circular perforations and optionally loaded with J-aggregates. Here, the period of the perforations is swept. The *white circles* represent the dispersion curve of the perforated silver film, whereas the *filled circles* represent the dispersion curve of the perforated film with the J-aggregates. Reprinted with permission from [40]. Copyright (2005) by the American Physical Society

these authors show a clear split in the dispersion curve at the exciton energy, see Fig. 6.3a. For a detailed study of the square-root dependency of the mode splitting and concentration (6.8) see for instance Balci et al. [39].

In addition to the Kretschmann and Otto [41] type of setups, adding periodic corrugations to the metal layer is another often-used practice to launch surface plasmon modes. For small perturbations, the SPP is largely the same as those that would exist on a perfectly flat film. For large periodic corrugations, however, the modes shift in frequency and also new modes can come into existence [42]. These new modes can resonantly enhance or suppress certain wavelengths. This additional degree of freedom can be exploited to tune the dispersion curve of the plasmon mode, instead of changing the angle of incidence for planar SPPs. An example is shown in Fig. 6.3b, which is the result of a measurement by Dintinger et al. [40]. Here, a silver film with periodic circular perforations was coated with J-aggregates, and the dispersion was measured as a function of the period of the perforations. Also here, the dispersion curve shows a distinct split, but the crucial difference is that the tuning parameter is the period of the holes instead of the angle of incidence. Strong coupling between J-aggregates and similar delocalized modes in periodic arrays of nanovoids was observed in [43], as well as strong coupling with localized resonances of the holes.

A noteworthy scientific achievement is that of Vasa et al. [44]. Here, the authors used a pump-probe scheme to monitor the dynamics of strong coupling between J-aggregates and plasmons in a grating-type structure. This experiment will be described into more detail in Sect. 6.4.

According to 6.9, the typically broad spectra of typical dye molecules seem to make an observation of strong coupling impossible. Nevertheless, in 2009 conclusive evidence was provided by Hakala et al. [45] of strong coupling of a dye molecule with a broad absorption spectrum, Rhodamine 6G, and SPPs. This result was later reproduced, not only for Rhodamine 6G [46–48], but also for other organic molecules with a broad absorption spectrum [49–52]. A more detailed treatment of the important result of [45] is provided in the next section. In that light, perhaps the classical result of 6.9 is better taken as a rule of thumb rather than a strict rule. Further (quantum) theoretical investigations are required to obtain a fully satisfactory description, matching the experimental observations [53].

Besides organic molecules, one could also choose to use quantum dots as the emitter in strong-coupling experiments, as for many cases quantum dots have desirable or even tunable properties. Indeed, strong coupling between SPPs and quantum dots has been shown [54, 55], but publications successfully showing strong coupling between SPPs and quantum dots are rather sparse and reaching the strong coupling condition, even with a high concentration of quantum dots [56], proves to be difficult.

We end this brief overview with one of the more recent results, which is the strong coupling of organic molecules with surface lattice resonances (SLR), sustained by periodic arrays of metal nanoparticles [57–59]. We will keep a more complete description of SLRs for Sect. 6.5, and will discuss the strong coupling between these lattice resonances and organic molecules in detail in Sects. 6.6.1 and 6.6.2. Furthermore, we refer the reader to [32, 60–62] for a detailed background on the topic of SLRs. The applicability of these periodic systems for lasing applications is discussed in Chap. 8.

6.4 Dynamics of Strong Coupling

The interpretation of a continuous energy exchange between the SPP mode and the emitter implies that dynamics, that is, the Rabi oscillations, are an important part of strong coupling. Therefore, a time-domain measurement showing this exchange is seemingly the most obvious way to investigate. However, due to the ultrafast time scales that are involved, on the order of tens of femtoseconds, a direct time-domain measurement of Rabi oscillations is far from trivial. Below, we will discuss one of the first studies of strong coupling between SPPs and a dye molecule in the time domain, using pump-probe spectroscopy. However, the dynamics of strong coupling with an SPP mode can also be altered by changing the interaction length, and therefore interaction time, that the SPP mode has with the dye. This, then, can be probed with frequency-domain methods, which we will discuss first.

6.4.1 Frequency Domain

In [45], strong coupling between propagating surface plasmon polaritons and Rhodamine 6G (R6G) molecules was studied. The dispersions of planar silver films with varying concentrations of R6G (mM range) were measured. With higher concentrations, a clear anticrossing in the dispersions was observed, first at the energy corresponding to the R6G absorption maximum (2.3 eV), and with the highest concentration, at 2.45 eV as well. The latter anticrossing corresponds to an R6G absorption shoulder, see Fig. 6.4 (the curves on the right). The energy splitting was found to be proportional to the square root of the R6G concentration, characteristic for strongly coupled systems (see Sect. 6.2).

The dynamics of the system was studied with plasmonic waveguides. The waveguides consisted of 5 μm wide Ag stripes. The SPPs were launched in a specific region on a waveguide, and after propagating a few micrometers, entered an area having R6G molecules on top of the waveguide, see the inset of Fig. 6.4. The luminescence spectra were measured at the end of this area. The interaction time between the SPPs and R6G molecules was adjusted by varying the length of the R6G area. By adjusting the length of the interaction area, one can induce a faster effective decay than the one obtained from the natural coherence times (or lengths) of the system.

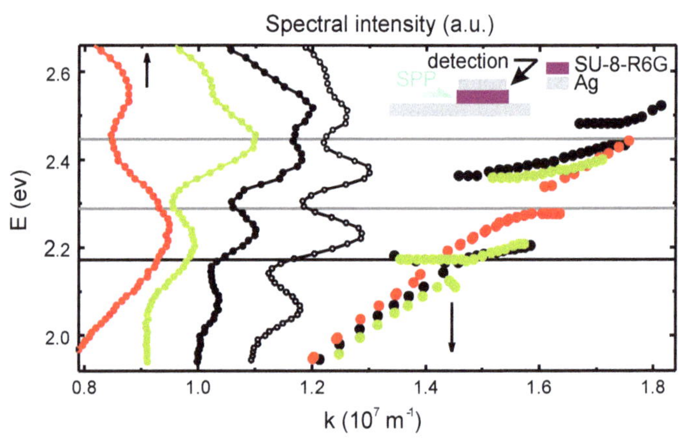

Fig. 6.4 The dispersion curves of the thin film samples (*right*). The *red*, *green*, and *black* dispersions correspond to samples having 4, 25 and 50 mM R6G concentration. On the *left*, luminescence spectra measured from the waveguide samples having different lengths of R6G deposited on top of a waveguide are shown. The *red*, *green* and *black curves* correspond to samples having 1, 2 and 5 μm long R6G areas. The curve with empty *black circles* is measured from a waveguide sample having a 5 μm long R6G area with a layer of silver deposited on top of this area (see the *inset*). The colour coding shows the cases where the dips in the luminescence spectra and the splits in the dispersions appear at same energies. The *black horizontal line* depicts the measured R6G emission and the *gray lines* the absorption maxima of R6G. Reprinted with permission from [45]. Copyright (2009) by the American Physical Society

This is analogous to optical microcavities, in which a lower cavity Q-factor reduces the roundtrips (and therefore the interaction time) of the cavity photon before exiting the cavity.

On the left of Fig. 6.4, one can see how the measured spectra evolve when the length of the R6G area is increased. For the 1 μm R6G area, the spectrum is dominated by two peaks, one corresponding to the initial peak wavelength of the propagating plasmons (2.55 eV) and the other to R6G emission (2.25 eV). This case corresponds to the weak coupling case, in which the decay of the mode is faster than the Rabi period. As the interaction time is increased with longer R6G areas, however, a dip in the luminescence spectra appears at 2.3 eV, corresponding to the R6G absorption maximum (green curve in Fig. 6.4). This dip was interpreted as a consequence of entering the strong coupling regime. Interestingly, for 5 μm interaction area, a second dip appears at 2.45 eV, corresponding to the R6G absorption shoulder (black curve in Fig. 6.4). As a separate test, to verify our reasoning, the SPP-matter interaction was also increased by fabricating a layer of silver on top of the interaction area (see empty black circles in Fig. 6.4). This decreases the mode volume and reduces the decay into radiative modes. Therefore, very pronounced strong-coupling features can be observed.

6.4.2 *Time Domain*

As said before, Rabi oscillations between the two uncoupled states can be interpreted as a periodic energy transfer between two oscillators. Due to ultra-small mode volumes in plasmonic systems, the splittings are typically very large, up to 10 % of the resonance energy. Consequently, the early studies focused on detecting this energy split. At visible wavelengths, a typical splitting in plasmonic systems on the order of 100 meV results in a periodic energy transfer rate of 30 fs. This poses considerable challenges on the direct probing of the dynamics in time domain. Recently, however, femtosecond-scale studies of dynamics in plasmon-exciton system have been carried out with an optical pump-probe setup, one of which we will discuss below [44].

The sample of [44] consisted of a one dimensional array of narrow (45 nm) Au nanoslits and J-aggregate molecules embedded in a polymer matrix, see Fig. 6.5a. Strong coupling between the periodic plasmonic structure and J-aggregates was evident from the anticrossing of the two upper (UP) and lower (LP) polariton modes in the measured white light reflectivity spectrum, see Fig. 6.5b. Ultrashort (15 fs) pulses were used to excite the system, which was followed by a similar (lower intensity) probe pulse. The differential transient reflectivity of the probe pulse was recorded as a function of the delay between the pump and the probe pulses. At 39° incident angle, the transient reflectivity of the UP is similar to the one obtained from a bare Au film overlaid with molecules, see Fig. 6.5c. The nonlinearity in this case is due to saturation of the exciton absorption induced by the pump beam. The transient reflectivity of the LP at 1.65 eV, however, exhibited temporal oscillations with a period of 27 fs.

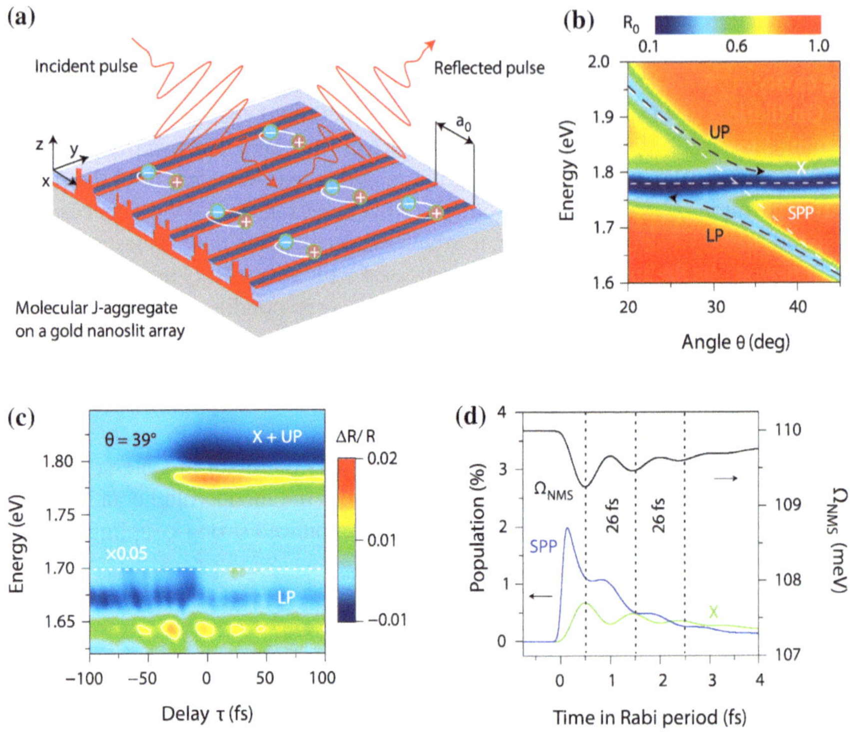

Fig. 6.5 **a** Schematic of the sample, **b** Angle-resolved reflectivity of the sample exhibiting a splitting of 110 meV, **c** Experimental transient reflectivity spectra. The spectra were obtained by using two almost collinearly propagating pulses with time delay τ (pulse length 15 fs, incidence angle 39°). Temporal oscillations are observed at LP resonance energy (1.65 eV). **d** The simulated time evolution of the pump-induced SPP and exciton population dynamics and the normal mode splitting. Reprinted by permission from Macmillan Publishers Ltd. from [44], Copyright (2013)

More recently [63], it was shown that in a similar system the dynamics were governed by two distinct contributions: a coherent resonant dipole-dipole interaction and an incoherent energy transfer due to the spontaneous emission/reabsorption of a photon. The interplay of the two pathways modified the radiative damping rate and may enable applications in coherent active plasmonic systems.

6.5 Surface Lattice Resonances

In the next section, we will discuss the strong coupling of emitters with so-called Surface Lattice Resonances (SLRs). Before we do so, we provide a primer on SLRs for the readers who might not be familiar with the topic. This discussion provides a basic understanding of the optical modes that can also support lasing as we shall see in Chap. 8. Surface lattice resonances are optical modes found in 2D lattice structures made from metallic nanoparticles. The spacing of the particles is similar to the wavelength of light in the dielectric media surrounding the particles. In contrast to metamaterials and metasurfaces where the periodicity is much smaller than the wavelength of light [64], diffraction plays a crucial role at the length scales involved in the SLR resonances.

Nanoparticle arrays are conceptually closely related to hole arrays in noble metal films. Hole arrays have subwavelength holes patterned in a metal film with spacings similar to the wavelength of light. The holes support localized plasmon resonances, analogously to the nanoparticles considered in this section, and the light diffracted by the structure couples into a combination of surface plasmon polaritons and quasi-cylindrical waves on the metal surface [65–67]. This is the main difference between the SLRs formed in nanoparticle arrays and the lattice resonances in hole arrays—in the SLRs the coupling between individual lattice sites has no contribution by surface plasmon polaritons [32].

Many of the dispersion properties of SLR modes can be understood by considering the implications of periodicity on the dispersion of plane waves, which we will consider first. Then, we refine our model by considering infinitely small point dipoles to describe the finite nanoparticles semi-analytically, using a coupled dipole approximation (CDA). Finally, we will briefly discuss the formation of band gaps in the dispersion, obtained through fully numerical finite-element calculations.

6.5.1 Empty Lattice Approximation

We start off our discussion with the propagation of light in a homogeneous medium. Given a refractive index n, light has the following dispersion relation:

$$k = \frac{nE}{\hbar c}, \tag{6.21}$$

where E is now the energy of the photons. We restrict our treatment to a space formed by k_x, k_y and E. In this space, propagating plane waves with wave vector $\vec{k} = (k_x, k_y, k_z)$ form a filled cone, with the edges defined by $k_x^2 + k_y^2 = (nE)^2/(\hbar c)^2$. Outside this cone, the modes become evanescent in the z-direction, as k_z becomes imaginary in order to fulfil 6.21. This is illustrated in Fig. 6.6. We are mainly interested in the optical modes in the plane of the lattice structure, so we can limit the treatment further by considering modes with $k_z = 0$. We are left with the surface of the cone.

Next, we assume that we have an infinite lattice in two dimensions with a periodicity of p_x in the x-direction and p_y in the y-direction. All the solutions to the wave equation in this periodic system are also periodic, with a fixed phase difference at the unit cell boundaries coming from the in-plane wave vector of the modes. All modes

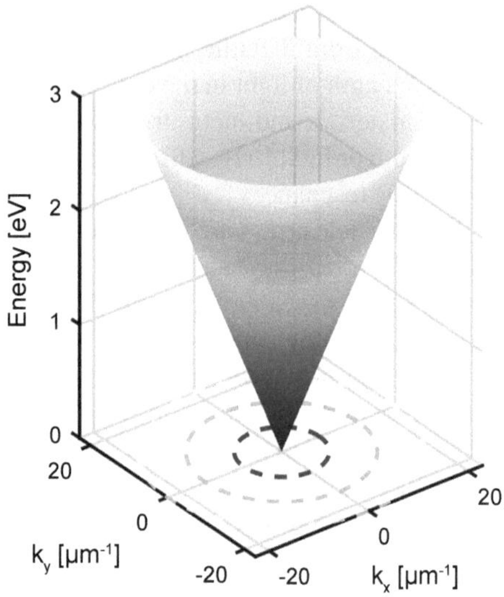

Fig. 6.6 Dispersion of light in k_x, k_y, E -space, $n = 1.5$. The modes inside the cone are plane waves, that propagate away from the (x,y)-plane. The modes outside the cone are evanescent in the z-direction due to an imaginary k_z. This causes the plane waves to decay instead of propagating freely. On the surface of the cone are the modes that can propagate in the (x,y)-plane

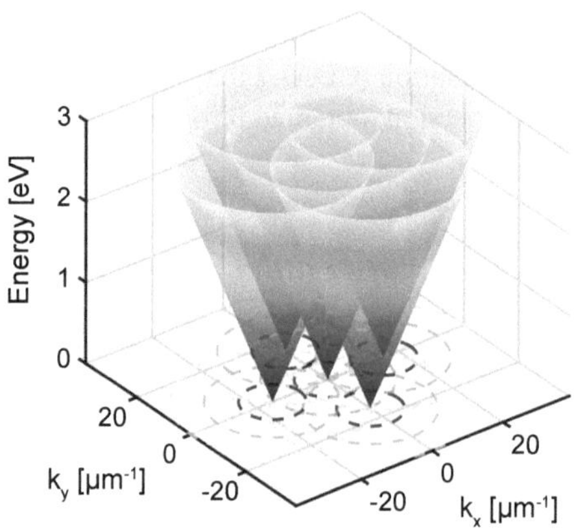

Fig. 6.7 Dispersion of light in a periodic structure in k_x, k_y, E-space. Only the first-order scattering is considered, which results in four extra cones in the dispersion, with the tips of the cones displaced from the origin by the reciprocal lattice vectors $\vec{G}_x = 2\pi/p_x\hat{x}$ and $\vec{G}_y = 2\pi/p_y\hat{y}$. The periodicity is assumed to be 400 nm in both x and y-directions, with $n = 1.5$

with the in-plane wave vector differing by an integer multiple of the reciprocal lattice vectors $\vec{G}_x = 2\pi/p_x\hat{x}$ or $\vec{G}_y = 2\pi/p_y\hat{y}$ are equivalent (Bloch waves) [68]. Therefore, the dispersion of the modes at lower wave numbers is 'copied' to the higher wave numbers. This is illustrated in Fig. 6.7 in terms of the cones introduced in the previous paragraph by plotting the dispersion in k_x, k_y and E-space.

Here we see that, by forcibly considering a periodic structure, we have artificially introduced a more complicated dispersion into the studied system. This was done by imposing a periodic boundary condition onto a homogeneous dielectric medium. In real systems, there usually will be a regularly arranged scattering element in order to define a periodic system, which also mediates the interactions between the different light cones where they intersect.

6.5.2 Lattice of Point Dipoles

In the case of sufficiently small nanoparticles, all higher-order multipoles can be neglected and the nanoparticles can be considered as individual point dipoles. We now consider what happens when we add point dipoles as the scatterers to the periodic system and drive them with light polarized linearly along the x-direction of the lattice. If we drive a point dipole with a linearly polarized electric field, the dipole will not radiate in the direction defined by the polarization of the driving field. This will make the effect of the periodicity in the x-direction negligible as long as the spacing between the dipoles is large enough to neglect higher-order multipoles induced by non-radiative (near-field) coupling. We will return to the higher-order multipoles in Sect. 6.5.3.

We will also limit the discussion to lattices with periodicities roughly equal to λ/n. What we are left with, under these assumptions, is shown in Fig. 6.8. There are two distinct directions where the light cones intersect that show qualitatively different behavior. The transverse magnetic modes (TM, Fig. 6.8b) can be obtained by looking at the $k_y = 0$ plane. For the TM mode the dispersion relation is $E = \pm\frac{\hbar c}{n}(k_x{}^2 + G_y{}^2)^{1/2}$, where $G_y = 2\pi/p_y$ is the momentum given by the lattice. The momentum given by the lattice is perpendicular to the in-plane wave vector of the incident light, thus giving the square root dependence for the total wave vector. When k_x is small when compared to G_y, the dispersion of the TM mode is approximately quadratic. Another type of dispersion is found for the transverse electric modes (TE, Fig. 6.8c), which are on the $k_x = 0$ plane. On this plane, the modes have a linear dispersion, that is $E = \frac{\hbar c}{n}(k_y \pm G_y)$, as the momentum gained from the lattice is in the same direction as the in-plane wave vector of the incident field. The Γ-point of the

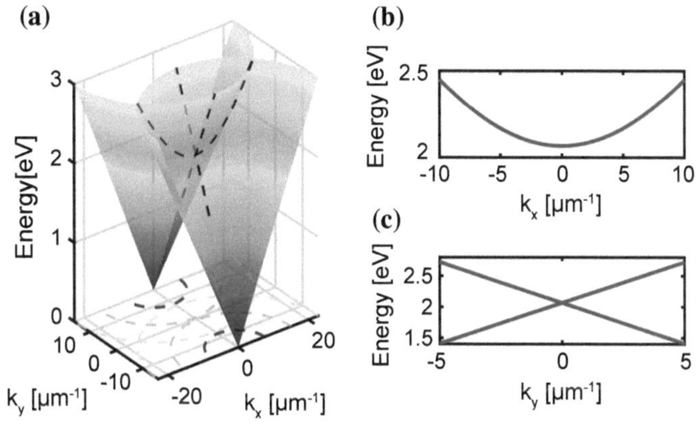

Fig. 6.8 Dispersion of light in a periodic structure in k_x, k_y, E-space (**a**). Only half of the (0, +1) and (0, −1) diffracted orders are shown for clarity. The crossing points of the two cones along the $k_x = 0$ and $k_y = 0$ are highlighted with *black dashed lines*. The TM modes ($k_y = 0$) are shown in (**b**), and the TE modes ($k_x = 0$) are shown in (**c**)

dispersion, where $k_x = 0$ and $k_y = 0$ is identical for the TE and TM modes. Figure 6.8 shows the case where there is no coupling between the light cones at their intersection points.

In order to calculate a better approximation of the real dispersions of the system, we turn to the coupled dipole approximation (CDA), also known as the discrete dipole approximation [69]. The CDA is typically used to approximate a single continuous (nano)particle by a discrete set of dipoles, where each dipole has a polarizability that depends on the material properties of the particle. However, it is computationally advantageous to apply the CDA to an array of nanoparticles as well. In that case, each discrete dipole represents a complete nanoparticle, with a polarizability that depends on both the material and the geometry of the particle [60].

The dipole moment $\vec{p}_i$ of the i:th nanoparticle can be written as

$$\vec{p}_i = \alpha_i(\omega)\vec{E}(\vec{r}_i) = \alpha_i(\omega)\left[\vec{E}_{ext}(\vec{r}_i) + \vec{E}_{d-d}(\vec{r}_i)\right] , \tag{6.22}$$

where $\alpha_i(\omega)$ is the frequency-dependent polarizability of the i:th nanoparticle and $\vec{E}(\vec{r}_i)$ is the total electric field at the location of the i:th particle. The latter in turn is subdivided into $\vec{E}_{ext}(\vec{r}_i)$, the external or driving electric field at the i:th particle, and $\vec{E}_{d-d}$, the electric field due to the all of the other nanoparticles in the system.

The $\vec{E}_{d-d}$ field at the i:th particle location is found by summing the fields due to the induced dipole moments at all other particle locations:

$$\vec{E}_{d-d} = \sum_{j\neq i} \frac{\mathrm{e}^{\mathrm{i}kR-\mathrm{i}\omega t}}{4\pi\varepsilon_0} \left\{ k^2 \frac{(\vec{e}_r \times \vec{p}_j) \times \vec{e}_r}{R} + \left[3\vec{e}_r(\vec{e}_r \cdot \vec{p}_j) - \vec{p}_j\right] \left(\frac{1}{R^3} - \frac{\mathrm{i}k}{R^2}\right) \right\} , \tag{6.23}$$

where $\vec{e}_r = (\vec{r}_i - \vec{r}_j)/R$ and $R = |\vec{r}_i - \vec{r}_j|$. The set of linear equations given by 6.22 is typically solved numerically. After solving, the extinction cross section σ_{ext} can be computed [69], which is closely related to the array transmission.

$$\sigma_{ext} = \frac{4\pi k}{|\vec{E}_{ext}|^2} \sum_{j=1}^{N} \mathrm{Im}(\vec{E}^*_{ext,j} \cdot \vec{p}_j) . \tag{6.24}$$

Equation 6.24 is generic, but we can further simplify 6.24 if we only take plane wave excitation at normal incidence into consideration. The benefit is that this allows us to separate the contribution of the array from that of the individual nanoparticles, providing further insight into the underlying mechanisms as well as greatly simplifying the calculation of σ_{ext}. Assuming an infinite array of identical nanoparticles, all dipole moments will have the same magnitude, phase and direction. In that case, the effects of the array can be summed into a term called the retarded dipole sum S [60].

$$S = \sum_{j\neq i} = \frac{(1 - \mathrm{i}kr_{ij})(3\cos^2\Theta_{ij} - 1)\mathrm{e}^{\mathrm{i}kr_{ij}}}{r_{ij}^3} + \frac{k^2 \sin^2\Theta_{ij}\mathrm{e}^{\mathrm{i}kr_{ij}}}{r_{ij}} , \tag{6.25}$$

where $\vec{r}_{ij} = \vec{r}_i - \vec{r}_j$ is the distance between the i:th and j:th nanoparticles in the lattice and the angle Θ_{ij} is the angle between the polarization of the incident electric field and the vector $\vec{r}_{ij}$. The location of the i:th particle in the sum in 6.25 can be chosen arbitrarily, for example the center of the array. With the aid of the retarded dipole sum, the polarizability of a single nanoparticle in an infinite array of identical nanoparticles can then be expressed as an effective polarizability α_{eff}, subject to the *external* driving field only. The effect of the neighboring particles has been fully taken into account by S. The expression for α_{eff} is [60]

$$\alpha_{eff} = (1/\alpha - S)^{-1} . \tag{6.26}$$

and the induced dipole moment of each particle can be expressed as $\vec{p} = \alpha_{eff}(\omega)\vec{E}_{ext}$. Finally, from the polarizability, one can now easily calculate the extinction cross section:

$$\sigma_{ext} = 4\pi k \mathrm{Im}(\alpha_{eff}) . \tag{6.27}$$

By evaluating the extinction cross section as a function of k_x, k_y, one obtains the dispersion curve, subject to the approximations made. For $k_x, k_y = 0$ (the Γ point), we calculate the extinction cross section with 6.27 for a 30 nm thick, 60 nm in diameter silver cylinder in an array with the same periodicity as in Fig. 6.8 (400 nm). The result of this calculation is shown in Fig. 6.9. The polarizability of the single cylinder is obtained numerically, see [57, 70] for details.

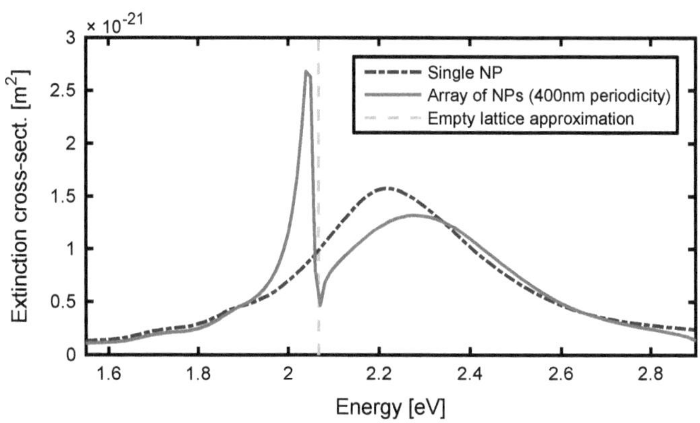

Fig. 6.9 Extinction cross sections for the single nanoparticle resonance of a 30 nm thick, 60 nm diameter silver disk (*dash-dotted line*) and the extinction cross section of an array with 400 nm periodicity in x and y composed of the silver disks (*solid line*). The refractive index is 1.5, and the empty lattice approximation for the location of the mode (Fig. 6.8) is shown as a *vertical dashed line*

From Fig. 6.9 one can see how making a lattice of metallic nanoparticles modifies their response. The sharp peak appearing on the low energy side of the diffracted orders (shown as the vertical dashed line) is called the surface lattice resonance (SLR). The SLR mode is a collective resonance, formed by the constructive interference of the scattered light from each nanoparticle in the plane of the lattice. One way to describe the SLR modes is to consider them as the resulting modes from strong coupling between the diffracted orders of the lattice, and the nanoparticle resonances [57].

In strong coupling, one typically obtains two new eigenmodes for the system. In the CDA simulations one can see a sharp peak, and the shifted residual of the nanoparticle resonance at the high energy side. This line-shape is of the Fano-type and characteristic for systems where a narrow resonance (now the diffractive order) couples strongly with a broad resonance (now the single nanoparticle resonance). In the following subsection we will proceed to discuss phenomena that originate from the existence of several diffractive orders in the system.

6.5.3 Band Gap Formation in SLR Dispersions

At $k = 0$, two diffracted orders are degenerate in energy, see Fig. 6.8c. The scattering from the nanoparticle provides a coupling between these diffracted orders, resulting to the formation of two hybrid modes, symmetric and antisymmetric, and a small photonic band gap. The latter mode becomes sub-radiant, or dark, due to its antisymmetry. The dark mode corresponds to a quadrupole excitation in the nanoparticle, which is by definition not captured in the CDA model discussed above. The existence of the dark mode is, however, apparent in experiments and in full simulations of the system using Maxwell's equations.

It has been experimentally observed that the band gap features depend on the nanoparticle shape [59]. So far, the nanoparticles have been described as simple dipoles, however, when the particle dimensions are increased, higher order resonances can also be excited. By using rod-shaped particles one can open a small (10–100 meV) band gap in the dispersion, where one of the band edges turns into a dark mode. For the dark mode it is not possible to couple with far-field radiation due to the mode having a zero net dipole moment. These modes can, however, be excited locally through the near-field components of the electromagnetic field of nearby emitters, for example. These modes also can be seen by exciting the structure at a slight angle to break the symmetry. An example is shown in Fig. 6.10, which was obtained through finite-element simulations, i.e., no dipole approximation was made.

Specifically, Fig. 6.10a shows the calculated dispersion (extinction) of a lattice structure of $450 \times 120 \times 38\,\text{nm}^3$ gold nanoparticles in a lattice with $p_x = 600\,\text{nm}$ and $p_y = 300\,\text{nm}$. The refractive index is 1.45 and the electric field polarization is along the y-axis. In the dispersion, the upper band at $k = 0$ is a bright mode, and the lower band is a dark mode. The corresponding charge distributions are shown

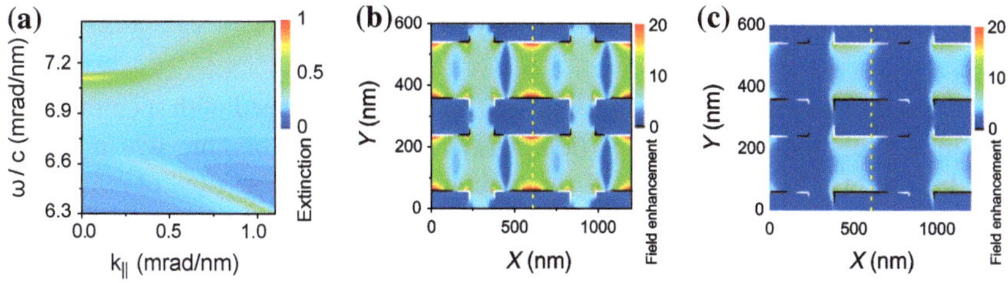

Fig. 6.10 **a** Numerically calculated dispersion for a lattice structure of $450 \times 120 \times 38\,\mathrm{nm}^3$ gold nanoparticles in a lattice with $p_x = 600$ nm and $p_y = 300$ nm. The polarization of the electric field is along the y-direction of the lattice. The electric field profiles in color scale and surface charge distribution (at an arbitrary phase) in *black* and *white* at the middle height of the particles are shown for the **b** upper band and **c** lower band edge SLR modes. The excitation was done at a small angle ($k_x = 0.15\,\mu\mathrm{m}^{-1}$) in order to see the quadrupolar mode in (**c**). Figure adapted from [62] (Creative Commons Attribution 3.0 License)

in Fig. 6.10b and c, for the upper and lower band, respectively. The field distributions were calculated with $k_x = 0.15\,\mu\mathrm{m}^{-1}$ in order to excite a mode close to the dark lower band edge at $k = 0$. Importantly, the dark mode has a quadrupolar charge distribution (two dipoles with a π phase shift), while the bright mode is composed of three dipoles. The middle one of the three dipoles has a π phase shift, but it is not strong enough to cancel the far-field radiation from the two other dipoles.

It is also possible to have a bright mode in the lower band and a dark mode in the upper band. This happens when the particle length is small enough compared to the periodicities [59]. In this case, the upper band edge is a quadrupole mode like above, but the charge distribution at the lower band edge corresponds to a single dipole. This single dipole is below the quadrupole in energy, and for this reason the lower band edge is visible in the dispersion.

6.6 Strong Coupling in Nanoparticle Arrays

Strong coupling between emitters and SLRs of nanoparticle arrays has been observed experimentally first in [57–59]. In those works, the SLR is scanned through the resonance of either only the emitter, or of both the emitter and the single nanoparticle. The method of scanning the SLR can vary. For example, the SLR can be changed in frequency by changing the angle between the (plane-wave) excitation and an array with a fixed periodicity. Another method is to change the period of the nanoparticles that make up the array while keeping the angle of incidence fixed, typically normal to the plane of the lattice. Below, we will discuss both types, studied by Väkeväinen et al. [57] and Shi et al. [59], starting with the latter method.

6.6.1 Spectral Transmittance Experiments

Strong coupling between SLRs and emitters was first established with spectral transmittance experiments on nanoparticle arrays. In [57] the organic dye molecule Rhodamine 6G (R6G) was used as the emitter. The arrays consisted of cylindrical silver nanoparticles of 65–90 nm in diameter and 30 nm in height. The nanoparticles were fabricated with electron beam lithography on a borosilicate microscope glass cover slip. The periodicity of the arrays h was varied between 275 and 495 nm in 10 nm steps, in order to scan the SLR frequency across the localized surface plasmon resonance (LSPR) of the nanoparticle and the R6G absorption maxima. Here, the SLR frequency is the resonant frequency at normal incidence ($k_x = k_y = 0$), which is inversely proportional to the lattice period. The R6G molecules were added on top of the arrays in a 50 nm thick matrix of poly(methyl methacrylate) (PMMA). The optical modes of the arrays were explored with transmission spectroscopy where the transmission spectra of the arrays were measured at normal incidence. The nanoparticle arrays were immersed in index matching oil and sandwiched between two glass slips to create a symmetrical index environment, which is crucial to obtain spectrally sharp SLRs [61].

Figure 6.11 displays the extinction spectra of a sample with only bare nanoparticle arrays, i.e., without any molecules. The measured extinction spectra are displayed together with the results of finite-difference time-domain (FDTD) simulations, and CDA calculations using 6.27. Here, we define the extinction as $(1 - T)h^2$, T being the transmission of the array. The dispersion diagrams show a clear avoided crossing between the LSPR and the diffracted orders (DOs) of the lattice (the optical mode in the empty lattice approximation, Sect. 6.5.1). This shows that SLRs are strongly coupled hybrid modes of the LSPR and DOs of the lattice. The dispersion was fitted with a coupled-modes model, with the uncoupled modes being the first and second-order DO, and the LSPR. This process retrieved the coupling coefficients between the individual modes. See for a more detailed discussion [26, 57]. From the fit, the coupling coefficients (splittings) between the LSPR and each DO were obtained, while the coupling coefficient between the DOs themselves was kept zero. This fit gave a splitting of 107 ± 1 meV between the LSPR and the first-order DO, and 142 $\pm$ 49 meV between the LSPR and the second-order (higher frequency) DO.

Figure 6.12 displays the dispersion of the optical modes for a sample which was measured without and with R6G molecules dispersed on top. Only the peak positions of extinction spectra are plotted such that the two dispersions (the one without emitters and the one with 200 mM concentration of R6G) can be plotted in the same graph. The sample was also measured with 50 and 100 mM R6G concentrations, in order to show that the magnitude of the splitting of the modes follows the square-root dependence on the molecular concentration (number of emitters, 6.8). The results of these measurements are shown in the inset in Fig. 6.12, proving the expected relation. Examples of extinction spectra that make up the underlying data set of Fig. 6.12 are displayed in Fig. 6.13, for different concentrations and array periodicities.

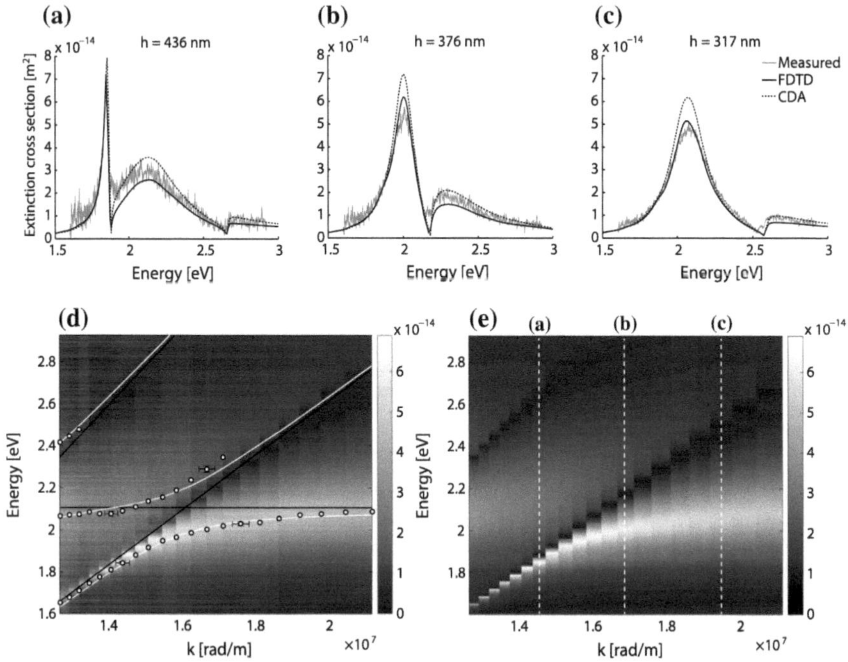

Fig. 6.11 Measured and calculated extinction spectra of nanoparticle arrays without emitters. **Panels a–c** depict the measurements, FDTD simulations, and the CDA calculations, at lattice periods for which the SLR is below, at, and above the LSPR frequency, respectively. In **d–e**, the spectra are shown as dispersion diagrams where k is the inverse lattice spacing $k = 2\pi/h$. **d** presents the measured extinction. The *diagonal black lines* depict the calculated DOs and the *horizontal black line* indicates the LSPR frequency. In addition, the maxima of the spectra are shown with *white dots* and the fit with a three-coupled-modes model is shown with the *white curved lines*. **e** presents the corresponding FDTD simulations. The k-values that correspond to the lattice spacing in **a–c** are marked with *vertical dashed lines*. Adapted with permission from [57]. Copyright (2013) American Chemical Society

To characterize the system, we must effectively describe it as five coupled modes: the LSPR, the two DOs, the R6G main absorption peak, and the R6G absorption shoulder. Even if direct couplings between the R6G and the DOs are ignored, the data can be fitted fairly well and the system stands as a strongly coupled mixture of the R6G, SLR, and LSPR modes. The resulting model has four parameters, two for the couplings between the R6G resonances and the LSPR, and two for the couplings between each of the two DOs and the LSPR. The coupling coefficients for the sample with a 200 mM R6G film on top are obtained with this truncated five-coupled-modes model. The model yield splittings of 150 ± 12 meV (LSPR and R6G main peak), 121 ± 12 meV (LSPR and R6G shoulder), 95 ± 6 meV (LSPR and first-order DO), 79 ± 7 meV (LSPR and second-order DO). Based on the FDTD simulations with

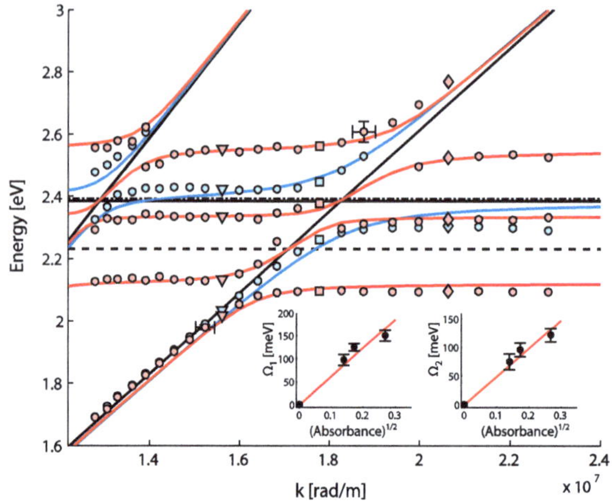

Fig. 6.12 Maxima of the measured extinction spectra without R6G, and with a 200 mM R6G film on top of the array, are plotted with *blue* and *red dots*, respectively. *Blue* and *red curved lines* represent the corresponding fits with the coupled-modes models. The *inset* presents the LSPR-R6G coupling coefficients obtained from the fits for four different concentrations (0, 25, 50 and 200 mM), showing the expected square-root dependence of the coupling strengths on the absorbance of the R6G film (number of emitters, 6.8). Here, Ω_1 denotes the coupling coefficient between the LSPR and the R6G main absorption peak, and Ω_2 denotes the coupling coefficient between the LSPR and the R6G absorption shoulder, respectively. Reprinted with permission from [57]. Copyright (2013) American Chemical Society

R6G molecules, it is estimated that approximately 3×10^4 molecules are involved in the strong coupling per one metallic nanoparticle.

Also a direct coupling between dye molecules and the SLR can be realized. In [59] the SLR was scanned across the absorption line of DiD dye molecule while the LSPR was at a significantly higher energy. This resulted to direct strong coupling between the SLR and the molecular excitation, see Fig. 6.14, with the size of the avoided crossing between the modes following the theoretically expected square-root dependence on the molecular concentration.

In summary, the R6G molecules were observed to strongly couple to the SLR modes in nanoparticle arrays, SLRs being hybridized modes of the LSPR and DOs. The SLRs can have long lifetimes and high quality factors (for plasmonic modes) as evidenced by the sharpness of the extinction peaks, whereas the LSPRs are broad. Nevertheless, the high local field enhancement that LSPRs provide may be the key to achieving strong coupling with emitters. Since the SLRs are delocalized modes within the array, reaching the strong coupling regime implies that emitters near distant nanoparticles are coherently coupled.

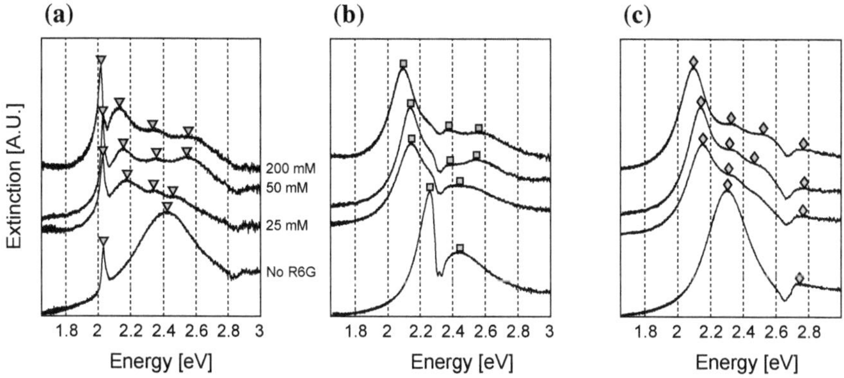

Fig. 6.13 Examples of the extinction spectra of nanoparticle arrays measured on the same sample as in Fig. 6.12. **a–c** correspond to 403, 354 and 305 nm lattice spacing, respectively. **a** presents the case when the SLR is lower in energy compared to the LSPR and the R6G absorption maxima. In **b**, the SLR is in between the R6G main absorption peak and the LSPR. In **c**, the SLR is higher in energy than the LSPR and the R6G resonances. The spectra at the *bottom* were measured without emitters and the spectra above correspond to the same nanoparticle arrays with R6G films of 25, 50 and 200 mM concentration, respectively. Extinction maxima are indicated with symbols on top of the curves. Triangles, squares and diamonds on the *bottom curve* (no R6G) and on the *top curve* (200 mM R6G layer) correspond to the data points plotted with the same symbols in Fig. 6.12. The spectra exhibit a clear evolution as a function of molecular concentration such that the strongly coupled hybrid modes (extinction peaks) emerge and become more prominent as they separate further away from each other. Adapted with permission from [57]. Copyright (2013) American Chemical Society

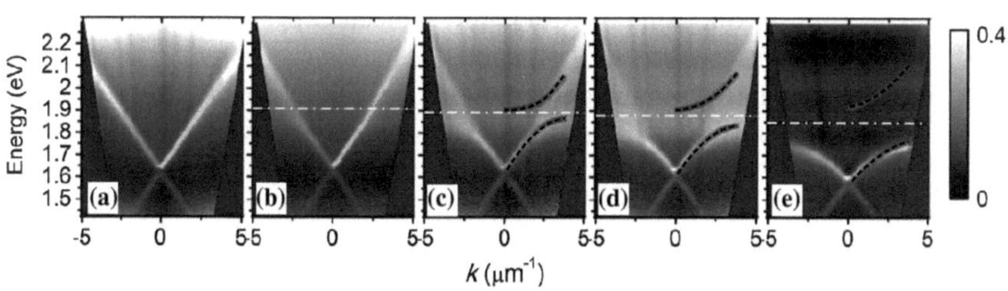

Fig. 6.14 **a** Extinction spectra of a nanoparticle array without emitters on *top*. **b–e** The same sample with 20, 200, 400, and 800 mM concentration of DiD molecules on top of the array, respectively. White areas correspond to maximum extinction. The *horizontal lines* correspond to the absorption maximum of the DiD film. Here, k is the in-plane component of the incident light momentum vector and therefore corresponds to the angle of incidence. Clear avoided crossings are observed for increasing molecular concentration. Reprinted with permission from [59]. Copyright (2014) by the American Physical Society

6.6.2 Spatial Coherence of Strongly Coupled Hybrid Modes

The hybridization of an initially delocalized plasmonic mode with a localized molecular excitation raises a natural question: what is the characteristic coherence length of the hybrid modes? In [71], the spatial coherence properties of a planar silver film overlaid with J-aggregate molecules were studied. The strongly coupled modes exhibited coherent emission of light from molecules that were separated by several micrometers. A different system, quantum dots on the silver film, was used as a weak coupling reference. In [59], the spatial coherence properties were studied for a system where silver nanoparticle arrays were coupling strongly to organic DiD molecules that were dispersed onto the array, similar to the system discussed in Sect. 6.6.1. Increasing concentrations of DiD were used, which allowed tracking the evolution of the spatial coherence, within the same system, while transiting from the weak to the strong-coupling regime.

In [59], a Young-type double slit was placed in the image plane of the sample, which leads to interference patterns in the far-field in case spatial coherence exists between the two separate parts of the sample (those imaged by the slits). Angle (k-space) and wavelength-resolved transmission spectra were collected with a spectrometer equipped with a CCD. Figure 6.15a–d shows how the interference pattern evolved as the concentration of the molecules was increased. Without DiD molecules, the interference pattern is created by the pure delocalized SLR mode. In this case, the sample acts as a spatially coherent light source which radiates through a double slit. The interference fringes are produced by the diffraction of the delocalized SLR mode from the double slit. A clear bending of the interference fringes is seen with increasing concentrations of DiD: this is a direct consequence of the emergence of new energy eigenstates for the strongly coupled system. Interestingly, the interference pattern persist even with the highest concentration studied (800 mM), even though the modes are very exciton-like (80 %, as obtained from a coupled oscillator model). This is in striking contrast with purely molecular excitations, which do not exhibit any interference pattern at all. As another control experiment, a spatial coherence image of a sample with randomly dispersed nanoparticles was measured, see Fig. 6.15e. The nanoparticle size, number and orientation as well as the molecular

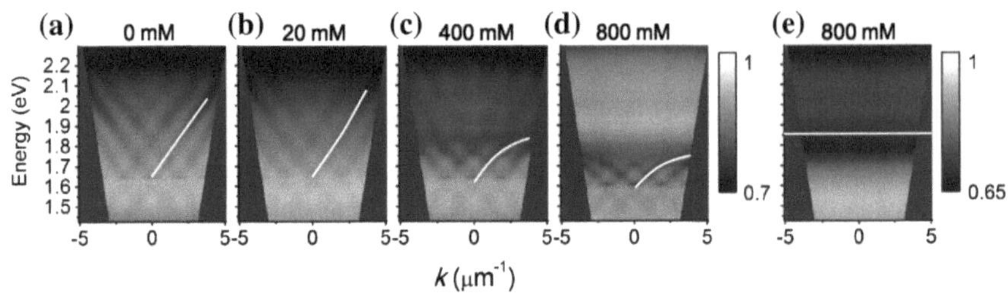

Fig. 6.15 **a–d** The spatial coherence images for a sample with DiD concentrations of 0, 20, 400, and 800 mM, respectively. The *white* (*black*) color corresponds to maximum (minimum) transmission. The *white lines* depict the measured dispersions. **e** A random sample with 800 mM DiD concentration. Only two transmission minima are seen: One at 1.85 eV which corresponds to absorption maximum of the DiD molecules (*white line*), and another at 2.25 eV, which corresponds to the single particle plasmon resonance. Reprinted with permission from [59]. Copyright (2014) by the American Physical Society

concentration are the same as in the sample in Fig. 6.15d. In this case, no interference fringes can be observed, highlighting the important role of the array periodicity for the emergence of long range spatial coherence. These results demonstrate how strong coupling can be be utilized to create nanosystems with designed properties: the long-range coherence (up to 10 μm) is achieved by hybridizing the initially localized molecular excitation with a delocalized SLR mode.

6.7 Outlook

In this book chapter, we have reviewed the theory and experiments on strong coupling between emitters and plasmonic modes. We focused on describing the dynamics and spatial-coherence studies on the topic. The second main theme was strong coupling in arrays of metal nanoparticles, where we also gave background information about how the modes of such periodic systems are formed.

While we certainly did not mention all the work on strong coupling in plasmonics (even the review [26] does not cite all of them despite more than two hundred references), we did discuss all the handful of articles reporting dynamics and spatial coherence studies on the subject. This obviously suggests that there is further work to do in these areas. In particular, exploring dynamical phenomena in the strong coupling regime in these systems is a challenge since the relevant time-scales are in the femtosecond range.

Another direction of future research is to go towards single emitter strong coupling. This has been studied theoretically in several works. Here follow some examples. The conclusion of [72] was that with present-day existing plasmonic nanostructures, one would need to work at 4 K to observe single-emitter strong coupling. However, the authors claim that possibly with modified structures which offer even more confined field hot-spots, room temperature observation could be possible. According to the predictions of [73], single-emitter vacuum Rabi splitting would be observable at room temperature for a quantum dot (with dipole moment $\mu/e \sim 0.3 - 0.5$ nm). Observable room temperature strong coupling was predicted in [74] where a dipole of similar strength was placed a few nanometers from the tip of a cigar-shaped nanorod. Such conclusions were obtained also in [75]. In [76], conical shaped nanoparticles were predicted to lead to single-emitter strong coupling at room temperature. It was found that the higher order multipoles of a nanoparticle can further enrich strong coupling phenomena and help in reaching strong coupling with small numbers of emitters [77]. In experiments, there has been progress towards ever smaller numbers of emitters showing strong coupling. In [78], silver nanoprisms and excitons in molecular J-aggregates were experimentally demonstrated to show strong coupling at room temperature when the estimated number of excitons contributing was as low as 70–85. Recently, interesting results on reaching strong coupling with average emitter concentrations that correspond to the single emitter limit have been reported [79]. If deterministic single-emitter strong coupling is achieved, the obvious next challenge is to observe the Jaynes-Cummings ladder. Even with many emitters, the Dicke model [35], that is the paradigm description of many emitters coupled with a field, predicts interesting quantum phenomena in the non-linear regime.

The regime of large numbers of photons, i.e., macroscopic mode population, provides many fruitful directions where strong coupling can play a role. Thresholdless lasing [27] may be possible if the interaction is so strong that all emission is channelled into a single mode. Furthermore, coherence due to strong coupling allows interference phenomena which may lead to lasing without population inversion [28]. In semiconductor systems excitons and photons may hybridize at strong coupling into polaritons, where the excitonic part provides effective photon-photon interactions. This has made it possible to create polariton condensates with inorganic [29] and organic [80, 81] materials, see Chap. 7 of this book. Similar condensates, photon fluids [30] as well as the analogues of photon condensates [82] may be possible in plasmonic systems, despite the high amount of losses [31]. The nanoparticle array structures as discussed in this chapter may be the ideal platform for such phenomena, and beyond, due to the possibility of designing the dispersion at will. In [31], the term *quantum plasmonic lattices* was coined for them; indeed first observations of lasing have already been achieved in such arrays [83], see Chap. 8 of this book.

The room temperature operation and on-chip nature of the plasmonic systems that have shown strong coupling promises applications. From the fundamental science point of view these systems are novel since the time-scales are extremely fast, and for instance the existence and nature of effective photon-photon interactions is largely unknown. This combined with the fantastic opportunities to design the mode structures and density of states by, for instance, nanoparticle lattices, promises rich physics to be discovered.

References

1. S.T. Thornton, J.B. Marion, *Classical dynamics of particles and systems* (Brooks Cole, 2003), 5th edn., chap. 12, pp. 468–511
2. Y. Kaluzny, P. Goy, M. Gross, J.M. Raimond, S. Haroche, Phys. Rev. Lett. **51**, 1175 (1983)
3. M. Brune, J.M. Raimond, P. Goy, L. Davidovich, S. Haroche, Phys. Rev. Lett. **59**, 1899 (1987)
4. M.G. Raizen, R.J. Thompson, R.J. Brecha, H.J. Kimble, H.J. Carmichael, Phys. Rev. Lett. **63**, 240 (1989)
5. Y. Zhu, D.J. Gauthler, S.E. Morin, Q. Wu, H.J. Carmichael, T.W. Mossberg, Phys. Rev. Lett. **64**(21), 2499 (1990)
6. G. Rempe, R.J. Thompson, R.J. Brecha, W.D. Lee, H.J. Kimble, Phys. Rev. Lett. **67**, 1727 (1991)
7. G. Rempe, H. Walther, N. Klein, Phys. Rev. Lett. **58**, 353 (1987)
8. R.J. Thompson, G. Rempe, H.J. Kimble, Phys. Rev. Lett. **68**, 1132 (1992)
9. C. Weisbuch, M. Nishioka, A. Ishikawa, Y. Arakawa, Phys. Rev. Lett. **69**(23), 3314 (1992)
10. R. Houdré, C. Weisbuch, R.P. Stanley, U. Oesterle, P. Pellandini, M. Ilegems, Phys. Rev. Lett. **73**, 2043 (1994)
11. R. Houdré, R.P. Stanley, U. Oesterle, M. Ilegems, C. Weisbuch, Phys. Rev. B **49**, 16761 (1994)
12. T.B. Norris, J.K. Rhee, C.Y. Sung, Y. Arakawa, M. Nishioka, C. Weisbuch, Phys. Rev. B **50**, 14663 (1994)

13. R. Houdré, J.L. Gibernon, P. Pellandini, R.P. Stanley, U. Oesterle, C. Weisbuch, J. O'Gorman, B. Roycroft, M. Ilegems, Phys. Rev. B **52**, 7810 (1995)
14. J.P. Reithmaier, G. Sek, A. Löffler, C. Hofmann, S. Kuhn, S. Reitzenstein, L.V. Keldysh, V.D. Kulakovskii, T.L. Reinecke, A. Forchel, Nature **432**(7014), 197 (2004)
15. T. Yoshie, A. Scherer, J. Hendrickson, G. Khitrova, H.M. Gibbs, G. Rupper, C. Ell, O.B. Shchekin, D.G. Deppe, Nature **432**(November), 9 (2004)
16. K. Hennessy, A. Badolato, M. Winger, D. Gerace, M. Atatüre, S. Gulde, S. Fält, E.L. Hu, A. Imamoğlu, Nature **445**, 896 (2007)
17. D.G. Lidzey, D.D.C. Bradley, M.S. Skolnick, T. Virgili, S. Walker, D.M. Whittaker, Nature **395**(6697), 53 (1998)
18. D.G. Lidzey, D.D.C. Bradley, A. Armitage, S. Walker, M.S. Skolnick, Science **288**, 1620 (2000)
19. P.A. Hobson, W.L. Barnes, D.G. Lidzey, G.A. Gehring, D.M. Whittaker, M.S. Skolnick, S. Walker, Appl. Phys. Lett. **81**(19), 3519 (2002)
20. R.J. Holmes, S.R. Forrest, Phys. Rev. Lett. **93**, 186404 (2004)
21. R.J. Holmes, S.R. Forrest, Org. Electron. **8**(2–3), 77 (2007). Coupled States of Excitons, Photons, and Plasmons in Organic Structures
22. D.G. Lidzey, D.D.C. Bradley, T. Virgili, A. Armitage, M.S. Skolnick, Phys. Rev. Lett. **82**(16), 3316 (1999)
23. I. Chiorescu, P. Bertet, K. Semba, Y. Nakamura, C.J.P.M. Harmans, J.E. Mooij, Nature **431**, 159 (2004)
24. A. Wallraff, D.I. Schuster, A. Blais, L. Frunzio, R.S. Huang, J. Majer, S. Kumar, S.M. Girvin, R.J. Schoelkopf, Nature **431**, 162 (2004)
25. T. Niemczyk, F. Deppe, H. Huebl, E.P. Menzel, F. Hocke, M.J. Schwarz, J.J. Garcia-Ripoll, D. Zueco, T. Hümmer, E. Solano, A. Marx, R. Gross, Nat. Phys. **6**, 772 (2010)
26. P. Törmä, W.L. Barnes, Rep. Prog. Phys. **78**(1), 013901 (2015). doi:10.1088/0034-4885/78/1/013901
27. P.R. Rice, H.J. Carmichael, Phys. Rev. A **50**, 4318 (1994)
28. A. Imamoğlu, J.E. Field, S.E. Harris, Phys. Rev. Lett. **66**, 1154 (1991)
29. J. Kasprzak, M. Richard, S. Kundermann, A. Baas, P. Jeambrun, J.M.J. Keeling, F.M. Marchetti, M.H. Szymanska, R. Andre, J.L. Staehli, V. Savona, P.B. Littlewood, B. Deveaud, L.S. Dang, Nature **443**, 409 (2006)
30. I. Carusotto, C. Ciuti, Rev. Mod. Phys. **85**, 299 (2013)
31. J.P. Martikainen, M.O.J. Heikkinen, P. Törmä, Phys. Rev. A **90**, 053604 (2014)
32. F.J. García de Abajo, Rev. Mod. Phys. **79**, 1267 (2007)
33. V. Agranovich, A. Malshukov, Opt. Commun. **11**(2), 169 (1974)
34. H. Raether, *Surface Plasmons* (Springer, 1988)
35. B. Garraway, Phil. Trans. R. Soc. A **369**, 1137 (2011)
36. F. Würthner, T.E. Kaiser, C.R. Saha-Möller, Angew. Chem. Int. Ed. **50**(15), 3376 (2011)
37. I. Pockrand, A. Brillante, D. Möbius, J. Chem. Phys. **77**, 6289 (1982)
38. J. Bellessa, C. Bonnand, J. Plenet, J. Mugnier, Phys. Rev. Lett. **93**(3), 036404 (2004)
39. S. Balci, C. Kocabas, S. Ates, E. Karademir, O. Salihoglu, A. Aydinli, Phys. Rev. B **86**(23) (2012)
40. J. Dintinger, S. Klein, F. Bustos, W.L. Barnes, T.W. Ebbesen, Phys. Rev. B **71**(3), 035424 (2005)
41. A. Otto, Zeitschrift für Physik **216**, 398 (1968)
42. F.J. Garcia-Vidal, L. Martin-Moreno, T.W. Ebbesen, L. Kuipers, Rev. Mod. Phys. **82**(1), 729 (2010)
43. Y. Sugawara, T.A. Kelf, J.J. Baumberg, M.E. Abdelsalam, P.N. Bartlett, Phys. Rev. Lett. **97**, 266808 (2006)
44. P. Vasa, W. Wang, R. Pomraenke, M. Lammers, M. Maiuri, C. Manzoni, G. Cerullo, C. Lienau, Nat. Photonics **7**(2), 128 (2013)
45. T.K. Hakala, J.J. Toppari, A. Kuzyk, M. Pettersson, H. Tikkanen, H. Kunttu, P. Törmä, Phys. Rev. Lett. **103**, 053602 (2009)

46. N.I. Cade, T. Ritman-Meer, D. Richards, Phys. Rev. B **79**, 241404 (2009)
47. R.J. Moerland, T.K. Hakala, A.I. Väkeväinen, A.P. Eskelinen, G. Sharma, J.J. Toppari, A. Kuzyk, M. Petterson, H. Kunttu, P. Törmä, Proc. SPIE **8070**, 80700D (2011)
48. R.J. Moerland, G. Sharma, A.I. Väkeväinen, A.P. Eskelinen, H.T. Rekola, P. Törmä, Proc. SPIE **8096**, 809606 (2011)
49. F. Valmorra, M. Bröll, S. Schwaiger, N. Welzel, D. Heitmann, S. Mendach, Appl. Phys. Lett. **99**, 051110 (2011)
50. S.V. Baieva, T.K. Hakala, J.J. Toppari, Nanoscale Res. Lett. **7**, 1 (2012)
51. M.A. Koponen, U. Hohenester, T.K. Hakala, J.J. Toppari, Phys. Rev. B **88**, 085425 (2013)
52. S. Baieva, J.A. Ihalainen, J.J. Toppari, J. Chem. Phys. **138**(4) (2013)
53. A. González-Tudela, P.A. Huidobro, L. Martín-Moreno, C. Tejedor, F.J. García-Vidal, Phys. Rev. Lett. **110**, 126801 (2013)
54. D.E. Gomez, K.C. Vernon, P. Mulvaney, T.J. Davis, Nano Lett. **10**(1), 274 (2010)
55. D.E. Gomez, K.C. Vernon, P. Mulvaney, T.J. Davis, Appl. Phys. Lett. **96**(7), 073108 (2010)
56. R.J. Moerland, H.T. Rekola, G. Sharma, A.P. Eskelinen, A.I. Väkeväinen, P. Törmä, Appl. Phys. Lett. **100**(22), 221111 (2012)
57. A.I. Väkeväinen, R.J. Moerland, H.T. Rekola, A.P. Eskelinen, J.P. Martikainen, D.H. Kim, P. Törmä, Nano Lett. **14**(4), 1721 (2014)
58. S. Rodriguez, J.G. Rivas, Opt. Express **21**(22), 27411 (2013)
59. L. Shi, T.K. Hakala, H.T. Rekola, J.P. Martikainen, R.J. Moerland, P. Törmä, Phys. Rev. Lett. **112**, 153002 (2014)
60. S. Zou, N. Janel, G.C. Schatz, J. Chem. Phys. **120**(23), 10871 (2004)
61. B. Auguié, W.L. Barnes, Phys. Rev. Lett. **101**, 143902 (2008)
62. S.R.K. Rodriguez, A. Abass, B. Maes, O.T.A. Janssen, G. Vecchi, J. Gómez Rivas, Phys. Rev. X **1**, 021019 (2011). doi:10.1103/PhysRevX.1.021019
63. W. Wang, P. Vasa, R. Pomraenke, R. Vogelgesang, A.D. Sio, E. Sommer, M. Maiuri, C. Manzoni, G. Cerullo, C. Lienau, ACS Nano **8**(1), 1056 (2014)
64. N. Meinzer, W.L. Barnes, I.R. Hooper, Nat. Photonics **8**(12), 889 (2014)
65. P. Lalanne, J. Hugonin, H. Liu, B. Wang, Surf. Sci. Rep. **64**(10), 453 (2009)
66. A.Y. Nikitin, F.J. García-Vidal, L. Martín-Moreno, Phys. Rev. Lett. **105**, 073902 (2010)
67. F. van Beijnum, C. Retif, C.B. Smiet, H. Liu, P. Lalanne, M.P. van Exter, Nature **492**(7429), 411 (2012)
68. J.D. Joannopoulos, S.G. Johnson, J.N. Winn, R.D. Meade, *Photonic Crystals: Molding the Flow of Light*, 2nd edn. (Princeton University Press, 2008)
69. B.T. Draine, P.J. Flatau, J. Opt. Soc. Am. A **11**(4), 1491 (1994)
70. P. Grahn, A. Shevchenko, M. Kaivola, New J. Phys. **14**(9), 093033 (2012)
71. S. Aberra Guebrou, C. Symonds, E. Homeyer, J.C. Plenet, Y.N. Gartstein, V.M. Agranovich, J. Bellessa, Phys. Rev. Lett. **108**, 066401 (2012)
72. T. Hümmer, F.J. García-Vidal, L. Martín-Moreno, D. Zueco, Phys. Rev. B **87**, 115419 (2013)
73. S. Savasta, R. Saija, A. Ridolfo, O. Di Stefano, P. Denti, F. Borghese, ACS Nano **4**, 6369 (2010)
74. A. Trügler, U. Hohenester, Phys. Rev. B **77**(11), 115403 (2008)
75. K. Słowik, R. Filter, J. Straubel, F. Lederer, C. Rockstuhl, Phys. Rev. B **88**, 195414 (2013)
76. S. D'Agostino, F. Alpeggiani, L.C. Andreani, Opt. Express **21**(23), 27602 (2013)
77. A. Delga, J. Feist, J. Bravo-Abad, F.J. Garcia-Vidal, Phys. Rev. Lett. **112**, 253601 (2014)
78. G. Zengin, M. Wersäll, S. Nilsson, T.J. Antosiewicz, M. Käll, T. Shegai, Phys. Rev. Lett. **114**, 157401 (2015)
79. R. Chikkaraddy, B. de Nijs, S. Barrow, O.A. Sherman, E. Rosta, A. Demetriadou, P. Fox, O. Hess, J.J. Baumberg, Nature **535**, 127 (2016)
80. J.D. Plumhof, T. Stöferle, L. Mai, U. Scherf, R.F. Mahrt, Nat. Mater. **13**, 247 (2014)
81. K.S. Daskalakis, S.A. Maier, R. Murray, S. Kena-Cohen, Nat. Mater. **13**, 271 (2014)
82. J. Klaers, J. Schmitt, F. Vewinger, M. Weitz, Nature **468**, 545 (2010)
83. W. Zhou, M. Dridi, J.Y. Suh, C.H. Kim, D.T. Co, M.R. Wasielewski, G.C. Schatz, T.W. Odom, Nature Nanotechnol. **8**, 506 (2013)

Chapter 7
Polariton Condensation in Organic Semiconductors

Konstantinos S. Daskalakis, Stefan A. Maier and Stéphane Kéna-Cohen

Abstract Because of their bosonic nature, exciton-polaritons can condense into a single macroscopic coherent state under the right conditions. This process is uniquely characterized by its out-of-equilibrium nature, which gives rise to some distinctions when compared to conventional Bose-Einstein condensation. This Chapter begins by describing condensation in general and then describes room-temperature demonstrations of polariton condensation in organic microcavities along with phenomena that have recently been observed in these systems such as long-range spatial coherence, spontaneous vortex formation and dynamic instabilities.

7.1 Introduction

In this Chapter, we will build on the description of exciton-polaritons introduced in the previous chapter to describe one of the most fascinating properties of these quasiparticles: their ability to condense in a single macroscopic coherent state. This phenomenon has been a rich playground for physics during the past decade, but mainly limited to low temperatures due to the small exciton binding energies typical of inorganic semiconductors [1]. In this Chapter, we will specifically focus on microcavities containing organic semiconductors, which readily show stable

K.S. Daskalakis
Department of Applied Physics, COMP Centre of Excellence,
Aalto University, PO Box 15100, Fi-00076 Aalto, Finland
e-mail: konstantinos.daskalakis@aalto.fi

S.A. Maier
Department of Physics, Imperial College London, London SW7 2AZ, UK
e-mail: s.maier@imperial.ac.uk

S. Kéna-Cohen (✉)
Department of Engineering Physics, Polytechnique Montréal, Montreal
QC H3C 3A7, Canada
e-mail: s.kena-cohen@polymtl.ca

S.I. Bozhevolnyi et al. (eds.), *Quantum Plasmonics*,
Springer Series in Solid-State Sciences 185,
DOI 10.1007/978-3-319-45820-5_7

excitons at room temperature and allow for simple fabrication [2]. The structures in this Chapter will use planar dielectric microcavities, as opposed to metallic ones, to reach the strong-coupling regime. The phenomenology that we will describe, however, gives a glimpse into that which could potentially be achieved in plasmonic structures such as those described in the previous chapter under the right conditions. From a practical standpoint, there has been substantial interest in organic polariton condensates due to their potential as coherent light sources with thresholds below those of conventional photon lasers.

In the first section we will give an overview of what mean by "condensation" and why nomenclature in the field has been such a matter of debate. In particular, within the context of this Chapter, we will make no distinction between the concepts of "polariton laser" or "polariton condensate". We will then describe the typical microcavity structure that has been used, the relaxation mechanisms that play an essential role in condensation phenomena and recent experiments on microcavities containing the organic 2,7-Bis[9,9-di(4-methylphenyl)-fluoren-2-yl]-9,9-di(4-methylphenyl)fluorene (TDAF).

7.2 What Is a Condensate?

The realization of atomic Bose-Einstein condensates (BEC) has been one of the greatest achievements of experimental physics. In three dimensions, this new state of matter emerges when the interparticle distance approaches the de Broglie wavelength $\lambda_{dB} = \sqrt{\frac{2\pi\hbar^2}{mkT}}$, where m is the mass of the particle. When this condition is satisfied, individual atomic wavefunctions overlap and combine to form a collective coherent state. Beyond the condensation threshold, each additional particle added to the system "condenses" into the collective ground state and therefore contributes to its macroscopic wavefunction [3]. The "textbook" observation of a weakly-interacting BEC in a dilute atomic gas was awarded the 2001 Nobel Prize in physics and required the development of techniques to cool atoms to nanokelvin temperatures. Yet, physical phenomena related to BEC extend beyond atomic gasses and can be observed in superconductivity (the condensation of Cooper pairs) and superfluid ^{4}He. It has been argued that the unifying feature behind such phenomena is the spontaneous appearance of off-diagonal long-range order (ODLRO) [4]. This condition, which requires that phase correlations between any two points within the condensate persist for large separations, has been used as a criterion to define condensation. It is quantified by the first-order correlation function $g^{(1)}(r, r') \equiv \langle \Psi^\dagger(r)\Psi(r')\rangle$, where $\Psi(r)$ is the annihilation operator for a particle at position r. It can be shown that the appearance of ODLRO is related to a breaking of the continuous U(1) phase symmetry of the condensate wavefunction.

Like their fundamental constituents, exciton-polaritons obey Bose statistics at low densities. At first glance, it is thus reasonable to expect polaritons to undergo

BEC below some critical temperature. Indeed, signatures of inorganic polariton condensation were observed in early reports and an exhaustive claim for polariton "BEC" was made in 2006 [5]. This work included measurements of the particle distribution function, long-range spatial coherence, spontaneous linear polarization, momentum-space narrowing and a linewidth collapse. A few important distinctions can be made between a photon laser and a polariton condensate. First, threshold is reached via bosonic relaxation–often called stimulated scattering–of uncondensed particles towards the ground state. This stimulated scattering is driven at least in part by the polariton matter component. This contrasts with a conventional laser where the final state is purely photonic in nature and relaxation is driven by the stimulated emission of photons. Second, energetic relaxation via the emission of phonons or from a thermal reservoir can lead to thermalisation of the polariton distribution. Finally, polaritons interact due to their matter component and this manifests itself explicitly as a blueshift of the condensate energy. Note that all of these differences are microscopic in nature and do not manifest themselves as drastically new physical phenomena. For example, thermalisation or blueshifts can also be observed for conventional lasing. The intricacies of polariton condensation have therefore been hotly debated for many years [6–8].

It is now generally agreed that polariton condensation should be distinguished from BEC due to its non-equilibrium nature [9]. In fact, in a two-dimensional system, there is no true ODLRO and the phase correlations should decay algebraically for large separations [10]. There is, however, a transition from a disordered phase to a quasi long-range ordered phase. Moreover, understanding the nature of the phase transition is complicated by the fact that typical exciton-polaritons systems are out of equilibrium, weakly anisotropic and of a finite size. These points have recently been reviewed in [11]. It has been argued that spontaneous breaking of the global U(1) symmetry should be used as defining feature of polariton condensation and that the achievement of quasi-equilibrium (as evidenced by e.g. a Boltzmann distribution) below threshold is rather unimportant [7]. In the end, it is clear that polaritons still allow for a broad range of fascinating phenomena to be observed such as superfluidity [12, 13], vortex formation [14], the Berezinskii-Kosterlitz-Thouless (BKT) phase transition [15], acoustic black holes [16] and more.

7.3 Planar Microcavity Structures

To date, organic polariton condensates have been reported in four different material sets: an organic single crystal, a thermally evaporated oligomer, a spin-coated polymer and a biologically produced fluorescent protein [17–20]. All of these structures use planar Fabry-Perot microcavities composed of distributed Bragg reflectors (DBRs) to enclose the active medium. In such a microcavity, shown schematically in Fig. 7.1a, the photonic resonance ω_{cav} is parabolic as a function of in-plane momentum $k_{//}$:

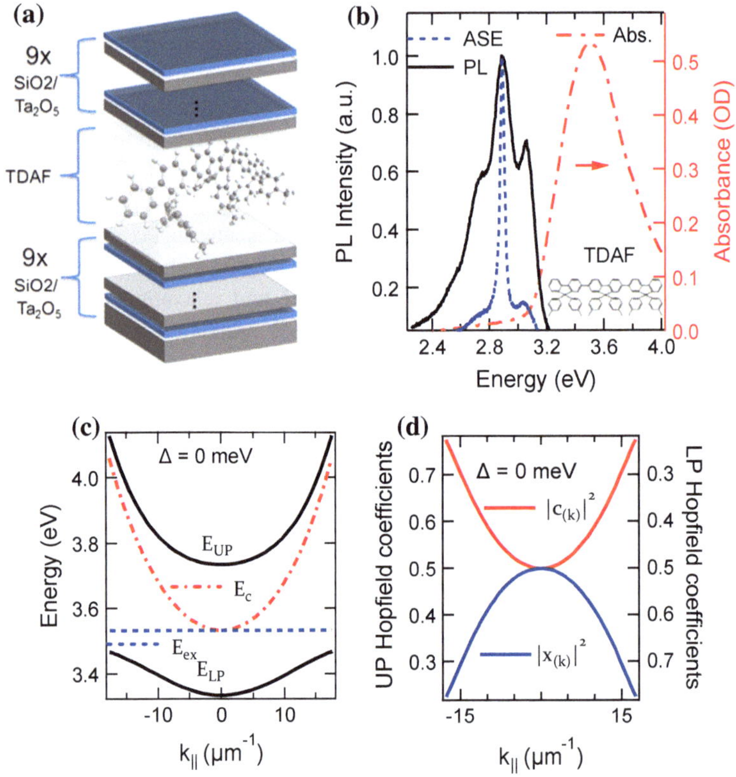

Fig. 7.1 **a** Schematic of a planar microcavity structure that will be used in experiments described later in the text. It consists of two dielectric mirrors of alternating tantalum pentoxide (Ta_2O_5) and silicon dioxide (SiO_2) pairs enclosing a single layer of TDAF. **b** The photoluminescence (*solid line*) of bare TDAF and its amplified spontaneous emission spectrum (ASE; *dashed line*) are shown on the left. Note the presence of vibronic replicas in the emission spectrum and that the ASE peak coincides with the emission maximum. The ASE spectrum was obtained by pumping a neat 120-nm-thick TDAF film with an elliptical spot long enough to achieve significant single-pass gain for the waveguided photons. The absorbance of a 60 nm TDAF film is shown on the right (*dash-dotted line*). It shows a strong inhomogeneously broadened exciton absorption band at ~3.5 eV. The molecular structure of TDAF is shown in the inset. **c** Calculated LP and UP dispersion relation for E_{ex} = 3.5 eV and V = 0.2 eV and zero detuning between the photon and exciton. The bare photon (E_c) and exciton (E_{ex}) dispersion relations are shown as dashed lines. The corresponding Hopfield coefficients for each polariton mode are shown in (**d**)

$$\omega_{cav}(k_{//}) = \sqrt{\omega_{cav}^2(0) + \frac{c^2}{n_{eff}^2} k_{//}^2}$$

where the cavity cut-off frequency $\omega_{cav}(0)$ and effective refractive index n_{eff} account for the finite penetration depth of the DBR, L_{DBR} [21]. Organic semiconductors typically possess broad, inhomgeneously broadened lineshapes. Figure 7.1b shows, for example, the absorbance, Stokes-shifted photoluminescence and amplified spontaneous emission (ASE) from a thin film of TDAF. The dispersion relation of a fictitious Fabry-Perot cavity and of an exciton with $E_{ex} = 3.5$ eV are shown schematically in Fig. 7.1c.

For large Rabi splittings, only the tail of the dielectric function affects the polariton dispersion [22]. These are typically homogeneous, of the form:

$$\varepsilon_r(\omega) = \varepsilon_b + \frac{Ne^2}{m\varepsilon_0}\frac{f}{\omega_{ex}^2 - \omega^2}$$

where ε_b is the background dielectric constant, N is the molecular density, f is the oscillator strength, ω_{ex} is the exciton frequency and m is the electron mass. In this case, the light-matter interaction strength is simply given by:

$$V(\theta) = \sqrt{\frac{Ne^2 f}{4m\varepsilon_0\varepsilon_B}\left(\frac{L_c}{L_c + L_{DBR}(\theta)}\right)}$$

where L_c and L_{DBR} is the DBR penetration depth defined in [21]. One then sees that the Rabi frequency is proportional to the square root of both the molecular density and the oscillator strength. The energies of the upper and lower polariton (UP and LP) branches and the Rabi splitting can be found by diagonalizing the usual light-matter Hamiltonian:

$$\hat{H} = \begin{pmatrix} E_{cav}(\theta) - i\gamma_{cav} & V(\theta) \\ V(\theta) & E_{ex} - i\gamma_{ex} \end{pmatrix}$$

where we have included losses via the cavity and exciton halfwidths at half maximum (HWHMs) γ_{cav} and γ_{ex}, respectively. Typical exciton HWHMs 25–250 meV, while the cavity HWHM can be as low as 1–5 meV. A typical dispersion relation showing both branches is given at zero detuning $\Delta/\hbar \equiv \omega_{ex} - \omega_{ph}$ in Fig. 7.1c for the case $V = 0.2$ eV. The corresponding Hopfield coefficients $|c(k)|^2$ and $|x(k)|^2$ for each polariton branch are shown in Fig. 7.1d. These are obtained by taking the magnitude of the eigenvector coefficients for each mode and give the photon $\left(|c(k)|^2\right)$ and exciton $\left(|x(k)|^2\right)$ content of each polariton [23].

Figure 7.2a shows experimental LP dispersions obtained from angle-resolved photoluminescence (PL) measurements performed on real TDAF cavities with the structure shown in Fig. 7.1a when the detuning is varied by changing the cavity length. We observe a flattening out of the parabolic dispersion relation for more positive detunings as well as a change in linewidth. Figure 7.2b shows the linewidth measured at normal incidence ($k_{//} = 0$) for various detunings. We see that it broadens as the mode approaches the broad exciton resonance as predicted by the light-matter Hamiltonian, but it also broadens for very negative detunings, where

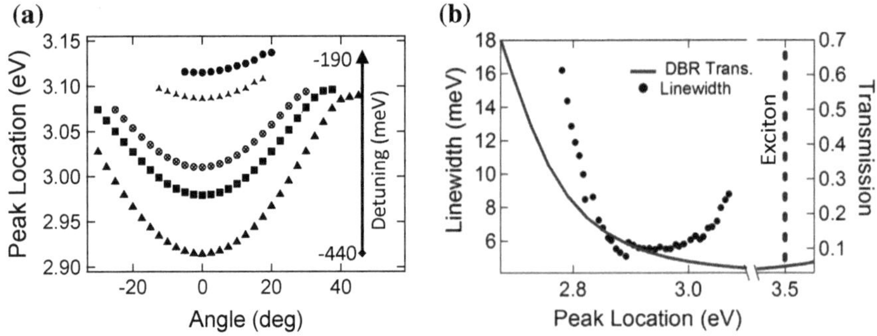

Fig. 7.2 **a** LP dispersion relations measured using angle-resolved PL performed on 7 negative-detuned microcavities. Note the flattening out of the LP dispersion away from the exciton resonance characteristic of polaritonic behaviour as the detuning is made more positive. **b** The LP linewidth measured at normal incidence for different detunings. Note the broadening on both sides due to the broad bare exciton resonance and the edge of the DBR stopband. The DBR transmission is shown as a *solid line*

the reduced reflectivity of the DBR near the edge of its stopband becomes important. The extracted light-matter interaction coefficient for these samples is $V(\theta) \sim 0.3$–0.35 eV throughout the range of angles shown [17].

7.4 Polariton Relaxation

Because of dissipation, principally via the cavity mirrors, polaritons are constantly lost through radiation. This, however, is a useful feature for measurements because any radiated photons preserve the population, phase and polarization of the original polaritons. The pump, which creates polaritons can be resonant, thus directly exciting polaritons, by choosing the appropriate wavevector $k_{//}$ and energy. In this case, it imparts a well-defined velocity and phase to the polaritons, which propagate within the plane of the microcavity. In condensation experiments, however, the microcavity is pumped non-resonantly (at high energies), often beyond the stopband of the DBRs. Under these excitation conditions, a "reservoir" of excitons is initially created as shown schematically in Fig. 7.3a. This can be understood by considering the system Hamiltonian, which mixes N_{exc} excitonic modes per unit surface (corresponding to the surface density of molecules within the excitation volume) with N_{phot} photonic modes per unit surface. Such a Hamiltonian will give N_{phot} "bright" states, which correspond to the polaritons and the remaining excitations will be "dark" because they are uncoupled to light [24]. In practice, N_{phot} is limited by the finite width of the DBR stopband, but we can provide a simple estimate by considering all of the photonic modes from k = 0 to the edge of the light cone $k_{//}^{\max} = \frac{2\pi}{\lambda}$. In this case, we find $N_{phot} = 2/\lambda^2$ where the factor of two accounts for polarization. Meanwhile, the volume density of a typical undoped organic film is 10^{21} cm^{-3} so for a typical thickness of d = 100 nm, $N_{\text{exc}} \sim 10^{16}$

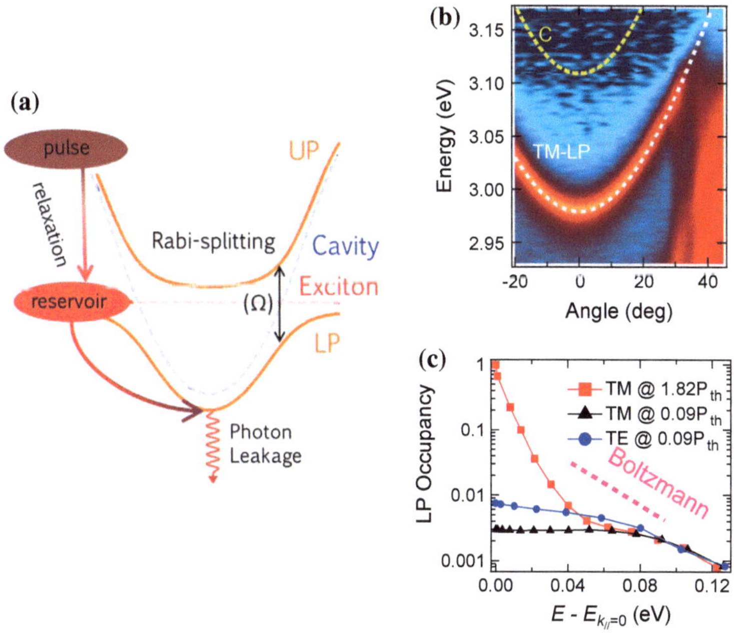

Fig. 7.3 **a** Schematic showing polariton emission under non-resonant pulsed excitation. For organic microcavities, the LP lifetime is typically as low as 150 fs and thus relaxation of polariton into the ground state is dominated by single scattering events. The *orange lines* show the UP and LP branches and the dashes-faded lines show the uncoupled exciton and cavity resonances. The impulsive pump mostly populates the reservoir, from which excitons can then scatter into polaritonic sates. **b** TM-polarized PL of a 120-nm-thick TDAF cavity. **c** The lower polariton (LP) population is shown on a logarithmic scale as a function of energy away from the polariton ground state. The k = 0 LP energy is shown as zero on the scale. The *pink dashed line* is a Boltzmann distribution $n(E)\propto e^{\left(-\frac{\Delta E}{k_B T}\right)}$, where k_B is Boltzmann's constant and T = 298 K

cm^{-2} and we find that only ~ 10^{-7} of the excitations are polaritonic, while the remaining correspond to "dark", molecule-like excitations [24].

The initially excited "dark" (or reservoir) excitons can then relax into polaritonic states by several mechanisms, as schematically depicted in Fig. 7.3a. First, organic molecules possess localized "molecular" vibrations with a varying range of energies (10–10^3 cm^{-1}). By emitting such phonons, excitons (or polaritons) can relax into lower-energy polaritonic states without any constraint on momentum conservation. The relaxation rate corresponding to this process is proportional to the exciton fraction $|x(k)|^2$ of the final state [25]. The correspondence between the polariton population, measured using photoluminescence, and the energy of Raman active vibrations was highlighted in [26]. Second, any pure (e.g. collisional) dephasing mechanism can also lead to relaxation into (from the reservoir) or along the LP branch [27]. Finally another mechanism for relaxation has been proposed due to radiation from the excited singlet exciton to vibronic replicas of the electronic ground state. It has been suggested that this type of transition can be treated perturbatively because the low-lying states are nearly unoccupied and lead to relaxation. As a result, any overlap between the Stokes-shifted PL of the organic material and the polariton modes can also be beneficial for populating the LP states [28, 29].

Figure 7.3b shows the angle-resolved PL intensity measured (TM polarisation) for a 120 nm-thick TDAF microcavity under pulsed, non-resonant optical excitation. One can see that in this case, no clear vibronic maxima along the LP are observed and the intensity decreases monotonically away from the branch minimum. From the measured linewidth, we find that the lower polariton lifetime is ~150 fs. Such a short lifetime suggests that the kinetics are dominated by single scattering events from the exciton reservoir, as opposed to multiple scattering along the dispersion.

By correcting for the photon fraction of the polariton modes, the emission intensity can be related to the occupation at the corresponding in-plane wavevector $k_{//}$. The resulting normalized polariton distribution function is shown in Fig. 7.3c. We find that the occupation is not thermalized near the bottom of the LP branch (where the LP is more photon like), but a fraction of the states 80 meV above the minimum appears to be thermalized with a characteristic temperature of 298 K. This clearly highlights the non-equilibrium nature of the system. The Figure also shows the distribution above the condensation threshold where this thermalized tail remains, but the low-*k* distribution is dramatically modified.

7.5 Condensate Formation

Using mean-field theory, it has been shown that polariton condensation can be reasonably well described by a Gross-Pitaeveskii equation (GPE) for the condensate wavefunction Ψ(r, t) coupled to an equation for the density of reservoir excitons n_R(r, t) [30, 31]:

$$i\hbar\frac{\partial\Psi(r,t)}{\partial t}=\left(-\frac{\hbar^2\nabla^2}{2m_{LP}}+\frac{i}{2}[Wn_R-\gamma_{LP}]+g|\Psi(r,t)|^2+\tilde{g}n_R\right)\Psi(r,t)$$

and

$$\frac{dn_R}{dt}=P(r,t)-\frac{n_R}{\tau_R}-k_b n_R^2-Wn_R|\Psi|^2$$

where m_{LP} is the LP effective mass, W is the effective scattering rate from the reservoir to $k = 0$, γ_{LP} is the LP decay rate at $k = 0$, g is the polariton-polariton

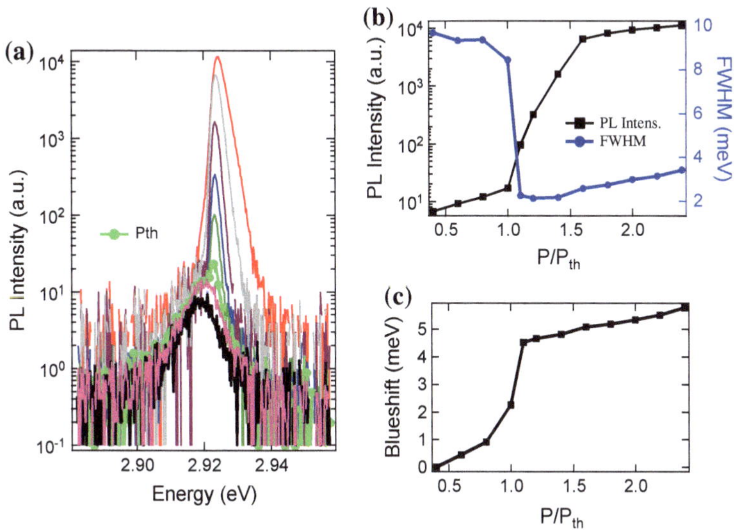

Fig. 7.4 **a** Photoluminescence measured at $k_{//} = 0$ for increasing pump fluence. **b** PL intensity and FWHM linewidth of LP as functions of pump fluence. **c** Shows the experimentally observed blueshift. In the pumped region, the blueshift is attributed principally to polariton–exciton scattering and as a result, a smaller blueshift is anticipated for the high-Q cavity. This is indeed observed in both the measurement and calculation. The clamping of the average reservoir exciton density can also be identified in the model

interaction coefficient and $\tilde{g}$ is the polariton-exciton interaction coefficient. In the second equation, $P(r, t)$ is the pump term, τ_r is the exciton lifetime and k_b is the bimolecular annihilation coefficient. In steady-state, the only non-zero solution to this pair of equations occurs when $Wn_R > \gamma_{LP}$, in other words a condensate forms when the net rate of scattering into the $k_{//} = 0$ condensate overcomes the decay rate of polaritons.

In TDAF microcavities, this threshold occurs at powers around ~30 μJ/cm^2 where we observe a superlinear increase in output intensity at $k_{//} = 0$ as shown in Fig. 7.4a, b. Beyond this point, we also observed a collapse of the distribution function towards the bottom of the LP dispersion as shown in Fig. 7.3c. These two effects are accompanied by a blueshift of the emission energy due to the interaction terms and a change from unpolarized to linearly polarized emission and a reduction in linewidth, indicative of an increase in temporal coherence [17]. From the above condition for threshold, if the polariton linewidth and reservoir occupation are known, the net scattering rate into the condensate can be calculated from the power dependence and we find $W = 2 \times 10^{-7}$ cm^{-3}s^{-1}.

The experiments shown in Fig. 7.4 were performed with a flat-top beam profile to exclude artifacts related to the beam profile from these measurements. Interactions give rise to two distinct regions in Fig. 7.4c. The first shows a gradual increase in the blueshift near threshold which nearly saturates around ~5 meV. This blueshift is due to an effective exciton-polariton repulsion due to a gradual saturation of the TDAF ground-state population. For structures with larger thresholds, this blueshift increases due to the need for stronger pumping. The fact that it becomes nearly constant at ~5 meV is related to a clamping of the exciton population at threshold just like in a conventional laser. This saturated part, however, has a small linear slope, which is due exclusively to polariton-polariton repulsion. From the slope, we can extract a mean-field polariton-polariton interaction coefficient $g = 10^{-6}$ meV μm^2. This interaction plays an essential role in the nonlinear behaviour of such condensates [9].

7.6 Condensate Coherence

BEC does not formally occur in 2D, but at the condition for quantum degeneracy $n\lambda^2 > 1$, where n is the particle density, there is a transition from a Gaussian decay of correlations to an exponential one with first-order correlation function $g^{(1)}(r) \approx e^{-r/l}$, where l is the correlation length. Furthermore, this correlation length $l \propto e^{n\lambda^2/2}$ grows exponentially and can readily span the entire system size [10].

To probe the decay of correlations, we have performed spatial coherence measurements by collecting the condensate emission using a NA = 0.42 objective and a Michelson interferometer with one arm on a piezoelectric delay stage to scan the phase and the other arm replaced by a retroreflector to invert one image [32]. Below threshold, we find that the emission is spatially incoherent and only a small autocorrelation fringe can be observed in the centre of the interferogram as shown in Fig. 7.5a. Beyond threshold, fringes emerge over the entire condensate area as shown in Fig. 7.5c. In this example, two centro-symmetric fork dislocations corresponding to a vortex can also be seen. The magnitude of the first-order correlation function can be obtained from such images by changing the phase delay and fitting the fringe visibility at each pixel location. For a small (27 × 25 μm) Gaussian pump spot, the value of $g^{(1)}$(r, r') averaged over a ~3 μm^2 area for separations $|r - r'|$ of 0, 4 and 8 μm is shown in Fig. 7.5d as a function of pump power. We find that in all cases, the correlation length increases dramatically beyond threshold and saturates at ~1.5 P_{th}. Moreover, significant phase correlations can be observed over the entire spot size.

In general, there are differences from spot to spot due to intrinsic disorder on the sample, but the most important phenomena affecting such coherence measurements are the shot-to-shot fluctuations between each realization of the condensate. Any phase fluctuations will artificially wash out the fringe visibility and magnitude of the first-order correlation function obtained using our extraction procedure. In particular, we find that although shot-to-shot fluctuations are observed for smaller

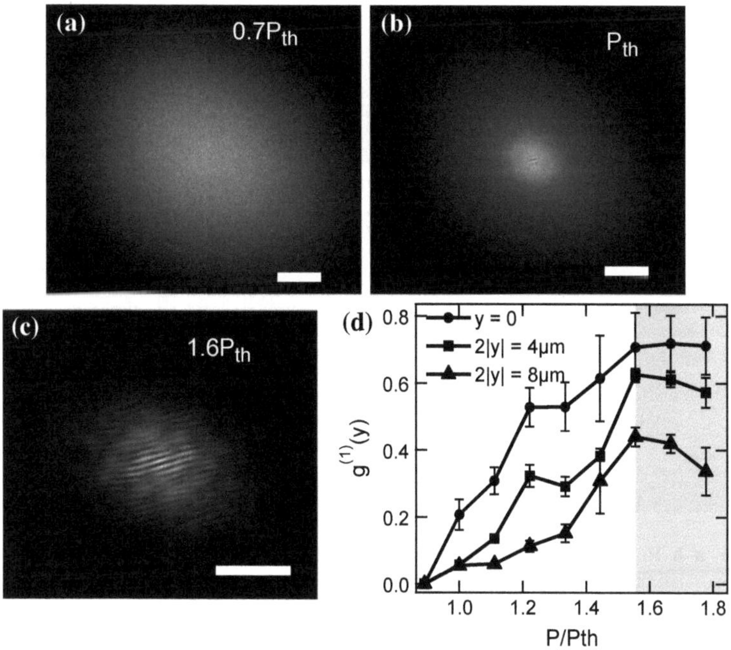

Fig. 7.5 **a–c** Condensate interferograms taken at **a** 0.7 P_{th}, **b** P_{th} and **c** 1.6 P_{th}. The interferometer used here is a Michelson interferometer in the retroreflector configuration. Below threshold, a small autocorrelation fringe can be seen. Note that or increasing pump power above the condensation threshold, fringes span over the entire condensate area indicative of macroscopic phase coherence. **d** Correlation function as a function of pump power. For a small (27 × 25 μm) Gaussian pump spot, we find a sharp increase in $g^{(1)}(r, r')$, followed by a plateau near ~1.5 P_{th} (*shaded part*). The value and error bars correspond to the mean and standard deviation obtained from averaging over a ~3 μm^2 area for 3 different areas on the sample. The scale bar corresponds to 5 μm

spot sizes, their role is dramatically increased for larger spot sizes and flat-top profiles as shown in Fig. 7.6. We find that in this case, vortices and strong phase gradients, which change location from shot-to-shot are often observed in single-shot interferograms. In real space, we find that the condensate breaks up into multiple domains. This behaviour is due to a type of dynamic instability supported by the GPE and its interaction with the reservoir. The behaviour of this instability was recently studied in the case of continuous wave pumping and a phase diagram for stability was calculated showing that stable regions only occur for very small pump spots or powers high above threshold [33]. In contrast, for the impulsive case measured here, the power dependence has been found to be opposite that of the continuous case. Using parameters for TDAF, numerical simulations can accurately reproduce the instabilities observed here for large spots or high pump powers [34]. In particular, this behaviour explains the artificial decrease in $g^{(1)}$ observed when the spot is made larger or the pump power is increased.

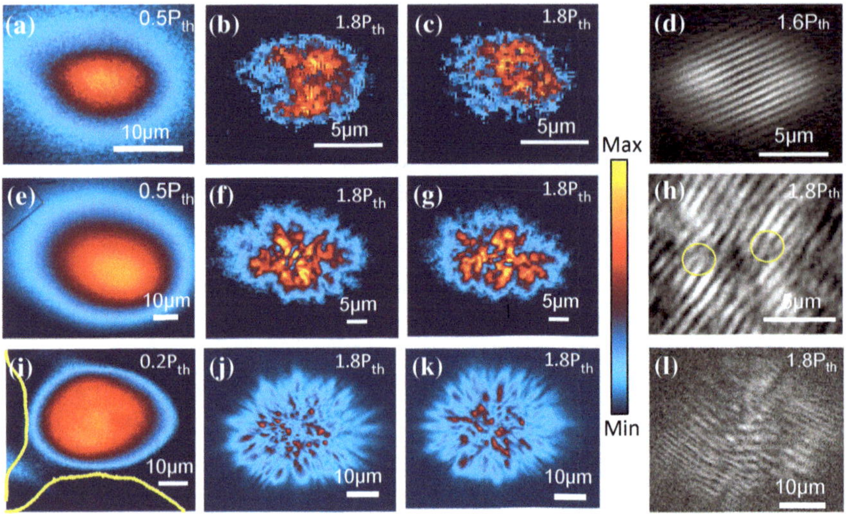

Fig. 7.6 **a–h** Real-space images and few-shots interferograms for a small (**a–d**) and a large (**e–h**) Gaussian pump. **i–l** Real-space **i–k** and 2-shot interferograms (**l**) for a flat-top pump. Below P_{th} real space images were integrated over >50 laser shots and above threshold over single laser shots. Note that for small Gaussian spots, shot-to-shot fluctuations do not disturb the interferograms. In contrast, larger Gaussian or flat-top pump single-shot interferograms show vortices and strong phase gradients, which change location from shot-to-shot

7.7 Conclusions

In conclusion, we have shown the range of behaviour that has been observed thus far in organic polariton condensates under non-resonant pumping: condensate formation, blueshifts, the spontaneous appearance of long-range spatial coherence, vortices and dynamic instabilities. Meanwhile, under resonant excitation ballistic flow and signatures of superfluidity have also been observed. Organic microcavities have shown to be a versatile playground for studying Gross-Pitaevskii physics at room temperature. Using simple microfabrication techniques, it is possible to further engineer the samples to imprint specific potentials profiles within or between condensates. In particular, plasmonic surface lattice resonances can be an attractive option to engineer both nonlinearities and spatial mode profiles given that they can possess quality factors Q > 700 and that this translates into lifetimes sufficient to achieve condensation. Finally, one strong motivation driving such research has been the dream of achieving electrically-pumped organic lasing by exploiting strong light-matter coupling. For this, however, thresholds will need to be lowered further and this can be done by increasing either the lower polariton lifetime or the scattering rate W from the reservoir into $k_{//} = 0$ polaritons.

References

1. H. Deng, H. Haug, Y. Yamamoto, Rev. Mod. Phys. **82**, 1489 (2010)
2. D.G. Lidzey, D.D.C. Bradley, M.S. Skolnick, T. Virgili, S. Walker, D.M. Whittaker, Nature **395**, 53 (1998)
3. C. Pethick, H. Smith, *Bose-Einstein Condensation in Dilute Gases*, 2nd edn. (Cambridge University Press, Cambridge, New York, 2008)
4. C.N. Yang, Rev. Mod. Phys. **34**, 694 (1962)
5. J. Kasprzak et al., Nature **443**, 409 (2006)
6. L.V. Butov, A.V. Kavokin, Nat. Photonics **6**, 2 (2012)
7. G. Malpuech, D. Solnyshkov, in *Exciton Polaritons in Microcavities*, ed. by S. D., V. Timofeev (Springer, Berlin, Heidelberg, 2012)
8. D. Snoke, in *Exciton Polaritons in Microcavities* (Springer, 2012), pp. 307
9. I. Carusotto, C. Ciuti, Rev. Mod. Phys. **85**, 299 (2013)
10. Z. Hadzibabic, J. Dalibard, Rivista del nuovo cimento **34** (2011)
11. J. Keeling, L.M. Sieberer, E. Altman, L. Chen, S. Diehl, J. Toner, arXiv preprint arXiv:1601. 04495 (2016)
12. A. Amo, J. Lefrere, S. Pigeon, C. Adrados, C. Ciuti, I. Carusotto, R. Houdre, E. Giacobino, A. Bramati, Nat. Phys. **5**, 805 (2009)
13. A. Amo et al., Science **332**, 1167 (2011)
14. K.G. Lagoudakis, M. Wouters, M. Richard, A. Baas, I. Carusotto, R. Andre, L.S. Dang, B. Deveaud-Pledran, Nat. Phys. **4**, 706 (2008)
15. G. Roumpos et al., Proc. Natl. Acad. Sci. **109**, 6467 (2012)
16. H.S. Nguyen, D. Gerace, I. Carusotto, D. Sanvitto, E. Galopin, A. Lemaître, I. Sagnes, J. Bloch, A. Amo, Phys. Rev. Lett. **114**, 036402 (2015)
17. K.S. Daskalakis, S.A. Maier, R. Murray, S. Kéna-Cohen, Nat. Mater. **13**, 271 (2014)
18. S. Kéna-Cohen, S.R. Forrest, Nat. Photonics **4**, 371 (2010)
19. J.D. Plumhof, T. Stöferle, L. Mai, U. Scherf, R.F. Mahrt, Nat. Mater. **13**, 247 (2014)
20. C.P. Dietrich, A. Steude, L. Tropf, M. Schubert, N.M. Kronenberg, K. Ostermann, S. Hofling, M.C. Gather, Dietrich et al., Sci. Adv. **8**, e1600666 (2016)
21. G. Panzarini et al., Phys. Solid State **41**, 1223 (1999)
22. R. Houdré, R.P. Stanley, M. Ilegems, Phys. Rev. A **53**, 2711 (1996)
23. A. Kavokin, *Microcavities* (Oxford University Press, Oxford, New York, 2007), Series on semiconductor science and technology, no 16
24. V.M. Agranovich, M. Litinskaia, D.G. Lidzey, Phys. Rev. B **67**, 085311 (2003)
25. M. Litinskaya, P. Reineker, V.M. Agranovich, J. Lumin. **110**, 364 (2004)
26. D.M. Coles, P. Michetti, C. Clark, W.C. Tsoi, A.M. Adawi, J.-S. Kim, D.G. Lidzey, Adv. Funct. Mater. **21**, 3691 (2011)
27. J. Feist, F.J. Garcia-Vidal, Phys. Rev. Lett. **114**, 196402 (2015)
28. L. Mazza, L. Fontanesi, G.C. La Rocca, Phys. Rev. B **80**, 235314 (2009)
29. L. Mazza, S. Kéna-Cohen, P. Michetti, G.C. La Rocca, Phys. Rev. B **88**, 075321 (2013)
30. M. Wouters, I. Carusotto, Phys. Rev. Lett. **99**, 140402 (2007)
31. M. Wouters, I. Carusotto, C. Ciuti, Phys. Rev. B **77**, 115340 (2008)
32. K.S. Daskalakis, S.A. Maier, S. Kéna-Cohen, Phys. Rev. Lett. **115**, 035301 (2015)
33. N. Bobrovska, E.A. Ostrovskaya, M. Matuszewski, Phys. Rev. B **90**, 205304 (2014)
34. N. Bobrovska, M. Matuszewski, K.S. Daskalakis, S.A. Maier, S. Kena-Cohen, arXiv preprint arXiv:1603.06897 (2016)

Chapter 8
Plasmon Particle Array Lasers

Y. Hadad, A.H. Schokker, F. van Riggelen, A. Alù and A.F. Koenderink

Abstract Diffractive arrays of strongly scattering noble metal particles coupled to a high-index slab of gain material can form the basis for plasmonic distributed feedback lasers. In this chapter, we discuss recent theoretical and experimental results describing the electromagnetic properties of these structures. Particularly, we investigate bandgap topology versus detuning between the plasmonic and Bragg resonances. We examine the complex dispersion relation, accounting for the fact that the particles are electrodynamic scatterers with radiation loss, that couple via a stratified medium system supporting guided modes. From the complex dispersion of this array we can deduce loss and outcoupling properties of the various Bloch modes, giving a handle on its lasing properties. From the experimental side, we show how to measure the dispersion relation using fluorescence microscopy, and systematically examine the array dispersion for realized plasmonic lasers as function of detuning between particle and lattice resonance. We conclude the chapter with a vision towards employing disordered, quasiperiodic and random plasmonic arrays to induce different optical responses, and experimentally demonstrate the exceptional robustness of lasing to disorder in these systems.

Y. Hadad (✉) · A. Alù
Department of Electrical and Computer Engineering, The University of Texas at Austin, UTA 7.215, 1616 Guadalupe Street, Austin, TX 78712, USA
e-mail: yakir@utexas.edu

A. Alù
e-mail: alu@mail.utexas.edu

A.H. Schokker · F. van Riggelen · A. Alù · A.F. Koenderink (✉)
Center for Nanophotonics, FOM Institute AMOLF, Science Park 104, 1098 XG Amsterdam, The Netherlands
e-mail: f.koenderink@amolf.nl

A.H. Schokker
e-mail: a.h.schokker@gmail.com

F. van Riggelen
e-mail: floor.vanriggelen@tno.nl

S.I. Bozhevolnyi et al. (eds.), *Quantum Plasmonics*,
Springer Series in Solid-State Sciences 185,
DOI 10.1007/978-3-319-45820-5_8

8.1 Introduction

Lasing takes place when gain and feedback are combined. In conventional laser systems the feedback mechanism is typically implemented in Fabry-Perot resonators with partially reflecting mirrors that enable a fraction of photons to escape the cavity after several roundtrips [1]. Alternatively, in distributed feedback lasers, the feedback is achieved through distributed resonances, such as Bragg resonances [2], or, when dealing with random systems, multiple scattering [3]. In general, similar to conventional lasers, the efficiency of the distributed feedback, together with the structure's Ohmic and radiation losses, plays an important role in determining the lasing threshold, i.e., the required gain above which the excited mode is self-sustained. Commonly, distributed feedback lasers are based on periodic dielectric gratings, and therefore they present inherently low loss, but also weak scattering cross sections, and thereby low feedback efficiency. For example, organic distributed feedback lasers have been widely studied since the mid-nineties for their ability to provide large-area lasing upon optical or electrical pumping, while being cheap to fabricate [4]. Such lasers generally consists of an organic gain medium that is deposited as a thin layer on a periodically corrugated dielectric surface, with a periodicity chosen such that it offers an in-plane Bragg diffraction condition within the gain window [2, 5]. A wide range of emission wavelengths can be chosen through the availability of a vast range of organic fluorophores and fluorescent polymers, while the relevant, usually weak, perturbative, corrugations can be realized, e.g., through optical lithography, or soft-imprint lithography [6, 7].

More recently a different class of lasers was proposed based on plasmonic materials. Plasmonics revolves around the fact that free electrons in metals can support collective resonances at optical frequencies [8]. This causes metal nanoparticles or nanostructured surfaces to provide highly enhanced and strongly localized electromagnetic fields upon irradiation, and to provide large spontaneous emission rate enhancements when coupled to nearby fluorescent sources [9–11]. This notion has triggered the development of several classes of plasmon lasers or 'spasers' [12]. Implementations include hybrid-mode nanowire lasers, in which light is tightly confined in the gap between a metal and II-VI gain medium, metal-insulator-metal lasers in III-V systems [13–15], and long-range surface plasmon systems coupled to organic gain media [16]. Here we focus on plasmonic distributed feedback lasers: when plasmonic particles are placed in two-dimensional diffractive periodic arrays, they not only can provide large field enhancement, but further provide control over emission directivity and brightness due to a hybridization of localized plasmonic resonances, and grating anomalies associated with the array [17–19]. In particular, these systems have been studied as substrates for Surface Enhanced Raman Scattering (SERS) [20], for sensing [21, 22] and for solid-state lighting [23]. Recently, several groups [24–26] have shown distributed feedback lasing in such plasmonic periodic systems. A main difference with conventional dielectric feedback lasers is that, while the dielectric perturbation is weak and non-resonant, for plasmonic systems the scattering strength per unit cell of the lattice can be made very strong, based on a resonance. One practical advantage is that strong scattering implies that any stop gap that arises in the band structure of the waveguide-particle system will be broader. In turn, broader stop gaps correspond to smaller Bragg scattering lengths [27] or, equivalently, much smaller required device sizes for lasing, and large robustness to disorder. However, this stronger coupling doesn't come without a price: plasmonic materials are lossier compared to dielectrics. Therefore a systematic study of these new systems is required.

This chapter gives an overview of recent theoretical and experimental efforts towards better understanding this subject. We begin with an overview of the typical experimental measurements in this field that support lasing effects, leading to observations of the underlying band structures. Next we present a theoretical dipole model description used in order to account for near- as well as far-field interactions between the particles as mediated by the waveguide structure. As opposed to a periodically corrugated dielectric waveguide with a weak periodic grating, the "nearly-free photon" numerical approach to calculate the band structure with very narrow stop gaps [28] is not valid here, and a more adequate model is developed here. This theoretical model is followed by a comparison with an experimental study of the band structure underlying lasing as the plasmon resonance is tuned to the lasing condition. Moreover, we demonstrate that, as the plasmon resonance crosses through the lasing condition, the loss characteristics of bands interchanges, and as a consequence also the stop gap edge at which lasing condition occurs moves from the low to the high end of the gap. We compare these findings with the theoretical electrodynamic point dipole model. The chapter concludes with a brief discussion of very recent results regarding quasiperiodic and quasi-random plasmonic systems, showing that the strong scattering offers unique opportunities for studies in the fields of aperiodic (dis)order.

8.2 Experiments on Plasmon Lattice Laser

8.2.1 Samples and Experimental Methods

In this section we review experimental methods to probe plasmon particle array distributed feedback lasers. All experiments to date report optically pumped, pulsed-excitation, laser action, typically measured either in a low-NA spectroscopy set up, or in a microscope. We have found it advantageous to use fluorescence microscopy techniques [25] so that in one instrument we can measure spectra, input-output curves, real space output and wave vector resolved output (Fig. 8.1). In essence, our method revolves around an infinity-corrected inverted fluorescence microscope with a very high-NA microscope objective (NA = 1.45, Nikon Plan Apo 100x). By focusing a pump laser (532 nm in our case) in the back aperture of the microscope objective, one can pump a reasonably-sized area on the sample, of

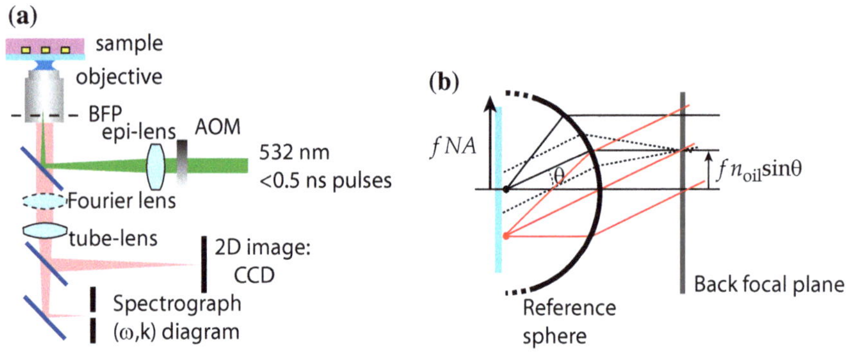

Fig. 8.1 **a** Diagram of an inverted fluorescence-microscope adapted to study plasmon lasers. Epi-illumination is provided by a 532 nm pulsed laser (0.5 ns, uJ energy). Collected light from single pump shots is collected either on a camera or spectrometer, where a Fourier lens can switch between real-space and k-space imaging. *Right* diagram explaining back-focal plane imaging and the Abbe sine condition. The objective lens is represented by its reference sphere of diameter *f* NA (with *f* the objective focal length). Physically the back focal plane is located near the baffle at the mounting thread of the objective

typically 20–100 μm across, the only difference with regular epi-illumination being that the excitation is pulsed (0.5 ns, μJ energy pulses in our case). Wide-field fluorescence, captured by the same objective and separated from the pump light by a dichroic beamsplitter, can subsequently be analyzed in various ways. As one typically uses low-repetition rate pulses, single-shot large area detection is called for, such as offered by current CCD cameras with effective sensitivities of about 5 photons per pixel count, at *effective* dynamic ranges of better than 2^{12}. First, by imaging fluorescence directly on a CCD camera, one can make real space images of sample output. Second, using a Bertrand, or conoscopic lens, one can also perform back-aperture imaging [29–31]. High-NA back-aperture images quantitatively map angular-emission profiles, with resolution in the order of 0.2°, over the entire angular acceptance range of the used objective (70°). Owing to the “Abbe-sine” rule these ‘Fourier’ images are direct maps of k|, parallel-momentum-space. While on a regular 2D CCD array (Andor, Clara Si CCD) one obtains wavelength-integrated intensity versus k_x and k_y, one can also relay the Fourier image to the entrance slit of an imaging spectrograph [32] (Shamrock 303i, with Si iVAC CCD). If this is equipped with a CCD camera, one can obtain direct maps of energy-momentum space for one chosen linecut through momentum space. The most important feature that this single-shot imaging approach lacks is the ability to assess temporal photon statistics, for which one would need continuous-wave operation, and time-correlated single photon counting.

8.2.2 *Input-Output Curves, Thresholds and Fourier Space*

In this section we discuss the typical features of the data one obtains when measuring plasmon lattice lasers, illustrated by the example of systems studied in our group that consist of particles in a wave guiding layer that supplies gain. It should be noted that several other works, notably of the group of Odom at Northwestern [24, 26, 33] have studied systems that share many of these broad features, yet also differ on important other aspects. These are the topic of Sect. 8.4.

As a typical example, we consider systems of Ag and Au particles in periodic square grids, fabricated using electron beam lithography and lift off, and deposited on glass. These are covered by a 400–450 nm thick layer of a high index polymer ($n = 1.6$ for SU8) doped with 0.25 wt% of the organic dye Rh6G. The thickness and index of the layer means that a single TE and a single TM waveguide mode are supported, with a mode index of around 1.54. With this particular choice, the gain window will be at around 590 nm, while the plasmon resonance for strongly scattering, i.e., large (100 nm diameter) metal disks will be further to the red. For lattice periodicities in the range of 360–400 nm, the kinematic 2nd order Bragg diffraction sweeps from 550 to 615 nm, through the gain window.

Figure 8.2 presents a typical below- and above threshold spectra, as well as an input output curve collected normal to the sample, on a lattice with a pitch of 380 nm. Typical findings in this system are a clear threshold at pulse energies of order 50 nJ or 1 m J/cm^2. At threshold, the spectrum under normal incidence sharpens to a peak of sub-nm width lasing peak that is to within a few percent accuracy at the kinematic 2nd order Bragg condition for the waveguide mode. This wavelength is hence tuneable by lattice pitch, and by the effective index associated to the diffractive resonance, which can be tuned by varying the waveguide or cladding index [26].

Broadly, this is the typical behavior for 2D distributed feedback lasing, also commonly observed in dielectric structures [2]. The threshold energies are comparable to those reported by Suh, Zhou and Yang [24, 26, 33], although they used a gain medium shifted to the near infrared (IR140 dye), and are quite comparable to thresholds of more optimized polymer DFB lasers [2, 4].

While the physics is similar to that of 2D dielectric DFB lasers, we propose that, if there are differences, they must be encoded in the underlying periodic-system band structure that depends on the strong, resonant scattering of the plasmon particles. Figure 8.2 shows a raw back focal plane image below threshold, as well as the corresponding photoluminescence enhancement below and above threshold. We remind the reader that back focal plane images form a map of parallel momentum $k_{||}$, or more precisely of $k_{||}/k_0 = n \sin\theta\,[\cos\varphi,\ \sin\varphi]$ where (θ, φ) are respectively the angle of emission relative to the sample normal, and the in-plane angle, and where n is substrate image. Viewed from the glass side, emission from any thin fluorescent layer at the air/glass interface will appear as a comparatively dim disk for $k_{||}/k_0 < 1$ surrounded by a bright ring for larger NA. For layers that support a waveguide mode, in fact most emission is expected to occur at $|k_{||}|/k_0$ equal to the

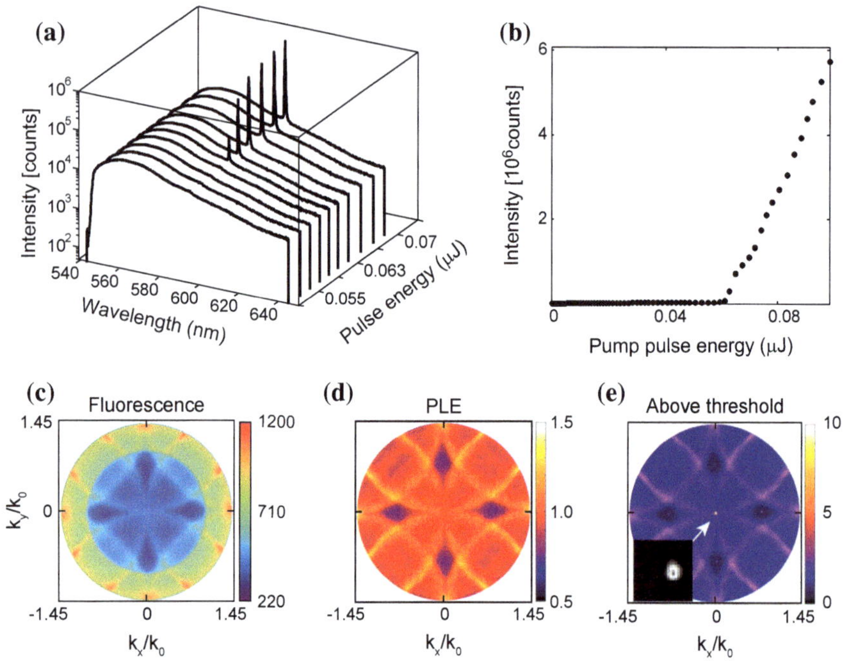

Fig. 8.2 **a** Emission spectra for various pump energies showing the typical emergence of a narrow laser line at the frequency of 2nd order Bragg diffraction of the gain-slab waveguide mode (pitch 380 nm). **b** Input-output curve plotting the integrated energy in just the spectral window of the laser peak versus pump power. **c** Panchromatic Fourier image just below threshold. (360 nm) pitch. **d**, **e** Photoluminescence enhancement for the same data, just *below* and just *above* threshold, obtained by dividing out the radial dependence for fluorescence in the same geometry without particles

waveguide mode index, which is by definition inaccessible to any objective. Due to the lattice periodicity, however, the waveguide dispersion folds back, causing the appearance of bright rings with radius of curvature 1.54, and centered at the reciprocal lattice vectors (or more precisely, at vectors $\mathbf{G}/k_0 = \lambda_0/d\ (m, n)$). Normalizing emission to that from an unpatterned slab, we find typical photoluminescence enhancements up to 1.5 times.

These enhancements are identical in mechanism to, but in magnitude far smaller than those, obtained by Lozano et al. [23], and Rodriguez et al. [34] in the framework of plasmon-lattice enhanced remote phosphors with optimized alignment of LSPR and diffractive waveguide coupling. As we cross the threshold, a bright and narrow beam emerges that has a donut profile. Polarization analysis evidences this lasing beam to be TE, i.e., s-polarized. From this we conclude that laser light is mainly outcoupled through the TE mode, which has a strong spatial overlap with the gain and with the particles. While the picture of band folding of the free photon/waveguide dispersion is generic to any periodic system, the distinct physics of the system is contained in fine features, such as the avoided crossings

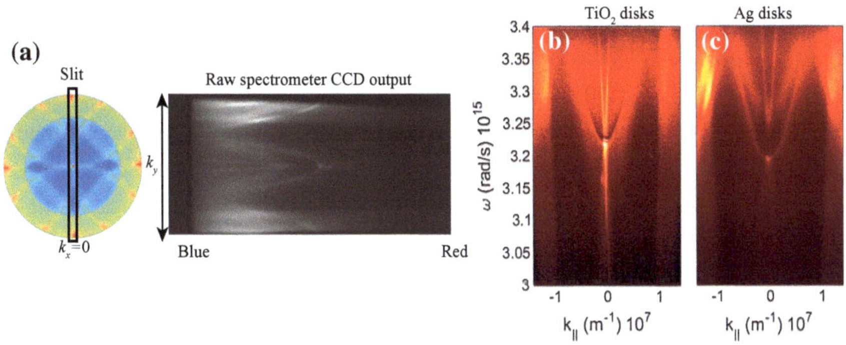

Fig. 8.3 **a** By relaying the Fourier image to the entrance slit of a spectrometer, one can disperse one slice of k-space over frequency. A band structure becomes apparent in the fluorescence. Panel **b** shows that for high-index dielectric particles (titiania disks of 150 nm diameter), the band structure hardly shows an avoided crossing. In contrast for silver disks (110 nm diameter) a distinct band anticrossing appears near the 2nd order Bragg condition (at 590 nm, set by the 380 nm pitch)

that are expected at Bragg conditions, i.e., precisely at the intersections of the free photon circles. Resolving such features requires spectral resolution at the same time as wave vector resolution. While results in Fig. 8.2 on a 2D CDD chip report the sum over all emission wavelengths, wavelength information can be obtained by dispersing one slice of k-space on a grating. Figure 8.3 show typical below-threshold spectrally resolved Fourier image. The folded waveguide dispersion appears as narrow bright features organized in four bands. Indeed, at second order Bragg diffraction in a square lattice four diffraction orders couple, namely diffraction along $2\pi/d(\pm 1, 0)$ and $2\pi/d(0, \pm 1)$ giving respectively straight lines crossing at the Γ-point, and two parabola. Due to strong scattering at the plasmon particles, these four bands display a splitting that strongly depends on the plasmonic nature of the particle. In particular we have found that even for very high-index dielectric particles (TiO_2, disks 150 nm in diameter were required [25] to obtain lasing) no observable stop gap is found. In contrast, for silver particles, stop gap widths up to 20 nm, or $\Delta\omega/\omega = 3\ \%$ occur. For Au particles, we have found only narrow stop gaps and high lasing thresholds, showing that more absorptive metals give weak scattering yet high loss.

8.3 Theory of Plasmon Lattices Coupled to Stratified Media

Now that the reader is familiarized with the general characteristics of 2D plasmon lattice lasers we turn to a theoretical model able to capture the main phenomena in an analytical framework. Starting from the nearly free photon approximation as intuition, we develop a point dipole model for the complex band structure.

8.3.1 Two-Dimensional Periodic Arrays, Folded Dispersion, and the "Nearly Free-Photon" Approximation

Plane waves are solutions of the wave-equation in infinite homogeneous media, governed by a conical dispersion $\omega = kc$, where $k = (k_x^2 + k_y^2)^{1/2}$ and c is the speed of light. If in the medium we introduce a periodic 2D array of identical particles, then coupling between the plane waves due to the particles will lead to new forms of solutions. These are termed Bloch or Floquet harmonics. The dispersion of these solutions is periodic in the (k_x, k_y) plane, the so-called reciprocal or momentum plane that is directly imaged by Fourier microscopy. For example, if the lattice in the physical domain is square with lattice constant d then in the k- space, the reciprocal lattice will also be square and with reciprocal lattice constant $2\pi/d$, as shown in Fig. 8.4a, b. The area defined by, $(k_x, k_y) \in [-\pi/\mathrm{d}, \pi/\mathrm{d}] \times [-\pi/\mathrm{d}, \pi/\mathrm{d}]$ in the (k_x, k_y) plane is termed the first Brillouin zone (BZ). In the limit of negligibly scattering particles the dispersion of the waves supported in the periodic array will consists of infinite replicas of the fundamental cone, as shown in Fig. 8.4 a. Also highlighted in Fig. 8.4 are two common cuts through this diagram. On one hand, Fig. 8.4b shows an equifrequency slice, similar to taking a constant energy (Fermi) surface in k-space for electrons. Figure 8.4c, instead, shows a cut through the dispersion relation at fixed $k_x = 0$, thus representing an ω versus k_y bandstructure.

If the interaction between the particles and the plane waves is weak but not negligible, then the new solutions can be described as a combination of just a few plane waves. In this case the new dispersion will be very close to the dispersion in case of negligibly scattering particles. Exceptions are frequency points where phase-matching between counter propagating plane waves takes place and thereby relatively efficient coupling between the plane waves occurs, namely, right at the

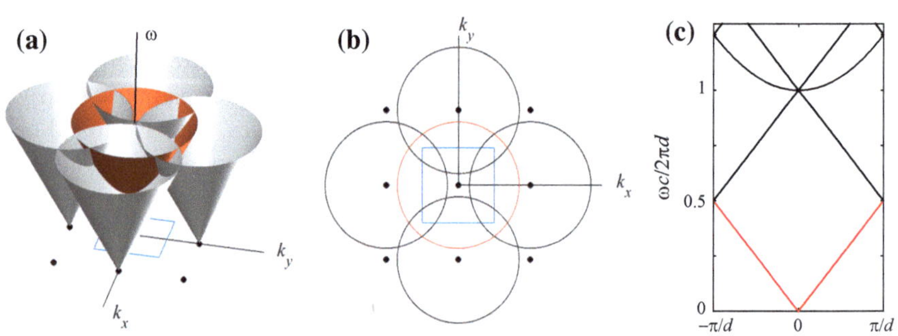

Fig. 8.4 Sketch of the nearly free photon dispersion in a 2D system **a** the conical dispesion relation $\omega = c|\mathrm{k}|$ is repeated every reciprocal lattice vector (*black dots*). The cone centered at $|\mathrm{k}| = 0$ is plotted in *orange*. **b** Constant frequency slice, showing the circular dispersion relation repeated every reciprocal lattice vector. The cut is taken at a frequency just below 2nd order Bragg diffraction—with increasing frequency the *circles grow*, and intersect at theorigin of k-space. **c** ω-k_y cut for $k_x = 0$, showing the well known band diagram in the first Brillouin zone (*blue box* in (**a**, **b**)

intersection points of different dispersion branches. In Fig. 8.4c this concerns the crossing of four cones at the origin in k-space, visible in the ω versus k_y slice as two intersecting lines, and a (degenerate) parabola. This points is the second order Bragg condition (the first being at the edge of the Brillouin zone). This description in terms of a folding of a dispersion relation is termed the "nearly free photon" approximation, and it is used to approximately describe the wave solutions in weak perturbation scenarios. This is similar also to the case of conventional dielectric-grating distributed feedback lasers. In this chapter, however, we study a rather different problem, namely that of a periodic lattice of plasmonic particles embedded in a stratified dielectric system that supports waveguide modes and provides gain. Plasmonic particles are strong scatterers, and therefore, although it may give the basic physics, the simplified, "nearly free photon" picture does not convey the full bandgap physics of such a system. In particular, distinct stop gaps are expected to open up. Indeed, if we compare the conceptual Fig. 8.4 with the measurements in Figs. 8.2 and 8.3 it is directly clear that there is a close correspondence, yet at the same time that theory must account for stop gaps, and the width of the bands.

8.3.2 Surface Lattice Resonances

Now we turn into a more detailed description of our system of interest, namely a particle lattice inside a 2D dielectric waveguide system. Consider a square lattice of plasmonic cylinders with diameter D and height H with lattice constant d positioned within a dielectric slab that supports a guided mode as illustrated in Fig. 8.5a (top and bottom). The guided slab mode wavenumbers k_m are related to the free-space wavelength λ_0 via the dispersion relation $k_m(\omega = 2\pi c/\lambda_0)$. A typical dispersion of the fundamental TE and TM modes supported by a slab with thickness of $h = 450$ nm that is made of relative dielectric constant $\epsilon_2 = 2.79$, and sandwiched between two half-spaces with relative dielectric constants $\epsilon_1 = 1$ and $\epsilon_3 = 2.25$ is shown in Fig. 8.5b. The periodic lattice exhibits Bragg resonances whenever the wavelength of the waveguide mode $\lambda_m = 2\pi/k_m$, rather than the vacuum wavelength, meets the Bragg condition

$$d = m\frac{\lambda_m}{2}, \; m = 1, 2, \ldots \tag{8.1}$$

Equation (8.1) describes the case of Bragg diffraction by rows of particles parallel to a Cartesian axis, for waves with wave vector perpendicular to it. By way of example, with $d = 400$ nm, a second-order ($m = 2$) Bragg resonance will take place at $\lambda^{TE}_{m=2} \approx 640$ nm for the TE mode with mode index around 1.55. A similar TM resonance will take place at a slightly shorter free-space wavelength, owing to the somewhat different mode index. Note that additional Bragg resonances will take

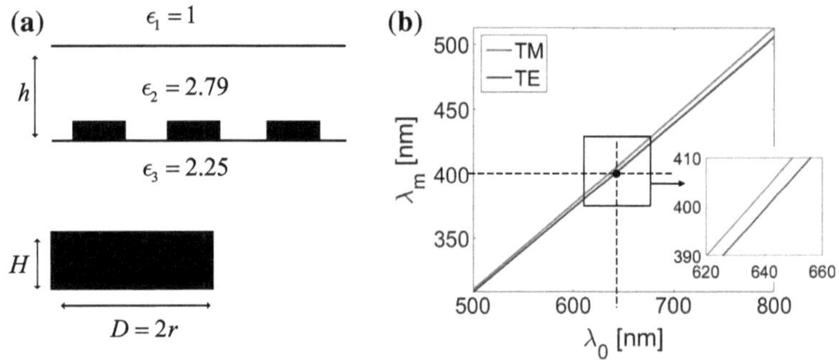

Fig. 8.5 **a** *Top* illustration of the slab and the plasmonic array. *Bottom* single plasmonic cylinder. **b** Typical dispersion of fundamental TE and TM modes in relatively thin slab

place for diffraction at differently oriented particle rows (i.e., other Miller indices *l, m*). As these fall outside the gain window of most reported laser studies, we ignore these.

8.3.3 Semi-analytical Approach: Polarizability and Lattice Sums

Ultimately, the lasing threshold for the Bloch modes that arise due to periodicity will be determined by the balance between the provided gain and the intrinsic loss. The net gain depends on the gain coefficient and the spatial overlap of the mode with the gain medium. The intrinsic propagation loss of the Bloch modes is due to Ohmic damping and radiative outcoupling. Therefore, dominant lasing will take place at or near frequencies where the lowest loss Bloch modes for feedback arise. In other words, to understand lasing we need to determine the dispersion relation of the plasmon-waveguide system, and seek the frequency regions in which the imaginary part of the complex dispersion (in the absence of gain) is minimal. Hence, in the following we calculate the passive array dispersion. We employ a dipolar model for the scatterers, taking into account Ohmic loss, as well as the coupling between the dipolar excitations in the array and the far field radiation. Thereby, we account for the fact that lasing will take place near the low loss points in the complex dispersion, while at the same time requiring that experimentally observable lasing needs outcoupling to radiation, and hence not a *fully* dark mode. Due to the strong scattering of the plasmonic particles, coupled mode theory and the "nearly free photon" assumption that are commonly used in the analysis of conventional dielectric distributed feedback lasers will lead to incomplete results for plasmonic arrays. Therefore we use the discrete dipole model that accounts for near as well as far field interactions between scatterers that are modelled as polarizable points that can carry a large dipole moment [35]. This method can give relatively accurate predictions particularly in the vicinity of the plasmonic particle resonance frequency. In this frequency range the single particle scattering behavior can be accurately modeled as dipolar, so that essential features are correctly captured.

Polarizability Model

The dipolar dynamics of each particle is encapsulated in its polarizability $\alpha(\omega)$. For a strongly and resonantly scattering particle in free space, such as in the case of a metal sphere described by a Drude model for its dielectric constant [36], the quasistatic polarizability is given by

$$\alpha_{\text{static}}(\omega) = \frac{V\omega_0^2}{\omega^2 - \omega_0^2 - i\omega\gamma} \tag{8.2}$$

(in CGS units, with ω angular frequency, ω_0 the particle resonance,γ an ohmic damping rate, and V an (effective) particle volume) To turn this static polarizability into that of a physical scatterer one must include radiation damping [35]

$$\frac{1}{\alpha} = \frac{1}{\alpha_{\text{static}}} - i\frac{2}{3}k^3 \tag{8.3}$$

with $k = n\omega/c$. As we are considering particles located inside an inhomogeneous medium, this "radiation correction" must be adapted, to account for the fact that both radiative damping and the resonance frequency are normalized. Moreover, although the particles we used in the experiments discussed in the following sections are electrically small they are not much smaller than the wavelength, thereby causing additional red-shifting in their resonance frequency due to phase retardation effects. While strictly speaking, the radiation damping term in (8.3) can be ab initio corrected using the Green's function of the stratified system [37, 38], here we adapt a more practical approach by fitting ω_0 and V in (8.2) and (8.3), to agree with full wave simulations of the scattering cross section shown in Fig. 8.6a below. Following this approximate approach, for metal disks of diameter $2r$ the fitted resonance frequency is given by $\omega_0 = 2\pi c/\lambda_{LSPR}$ with $\lambda_{LSPR} = 334 \times 10^{-9} + 3.6 \times 2r[m]$ and the damping rate by $\gamma \approx 0.05\omega_0$, valid for particle diameters in the range 40–110 nm.

Lattice Sum: Bloch Wave Assumption and Dipole Polarization

The structure is shift invariant in both x and y, and to derive a dispersion relation for the particle-waveguide system, we can hence assume that the dipolar excitation on the lattice has a Bloch form with dipolar moment

$$\mathbf{p}_{mn} = \mathbf{p}_{00}e^{id(nk_x + mk_y)} \tag{8.4}$$

where m, n denote the index real-space particle sites and (k_x, k_y) is the wavevector of the excited collective plasmonic mode. In this work we focus on polarization of $\mathbf{p}_{mn}$ only in the layer plane, containing components only in the x and y directions. This follows from the fact that the reported experiments work with flat silver disks,

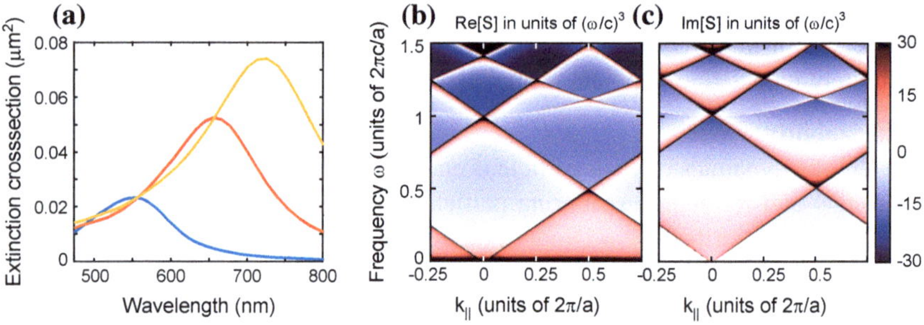

Fig. 8.6 **a** Extinction cross section of silver disks of 30 nm high and diameters 60, 90 and 110 nm, in SU8 (n = 1.65) on glass (n = 1.5) calculated using FDTD, assuming normal incidence from the glass side, and using tabulated optical data for silver [39], **b**, **c** Illustration of the lattice sum G_{xx} (summation in (8.5) for free space [35]. The real and imaginary part both show strong resonances exactly at the folded free dispersion, which gives rise to surface lattice resonances in the lattice polarizability

which are essentially only polarizable in the plane. Furthermore, second order Bragg resonance that corresponds to the square lattice pitch d has to take place in two orthogonal directions, parallel to the lattice primitive vectors.

Generally, the response of any particle in the lattice will be given by its polarizability and the field that it experiences due to incident field, plus the field of all the particles in the lattice. If we denote with $G(\omega, \mathbf{r}', \mathbf{r})$ the electric field dyadic Green function (meaning $G(\omega, \mathbf{r}', \mathbf{r}) \cdot \mathbf{p}$ specifies the field at $\mathbf{r}'$ due to a dipole moment $\mathbf{p}$ located at $\mathbf{r}$), the lattice response is

$$\mathbf{p}_{00} = \alpha \left[\mathbf{E}_{in} + \sum_{m,n \neq 0} G(\omega, \mathbf{r}_{00}, \mathbf{r}_{mn}) \mathbf{p}_{00} e^{id(mk_x + nk_y)} \right],$$

or in other words [35]

$$\mathbf{p}_{00} = \left[\alpha^{-1} - \sum_{m,n \neq 0} G(\omega, \mathbf{r}_{00}, \mathbf{r}_{mn}) e^{id(mk_x + nk_y)} \right]^{-1} \mathbf{E}_{in}.$$

In this expression, the summation of the single-dipole Green function over the real space lattice is known as *lattice sum*, and it contains all the physics of dipolar near- and far field coupling, to all multiple scattering orders. We note that if one looks for a dispersion relation, one considers the structure in absence of any driving, setting $\mathbf{E}_{in} = 0$ and looking for poles in the prefactor, which plays the role of a polarizability renormalized by interactions in the lattice. In our work we are interested in lasing near 2nd order Bragg diffraction only. In that case, since the x and y direction are equivalent, we can focus on the $k_x = 0$ slice of wave vector space, and can set the dipole polarization to $\hat{x}$. In this case, the modal matrix problem reduces to the simplified scalar equation

$$\Delta(\omega, k_x, k_y) = \alpha(\omega)^{-1} - \sum{}' G_{xx}(\omega, \mathbf{r}_{00}, \mathbf{r}_{mn})e^{id(mk_x+nk_y)}. \tag{8.5}$$

In (8.5) the symbol *primed-sum* $\sum{}'$ denotes summation over all indices except for $(m, n) = (0, 0)$, and G_{xx} is the xx component of the electric Green's function tensor.

Figure 8.6b shows G_{xx} for the case of a lattice in free space. Clearly, the lattice sum is strongly structured, with diffractive resonances occur right at the nearly free photon dispersion relation. Indeed, these are the surface lattice resonances on which lasing was reported by Zhou et al. [24], who operated at exact index matching between substrate, superstrate and gain slab. We also refer to Chap. 7 in this book for an overview of results one obtains in extinction and emission on basis of the same formalism.

Approximation of the Green's Function in Layered Media

Taking the full spectral content of the Green's function into account for the three-layer system of a substrate, a gain medium, and a superstrate, in the infinite summation in (8.5) is numerically challenging, and not essential. In order to simplify the analysis, we take the following physical considerations into account that are valid for systems in which the gain originates from a high index slab, as in our work [25]:

(a) Particle-particle interaction will mainly arise through the waveguide modes.
(b) As the TE and TM mode indices are very close in the organic gain systems typically studied [24–26, 33, 40] significant TE-TM coupling is expected.

Due to these physical reasons we may replace the full Green's function G_{xx} by its modal part, G^m_{xx} including both TE and TM mode contributions, i.e. $G^m_{xx} = G^{TE}_{xx} + G^{TM}_{xx}$, where the TE and TM contributions are separately given by

$$G^{TE}_{xx} = 2A_{TE}\left[H_0^{(1)}(k_{TE}\rho) + \frac{\partial^2_{x'}H_0^{(1)}(k_{TE}\rho)}{(k_{TE})^2}\right] \tag{8.6}$$

$$G^{TM}_{xx} = -2A_{TM}\left[\frac{\partial^2_{x'}H_0^{(1)}(k_{TM}\rho)}{k^2_{TM}}\right] \tag{8.7}$$

where $\rho = \sqrt{(x-x')^2 + (y-y')^2}$, and k_{TE}, k_{TM} are the wavenumbers in the transverse direction of the guided slab mode in the absence of the array, and are given by a solution of the corresponding mode transcendental equation. The amplitudes A_{TE}, A_{TM} are given by

$$A_X = \frac{k_0^3}{4\pi\epsilon_0}\frac{i}{2\eta_0}2\pi\xi_X g(z, z, \xi_X), \quad X = TE, TM \tag{8.8}$$

where $\xi_X = k_X/k_0$, and g is the 1D Green's function given in Appendix A. The infinite summation in (8.5) is slowly converging due to the inverse square root dependence of the Hankel function with respect to its argument. However, its convergence can be significantly accelerated by using the so called Ewald summation technique, adapted to the problem of interest. More details can be found in Appendix B. Note that taking nearest neighbor interactions only would be very inaccurate, since the propagators that couple the particles are dominated by slowly decaying slab modes that decay as $1/\sqrt{\rho}$. While far-field interactions are critical, since the particles are separated by roughly one guided wavelength from each other, the singular $1/r^3$, $1/r^2$ terms of the Green's function are practically irrelevant.

8.3.4 *Theoretical Model—Results*

Coupling of Collective Plasmon Resonances to Far-Field Radiation

The solution of (8.5) provides the complex dispersion of the collective plasmonic excitation of the array. In other words, at each frequency-wavenumber combination where (8.6) approaches zero, the lattice shows a strong response at weak, or zero, excitation. We therefore evaluate the complex dispersion relation by solving (8.6) at real frequencies, yet complex wave vector. The lowest loss mode correspond to zeros with the smallest imaginary part of the complex dispersion. In addition to evaluating the complex dispersion of Bloch modes, we also evaluate the outcoupling of Bloch modes. In analogy with conventional laser systems, a resonator based on perfect, rather than partially reflecting, mirrors will yield field amplification but no outcoupled laser light. In order to obtain the outcoupling efficiency for a given combination of real ω and k_y we invoke reciprocity. The excited dipolar moment p_{00} due to an impinging x-polarized plane wave with amplitude E_0 at ω and with $(0, k_y)$ is given by $p_{00} = E_0/\Delta(\omega, 0, k_y)$. By reciprocity, also the radiated field at given dipole moment p_{00} at ω and with $(0, k_y)$ will be proportional to $1/\Delta(\omega, 0, k_y)$. Figure 8.7 reports this quantity, i.e., the outcoupling at fixed dipole strength, as a function of real frequency ω and parallel momentum, as greyscale images. Here, white (black) represents efficient (poor) coupling. Panels (a)–(c) correspond to three frequency detuning cases $\lambda_B^{TE} < \lambda_{LSPR}$, $\lambda_B^{TE} \approx \lambda_{LSPR}$, and $\lambda_B^{TE} > \lambda_{LSPR}$, respectively, where λ_{LSPR} is the wavelength corresponds to the plasmonic particle resonance frequency, and λ_B^{TE} is the free space wavelength at which the 2nd order TE mode Bragg resonance takes place according to (8.1). The most notable features in Fig. 8.7 are that the repeated zone scheme dispersion becomes apparent as sharp features. However, clear stop gaps open up between bands. Moreover, it should be noted that some bands appear dark exactly at or near $k_y = 0$, indicating that outcoupling is forbidden, for instance by symmetry. Finally, the zero-detuning case shows a markedly distinct behavior.

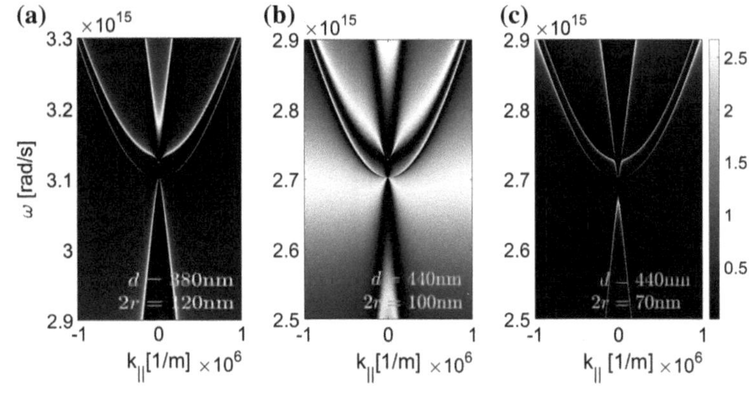

Fig. 8.7 Coupling between x-polarized dipolar excitation and x-polarized far-field. **a** $\lambda_B^{TE} < \lambda_{LSPR}$, **b** $\lambda_B^{TE} \approx \lambda_{LSPR}$, **c** $\lambda_B^{TE} > \lambda_{LSPR}$

Mixed TE-TM and the Effect of Resonance Detuning

In Fig. 8.8 we zoom-in on the frequency-wavenumber range where the TE-Bragg condition is satisfied and plot the outcoupling efficiency of Fig. 8.7 as grayscale alongside the complex dispersion of the collective plasmonic excitation, obtained as a solution of (8.5). The three panels, a b and c, of the figure correspond to the three detuning cases $\lambda_B^{TE} < \lambda_{LSPR}$, $\lambda_B^{TE} = \lambda_{LSPR}$, and $\lambda_B^{TE} > \lambda_{LSPR}$, respectively. The curves (blue only or blue and brown) in each figure represent the complex dispersion. Shown in the left (right) side of each of the panels is the dispersion of the imaginary (real) part of $k_{||} = k_y$ (recall $k_x = 0$). There are additional dispersion branches with much higher imaginary part that are not shown, as only the lowest-loss branches are important for lasing. In the left panels the frequencies that correspond to the TE and TM Bragg resonances are also marked with dashed lines. Distributed resonance and signal amplification in the presence of gain will take place at the points of minimum imaginary part. In all three detuning cases considered in Fig. 8.8 there are two frequencies at which the imaginary part of $k_{||}$ has a minimum. Having two, and not one, condition with minimal loss is a direct consequence of TE-TM coupling. If the two polarization families were completely decoupled, each would separately give rise to a stop band, where one of the two stop band edge would correspond to minimal loss (mode with nodes at the scatterers), and one would correspond to large loss (mode with antinodes at the scatterers). In the lasers studied in experiments, however, the TE and TM modes are very close (waveguide systems) or even identical (surface lattice resonance of Zhou [24]) in dispersion. Scattering by the particles can couple TE and TM modes with in-plane momentum at right angles to each other.

The minimal loss points occur at or near $k_{||} = 0$, corresponding to 'distributed' standing-wave resonances, and thereby will be amplified if sufficient gain is supplied. To observe lasing, at the same time outcoupling is required. For that, the real

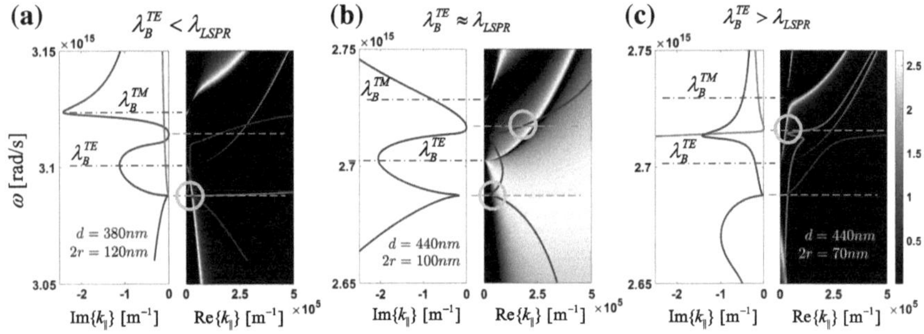

Fig. 8.8 Lasing versus detuning. **a** $\lambda_B^{TE} < \lambda_{LSPR}$. *Left* dispersion of imaginary part of k_y, distributed resonances takes place at the minimum points. *Right* dispersion of the real part of k_y. *Greyscale image* coupling between the excited mode to radiation. **b** as (**a**) but with $\lambda_B^{TE} = \lambda_{LSPR}$. **c** as (**a**) but with $\lambda_B^{TE} > \lambda_{LSPR}$

$k_{||}$ that corresponds to the point of minimal $\mathrm{Im}\{k_{||}\}$ should co-locate with strong outcoupling, i.e., a whiter region in the underlying grayscale map. Following this idea, in Fig. 8.4a, $\lambda_B^{TE} < \lambda_{LSPR}$, the only point where we have simultaneously a minimum imaginary part of $k_{||}$ and a corresponding real part of $k_{||}$ in a high coupling region in the greyscale map, is right above the TE Bragg condition. In the second case (8.4b), $\lambda_B^{TE} = \lambda_{LSPR}$, we find two points, one above and one below the TE Bragg condition, where we have low imaginary part of $k_{||}$ and simultaneously the real part of $k_{||}$ in a regime of good outcoupling. Finally, in the last case (8.8c), $\lambda_B^{TE} > \lambda_{LSPR}$, we see that laser output is expected only just below the TE Bragg condition. These observations implies that for different detuning we would expect to see lasing from different frequency regions, either only below, both above and below, or only above the TE Bragg resonance frequency.

8.3.5 Stop Gap and Band Crossing

With the theory in hand, we return to the magnitude and the topology of the band structure and lasing observable in recent experiments [24–26]. As the bandwidth of plasmon resonances is quite wide relative to the gain bandwidth even of organic dyes, it is not easy to measure band structures for all different detuning cases $\lambda_B^{TE} > \lambda_{LSPR}$ and $\lambda_B^{TE} < \lambda_{LSPR}$ without also changing the gain medium. To this end, we combine results from two sets of samples, one using Rh6G and a pitch of 380 nm to obtain lasing $\lambda_B^{TE} =$ 590 nm, well to the red of the plasmon resonance (near 700 nm for 110 nm diameter disks), and one using a different dye (Rh700) and pitch (460 nm) to obtain lasing at 710 nm, well to the red of the plasmon resonance (occurring at 650 nm, for smaller, 80 nm diameter disks). Figure 8.9 shows

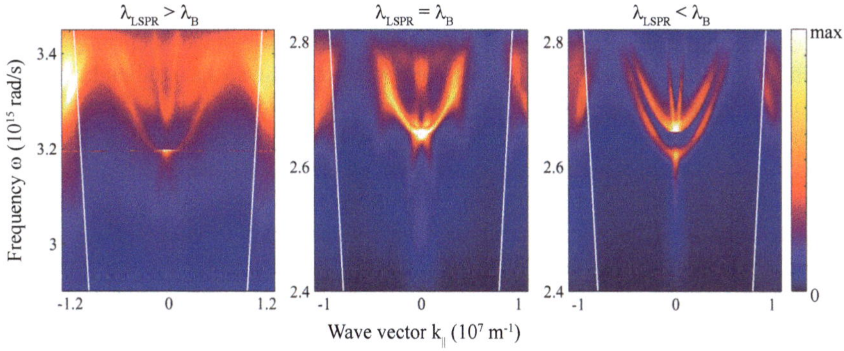

Fig. 8.9 Measured band diagrams for three distinct cases, namely when the plasmon resonance is red detuned from the lasing wavelength as given by the Bragg condition, when the two are on resonance, and when the localized plasmon resonance is *blue*-detuned from the lasing condition. The first case is achieved using Rh6G as gain medium, taking particles of 110 nm diameter at 380 nm pitch (590 nm lasing wavelength). The other two use a Rh6G-Rh700 dye mixture as gain medium and a pitch of 460 nm (lasing at 710 nm). The plasmon is tuned by working with large (117 nm diameter) resp. small (80 nm diameter) particles. These diagrams are taken just at (*left*) or *below* the lasing threshold. Lasing occurs on the lower band edge in the *left diagram* ($\omega = 3.19\ 10^{15}$ rad/s, note CCD over-exposure artefact), and on the upper band edge ($2.67\ 10^{15}$ rad/s) in the *right diagram*. Color scale is linear in intensity

representative measured band structures. Commensurate with the theory predictions, for these large detunings the band structure can be clearly interpreted as a perturbed free photon dispersion, with noticeable stop gaps. For the on-resonance case (replacing the 85 nm by 120 nm disks), the dispersion is markedly different. Importantly, we find that lasing occurs on the red edge of the measured stop gap in the case $\lambda_B^{TE} < \lambda_{LSPR}$, while it occurs on the blue edge in the case $\lambda_B^{TE} > \lambda_{LSPR}$. This stands in good agreement with the predictions of the dipole model.

8.4 Open Questions for Periodic Plasmon Lasers

In this section we provide a discussion of open questions, and crucial differences between the various experimental reports of plasmon array lasers [24, 25, 33]. The generic conclusion common to all works in this field is that plasmon particles allow distributed feedback lasing at thresholds of about 10 m J/cm^2, quite comparable to organic DFB lasers. Two limiting factors for this threshold are: first, only about 10 % of pump light is absorbed and, second, that feedback via multiple scattering suffers from absorption loss. Indeed, for Au particles, thresholds are about 20 fold-higher. As regards which plasmonic properties make these DFB lasers from their dielectric counterparts, a few distinct physical mechanisms have been proposed. First, the effect of stronger scattering per particle than in the case of dielectric systems is responsible for the large stop gap, which indicates that only few lattice spacings (in the order of 30) are needed for strong feedback [25]. Second, strong local field enhancement and Purcell factors have been proposed as facilitating lasing action [24, 26, 33]. Here we note that so far studies have been performed in two very distinct limits. On one hand, Schokker et al. [25] have studied lasing action in gain media with intrinsically high quantum efficiency (>90 %). For these cases it is expected that local field enhancement and Purcell factors have no strong role. Indeed, Purcell factors near plasmon particles are typically enhanced only in very small volumes, estimated to encompass no more than a few percent of the unit cell. A very recent stochastic superresolution-imaging map of LDOS in periodic plasmon arrays confirms this estimate [41]. For media with high internal quantum efficiency, this means that >95 % of fluorophores participate in lasing without actually enjoying particularly strong local field enhancements, yet benefitting from feedback by strong scattering. On the other hand, in the work of Suh and Zhou, [24, 33] very low internal quantum efficiency fluorophores were used (<10 %). Since Purcell factors can increase the internal efficiency of low efficiency fluorophores, lasing action is expected to occur preferentially on basis of gain medium close to the metal. Indeed, Suh et al. and Zhou et al. [24, 33] have performed ultrafast dynamics measurements to evidence that large Purcell factors of order 300 play an important role in bow-tie lattice lasers. To systematically elucidate the relative importance of scattering, plasmonic absorption, and Purcell enhancement, it would be very useful to perform measurements with systems in which the internal quantum efficiency of the gain medium could be tuned at will, or in which the spatial distribution of the pump beam *within* the unit cell can be controlled so as to preferentially excite emitters close to, or far from, the metal [42]. Moreover, from the photonic side, it would be ideal to perform measurements on a series of samples in which physical particle size is kept constant, yet polarizability and albedo are varied. Unfortunately, this is difficult to realize in practice. Finally, we note that in these systems the gain medium and laser output will show rich spatiotemporal dynamics. Dynamic modelling and measurements are hence required.

8.5 Scattering, Aperiodic and Finite Lasers

The fact that plasmon lattice lasers have comparatively wide stop gaps, means that the feedback needs only a few lattice constants (of order $(\Delta\omega/\omega)^{-1}$) as compared to their dielectric counterparts. On this basis, one would expect plasmon lattice lasers to be very robust against finite-size effects, random removal of particles, or shuffling of particles. Indeed, in our group we have found that oligomers of below 10 by 10 lattice constants already provide lasing output at the 2nd order Bragg diffraction condition. A particularly interesting finding [40] is that the lasing persists for extreme perturbations of the lattice in terms of random particle removal. We have studied structures for which as many as 95 % of particles were removed from the lattice, while the remaining 5 % were left in place. Remarkably, for this type of randomized plasmonic structure the threshold remains essentially unaffected when up to 80 % of particles are removed, while the slope efficiency is about twice better for lattices with disorder, than for lattices without disorder (Fig. 8.10). The lasing wavelength remains pinned to the 2nd order Bragg diffraction wavelength of the system. When even more particles are removed, the threshold does increase strongly, and laser output becomes weak.

One of the main reasons we identify for this robustness is that the planar waveguiding geometry is itself very amenable to amplification—indeed slab waveguides have been widely studied for Amplified Spontaneous Emission (ASE) sources [1]. Since scattering per particle is very strong compared to the scattering per unit cell in dielectric DFB lasers (taking the ratio in stop gap widths as measure, suggests a factor 50), it stands to reason that sufficient feedback remains when many scatterers can be removed. A requirement is that one keeps the remaining particles on the lattice sites, so that the Fourier transform of the lattice remains strongly peaked. An exciting idea for future experiments is to explore random lasing and “designer-disorder” lasing [3, 43–46]. Compared to other 2D systems, this lasing sample system is very easy to build and operate.

As an example highlighting this research direction, we have also studied lasing in a variety of quasiperiodic and aperiodic structures. Figure 8.11 shows lasing output in Fourier space for one such structure, a 2D Fibonacci lattice. This lattice is created from a square lattice (again, pitch 380 nm), by removing particle according to a deterministic, Fibonacci-sequence based scheme. The lasing wavelength remains as set by the 2nd order Bragg diffraction condition of the underlying lattice. The lasing output, however, now quantitatively matches the Fourier transform of the Fibonacci lattice. The interpretation is that this is exactly the pattern expected when considers diffraction of the $k_{||} = 0$ lasing beam by the quasiperiodic structure upon outcoupling. The practical importance of this example is that starting from the

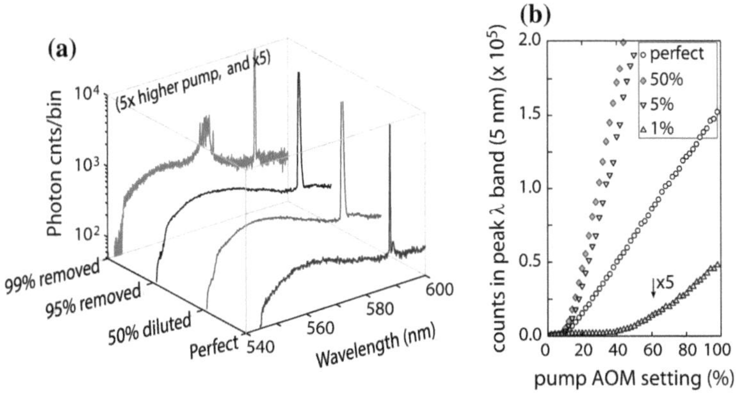

Fig. 8.10 **a** 2D plasmon arrays (pitch 380 nm, Rh6G gain medium) with particles randomly removed still show distinct lasing at the 2nd order Bragg condition (590 nm), even for extremely large dilution of the lattice. When just 1 % of the particles is left, lasing persists. In this regime also random-lasing type emission spikes appear **b** Input-output curves

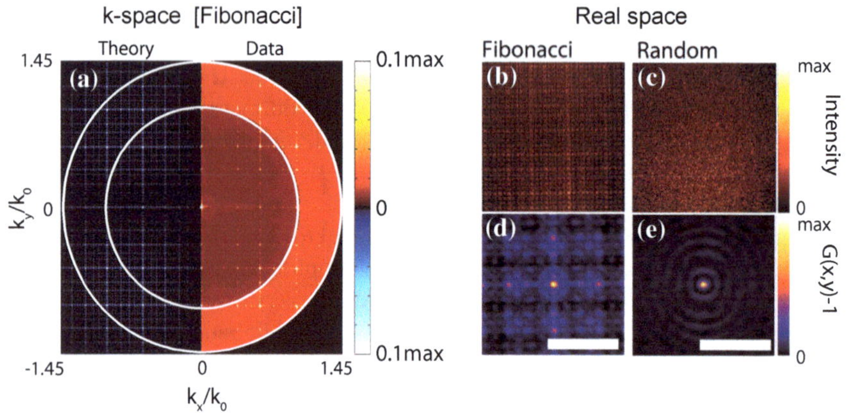

Fig. 8.11 **a** Fourier space output of a 2D Fibonacci plasmon array (d = 380 nm pitch, lasing at 590 nm, as does the underlying square lattice). The *left* part of this *panel* shows the structure factor (absolute value of the structures Fourier transform) of the array. The color scale is oversaturated. **b** Real space image of the sample plane *above* threshold for a Fibonacci lattice, and a randomized laser (50 % of particles randomly removed) showing *speckle* (20 μm field of view). Speckle is a direct evidence of spatial coherence (**d**, **e**) Autocorrelation of the speckle pattern (3 μm *scale bar*)

periodic lattice, one can impose amplitude masks (in this case the binary choice of keeping or removing partcles) or possibly also phase masks (varying particle resonance by size) directly in the plasmon structure that directly function as beam shaper. Further observations include the occurrence of many new lasing conditions at strong quasi-Bragg conditions, i.e., peaks in the structure factor.

Beyond the k-space output, in the field of random lasers and correlated disorder also real-space characteristics are important, or more concretely the spatial distribution and correlations in speckle. In our experiments, real space images of the laser structures invariably show a sharp transition from featureless, uncorrelated, Poisson noise for the incoherent fluorescence measured when pumped below threshold, to a distinct speckle pattern as threshold is crossed (Fig. 8.11b). For lattices disordered by random particle removal, this speckle pattern simply follows the laws established for uncorrelated scattering. These state that intensities follow Rayleigh statistics, and the spatial speckle autocorrelation has a sinc-squared shape of diffraction limited width [47]. Once the arrangement is correlated, however, as in the case of the Fibonacci structure, the autocorrelation shows a plethora of additional features. Aside from providing an additional handle on unraveling the optics of deterministic aperiodic structures, we note that this observation could also have a use. Indeed, several forms of 'speckle'-based microscopy are now developed [48, 49] where images are retrieved after illumination with speckle patterns, where the sharp autocorrelation of speckle essentially plays the role of point-spread function. Similar to traditional point-spread function engineering (for instance, by super-oscillatory lenses), deterministic aperiodic lattices provide control over the speckle autocorrelation.

8.6 Conclusions

This chapter has discussed experimental as well as theoretical aspects of plasmonic distributed feedback lasers. The discussion has been done to draw a direct comparison with dielectric gratings DFL, as well as between different approaches of plasmonic DFL. Plasmonic particles, particularly at their resonance are highly scattering elements thereby reducing significantly the finite array size required to establish significant 2nd order Bragg resonance. But this doesn't come without a price to pay in the form of larger losses compared with dielectric gratings structures. Therefore, the lasing threshold obtained with Ag particles (that are relatively low loss) is only similar to that achievable with organic lasers and not much weaker as one may expect just in light of their high scattering.

The interplay between the Bragg resonance frequency and the plasmonic resonance is a unique character of plasmonic DFL, that modifies the region in the band diagram where the lasing takes place. Another interesting consequence of the strong coupling between the plasmonic particles is that even when randomizing the array and removing 95 % of its particles lasing continues to take place with a moderate change in the lasing threshold.

Acknowledgements This work was supported by the Royal Dutch Academy of Sciences (KNAW), the U.S. Air Force Office of Scientific Research and the Welch Foundation with grant No. F-1802.

Appendix A: 1D Green's Function

First we define the normalized longitudinal wavenumbers $\zeta_i^X = \sqrt{\epsilon_{ri} - \xi_X^2}$, with $X = TE/TM$ and subject to the radiation condition $\mathrm{Im}\{\zeta_i^X\} \geq 0$. Then, the 1D Green's function used in (8.8) is given by

$$g(\omega, z, z') = \frac{1}{2}\frac{Z_2^X}{D_X}\left(e^{ik_z^X|z-z'|} + R_1^X e^{ik_z^X(2h-(z+z'))} + R_3^X e^{ik_z^X(z+z')} + R_1^X R_3^X e^{ik_z^X(2h-|z-z'|)}\right) \tag{8.9}$$

where h is the SU8 layer thickness and $k_z^X = k_0\zeta_2^X$, and

$$R_i^X = \frac{Z_i^X - Z_2^X}{Z_i^X + Z_2^X}, \quad i = 1, 3 \tag{8.10}$$

$$Z_i^{TM} = \eta_0 \frac{\zeta_1^{TM}}{\epsilon_{ri}}, \quad Z_i^{TE} = \frac{\eta_0}{\zeta_i^{TE}} \quad i = 1, 2, 3 \tag{8.11}$$

and

$$D_X = \frac{d}{d\xi}\left(1 - R_1^X R_3^X e^{2ik_0\zeta_2^X h}\right)\Big|_{\xi_X = k_X/k_0} \quad (8.12)$$

Appendix B: Ewald Summation

The convergence of the infinite summation in (8.5) can be significantly accelerated by using the Ewald summation technique [50–52]. First, we write

$$C(\omega, k_x, k_y) = 2A_{TE}\left(S(k_{TE}) + \frac{S_{xx}(k_{TE})}{k_{TE}^2}\right) - 2A_{TM}\frac{S_{xx}(k_{TM})}{k_{TM}^2} \quad (8.13)$$

with $k_{TE} = k_0\xi_{TE}$, and $k_{TM} = k_0\xi_{TM}$, and

$$S(k) = \lim_{x'y' \to 0} \sum{}' H_0^{(1)}(kR_{mn})e^{id(mk_x + n_k y)}, \quad (8.14)$$

$$S_{xx}(k) = \partial_{x'x'} S(k) \quad (8.15)$$

Where $R_{mn} = \sqrt{(x' - md)^2 + (y' - nd)^2}$. The primed summation sign in (8.14) is used to exclude the $(m, n) = (0, 0)$ term from the infinite two dimensional summation. The summation can also be written as

$$S(k) = \lim_{x'y' \to 0} \sum H_0^{(1)}(kR_{mn})e^{id(mk_x + n_k y)} - H_0^{(1)}(k\rho'), \quad (8.16)$$

where $\rho' = \sqrt{x'^2 + y'^2}$. The unprimed summation is used for infinite summation $(m, n) \in (-\infty, \infty) \times (-\infty, \infty)$. Next we replace the Hankel function by one of its integral representations

$$H_0^{(1)}(kR_{mn}) = -\frac{2i}{\pi}\int_0^\infty \frac{du}{u} e^{\left(k^2/4u^2 - R_{mn}^2 u^2\right)} \quad (8.17)$$

Note that since $R_{mn}^2 > 0$, and assuming that $k^2 > 0$, to formally guarantee convergence of the integral representation in (8.17), we have to require that u goes to infinity along the line $arg u = -\pi/4$. However, once we use this representation and derive an alternative, rapidly converging representation for the summation, we may apply Cauchy theorem and calculate the required integrals along a more convenient path. The semi-infinite integration path above is decomposed into two intervals, $0 \to E$, and $E \to \infty$, where E is an arbitrarily chosen constant picked as a trade-off between fast convergence of S_1 and S_2. We define

$$S_1 = \sum - \frac{2i}{\pi} \int_0^E \frac{du}{u} e^{\left(k^2/4u^2 - R_{mn}^2 u^2\right)} e^{id(mk_x + nk_y)} \tag{8.18}$$

$$S_2 = \sum{}' \frac{2i}{\pi} \int_E^\infty \frac{du}{u} e^{\left(k^2/4u^2 - R_{mn}^2 u^2\right)} e^{id(mk_x + nk_y)} \tag{8.19}$$

$$C = \frac{2i}{\pi} \int_0^E \frac{du}{u} e^{\left(k^2/4u^2 - \rho'^2 u^2\right)} \tag{8.20}$$

such that $S = S_1 + S_2 + C$. Note that as long as $E \gg k/2$, the integration in the summands of S_2 yields a Gaussian decay of the summands with respect to the summation indexes hence the summation over this part of the integral convergence rapidly. Similarly, the integration required to calculate C converges rapidly. The only issue left is the slow convergence of S_1 which is similar to the poor convergence of the original series. In this case, however, we are able to apply Poisson summation to accelerate the convergence. We obtain

$$S_1 = \frac{4i}{d^2} \sum_{p,q} \frac{e^{k_{zpq}^2/4E^2}}{k_{zpq}^2} \tag{8.21}$$

where $\mathbf{k}_{\rho pq} = (k_x, k_y) - 2\pi/d(p, q)$, and $k_{zpq}^2 = k^2 - \mathbf{k}_{\rho pq} \cdot \mathbf{k}_{\rho pq}$, $p, q \in (-\infty, \infty) \times (-\infty, \infty)$. The convergence of the summation S_1 in its new representation is Gaussian, therefore, practically only a few terms are required.

Finally, we have $S_{xx} = S_{1xx} + S_{2xx} + C_{xx}$ where

$$S_{1xx} = -\frac{4i}{d^2} \sum_{p,q} \frac{e^{k_{zpq}^2/4E^2}}{k_{zpq}^2} \left(k_x - \frac{2\pi}{d} p\right)^2 \tag{8.22}$$

$$S_{2xx} = \sum{}' \frac{4i}{\pi} \int_E^\infty du(1 - 2m^2 d^2 u^2) u \times e^{\left(k^2/4u^2 - R_{mn}^2 u^2\right)} e^{id(mk_x + nk_y)} \tag{8.23}$$

$$C_{xx} = -\frac{4i}{\pi} \int_0^E du u e^{\left(k^2/4u^2 - \rho'^2 u^2\right)} \tag{8.24}$$

References

1. A.E. Siegman, *Lasers* (University Science Books, 1986)
2. G.A. Turnbull, P. Andrew, M.J. Jory, W.L. Barnes, I.D.W. Samuel, Relationship between photonic band structure and emission characteristics of a polymer distributed feedback laser. Phys. Rev. B **64**, 125122 (2001)
3. D.S. Wiersma, The physics and applications of random lasers. Nat. Phys. **4**, 359–367 (2008)
4. I.D.W. Samuel, G.A. Turnbull, Organic semiconductor lasers. Chem. Rev. **107**, 1272 (2007)
5. G. Heliotis, R. Xia, G.A. Turnbull, P. Andrews, I.D.W. Barnes, D.D.C. Bradley, Emission characteristics and performance comparison of polyfluorene lasers with one- and two-dimensional distributed feedback. Adv. Funct. Mater. **14**, 91–97 (2004)
6. P. Del Carro, A. Camposeo, R. Stabile, E. Mele, L. Persano, R. Cingolani, D. Pisignano, Near-infrared imprinted distributed feedback lasers. Appl. Phys. Lett. **89**, 201105 (2006)
7. E. Mele, A. Camposeo, R. Stabile, P. Del Carro, F. Di Benedetto, L. Persano, R. Cingolani, D. Pisignano, Polymeric distributed feedback lasers by room-temperature nanoimprint lithography. Appl. Phys. Lett. **89**, 131109 (2006)
8. S.A. Maier, *Plasmonics: Fundamentals and Applications* (Springer, 2007)
9. G.M. Akselrod, C. Argyropoulos, T.B. Hoang, C. Cirac, C. Fang, J. Huang, D.R. Smith, M.H. Mikkelsen, Probing the mechanisms of large Purcell enhancement in plasmonic nanoantennas. Nat. Photonics **8**, 835 (2014)
10. C. Belacel, B. Habert, F. Bigourdan, F. Marquier, J.-P. Hugonin, S. Michaelis de Vasconcellos, X. Lafosse, L. Coolen, C. Schwob, C. Javaux, B. Dubertret, J.-J. Greffet, P. Senellart, A. Maitre, Controlling spontaneous emission with plasmonic optical patch antennas. Nano Lett. **13**, 1516 (2013)
11. E.J.R. Vesseur, J.F. Garcia de Abajo, A. Polman, Broadband purcell enhancement in plasmonic ring cavities. Phys. Rev. B **82**, 165419 (2010)
12. D.J. Bergman, M.I. Stockman, Surface plasmon amplification by stimulated emission of radiation: quantum generation of coherent plasmons in nanosystems. Phys. Rev. Lett. **90**, 027402 (2003)
13. M.T. Hill, Y.-S. Oei, B. Smalbrugge, Y. Zhu, T. De Vries, P.J. Van Veldhoven, F.W.M. Van Otten, T.J. Eijkemans, J.P. Turkiewicz, H. De Waardt, E.J. Geluk, S.-H. Kwon, Y.-H. Lee, R. Notzel, M.K. Smit, Lasing in metallic-coated nanocavities. Nat. Photonics **1**, 589 (2007)
14. M.T. Hill, M. Marell, E.S.P. Leong, B. Smalbrugge, Y. Zhu, M. Sun, P.J. van Veldhoven, E. J. Geluk, F. Karouta, Y.-S. Oei, R. Notzel, C.-Z. Ning, M.K. Smit, Lasing in metal-insulator-metal sub-wavelength plasmonic waveguides. Opt. Express **17**, 11107 (2009)
15. M. Khajavikhan, A. Simic, M. Katz, J.H. Lee, B. Slutsky, A. Mizrahi, V. Lomakin, Y. Fainman, Thresholdless nanoscale coaxial lasers. Nature **482**, 204 (2012)
16. P. Berini, I. De Leon, Surface plasmon-polariton amplifiers and lasers. Nat. Photonics **6**, 16 (2012)
17. S. Murai, M.A. Verschuuren, G. Lozano, G. Pirruccio, S.R.K. Rodriguez, J.G. Rivas, Hybrid plasmonic-photonic modes in diffractive arrays of nanoparticles coupled to light-emitting optical waveguides. Opt. Express **21**, 4250 (2013)
18. G. Vecchi, V. Giannini, J.G. Rivas, Shaping the fluorescent emission by lattice resonances in plasmonic crystals of nanoantennas. Phys. Rev. Lett. **102**, 146807 (2009)
19. A.G. Nikitin, A.V. Kabashin, H. Dallaporta, Plasmonic resonances in diffractive arrays of gold nanoantennas: near and far field effects. Opt. Express **20**, 27941 (2012)
20. P.L. Stiles, J.A. Dieringer, N.C. Shah, R.R. Van Duyne, Surface-enhanced Raman spectroscopy. Annu. Rev. Anal. Chem. **1**, 601 (2008)
21. M.E. Stewart, C.R. Anderton, L.B. Thompson, J. Maria, S.K. Gray, J.A. Rogers, R.G. Nuzzo, Nanostructured plasmonic sensors. Chem. Rev. **108**, 494 (2008)
22. J.N. Anker, W.P. Hall, O. Lyandres, N.C. Shah, J. Zhao, R.P. Van Duyne, Biosensing with plasmonic nanosensors. Nat. Mater. **7**, 442 (2008)

23. G. Lozano, D.J. Louwers, S.R.K. Rodriguez, S. Murai, O.T.A. Jansen, M.A. Verschuuren, J. G. Rivas, Plasmonics for solid-state lighting: enhanced excitation and directional emission of highly efficient light sources. Light Sci. Appl. **2**, e66 (2013)
24. W. Zhou, M. Dridi, J.Y. Suh, C.H. Kim, D.T. Co, M.R. Wasielewski, G.C. Schatz, T.W. Odom, Lasing action in strongly coupled plasmonic nanocavity arrays. Nat. Nanotech. **8**, 506–511 (2013)
25. A.H. Schokker, A.F. Koenderink, Lasing at the band edges of plasmonic lattices. Phys. Rev. B **90**(155452), 1–10 (2014)
26. A. Yang, T.B. Hoang, M. Dridi, C. Deeb, M.H. Mikkelsen, G.C. Schatz, T.W. Odom, Real-time tunable lasing from plasmonic nanocavity arrays. Nat. Commun. **6**, 1–7 (2015)
27. K.W.-K. Shung, Y.C. Tsai, Surface effects and band measurements in photonic crystals. Phys. Rev. B **48**, 11265
28. H. Kogelnik, C. Shank, Stimulated emission in a periodic structure. Appl. Phys. Lett. **18**, 152 (1971)
29. I. Sersic, C. Tuambilangana, A.F. Koenderink, Fourier microscopy of single plasmonic scatterers. New J. Phys. **13**(083019), 1–14 (2011)
30. Y. Fontana, G. Grzela, E.P.A.M. Bakkers, G.J. Rivas, Mapping the directional emission of quasi-two-dimensional photonic crystals of semiconductor nanowires using Fourier microscopy. Phys. Rev. B **86**, 245303 (2012)
31. J.A. Kurvits, M. Jiang, R. Zia, Comparative analysis of imaging configurations and objectives for Fourier microscopy. J. Opt. Soc. Am. A **32**, 2082–2092 (2015)
32. T.H. Taminiau, S. Karaveli, N.F. van Hulst, R. Zia, Quantifying the magnetic nature of light emission. Nat. Commun. **3**, 979 (2012)
33. J.Y. Suh, C.H. Kim, W. Zhou, M.D. Huntington, D.T. Co, M.R. Wasielewski, T.W. Odom, Plasmonic bowtie nanolaser arrays. Nano Lett. **12**, 5769–5774 (2012)
34. S.R.K. Rodriguez, S. Murai, M.A. Verschuuren, G.J. Rivas, Light-emitting waveguide plasmon polaritons. Phys. Rev. Lett. **109**, 166803 (2012)
35. F.J. García de Abajo, Colloquium: light scattering by particle and hole arrays. Rev. Mod. Phys. **79**, 1267 (2007)
36. M. Agio, A. Alù, *Optical Antennas* (Cambridge University Press, 2013)
37. L. Novotny, B. Hecht, *Principles of Nano-Optics* (Cambridge University Press, 2006)
38. Y. Hadad, B. Z. Steinberg, Electrodynamic synergy of micro-properties and macro-structure in particle arrays, 2010 URSI international symposium on electromagnetic theory. Berlin, Germany (2010)
39. M.J. Weber, *Handbook of Optical Constants of Materials* (CRC Press, 2003)
40. A.H. Schokker, A.F. Koenderink, Statistics of randomized plasmonic lattice lasers. ACS Photonics **2**, 1289–1297 (2015)
41. K. Guo, M.A. Verschuuren, A.F. Koenderink, Superresolution imaging of local density of states in plasmon lattices. Optica (2016)
42. G. Pirruccio, M. Ramezani, S.R.K. Rodriguez, Coherent control of the optical absorption in a plasmonic lattice coupled to a luminescent layer. Phys. Rev. Lett. (2016)
43. M.G. Conley, M. Burresi, F. Pratesi, K. Vynck, D.S. Wiersma, Light transport and localization in two-dimensional correlated disorder. Phys. Rev. Lett. **112**, 143901 (2014)
44. N. Bachelard, J. Andreasen, S. Gigan, P. Sebbah, Taming random lasers through active spatial control of the pump. Phys. Rev. Lett. **109**, 033903 (2012)
45. H. Noh, J.K.-K. Yang, S.V. Boriskina, M.J. Rooks, G.S. Solomon, L. Dal Negro, H. Cao, Lasing in Thue-Morse structures with optimized aperiodicity. Appl. Phys. Lett. **98**, 201109 (2011)
46. L. Dal Negro, S.V. Boriskina, Deterministic aperiodic nanostructures for photonics and plasmonics applications. Laser Photon. Rev. **6**, 178–218 (2011)
47. Dainty, J. C., editor: *Laser Speckle and Related Phenomena*. Topics in Applied Physics. Springer (1975)

48. E. Mudry, K. Belkebir, J. Girard, J. Savatier, E. Le Moal, C. Nicoletti, M. Allain, A. Sentenac, Structured illumination microscopy using unknown speckle patterns. Nat. Photonics **5**, 312–315 (2012)
49. H. Yilmaz, E.G. van Putten, J. Bertolotti, A. Lagendijk, W.L. Vos, A.P. Mosk, Speckle correlation resolution enhancement of wide-field fluorescence imaging. Optica **5**, 424–429 (2015)
50. F. Capolino, D.R. Wilton, W.A. Johnson, Efficient computation of the 2-D green's function for 1-D periodic structures using the Ewald method. IEEE Trans. Ant. Prop. **53**(9), 2977–2984 (2005)
51. K.E. Jordan, G.R. Richter, P. Sheng, An efficient numerical evaluation of the green's function for the Helmholtz operator on periodic structure. J. Comput. Phys. **63**, 222–235 (1986)
52. S. Steshenko, F. Capolino, P. Alitalo, S. Tretyakov, Effective model and investigation of the near field enhancement and subwavelength imaging properties of multilayer arrays of plasmonic nanospheres. Phys. Rev. E **84**, 016607 (2011)

Chapter 9
Surface Plasmon Enhanced Schottky Detectors

Pierre Berini

Abstract Surface plasmon Schottky detectors combine a structured metal contact that supports surface plasmons with a semiconductor, forming a rectifying metal-semiconductor junction. Internal photoemission occurs in such junctions via the excitation of hot carriers in the metal due to the absorption of surface plasmons therein, leading to photocurrent collected in the semiconductor. The cut-off wavelength of such detectors is determined by the Schottky barrier height, enabling detection below the bandgap of the semiconductor. The metal contact can be structured as a waveguide, grating or antenna on which surface plasmons are supported. Surface plasmon sub-wavelength confinement and field enhancement lead to significant enhancement of the internal photoelectric effect. The operating principles behind surface plasmon detectors based on internal photoemission are reviewed, the literature on the topic is surveyed, and avenues that appear promising are highlighted.

9.1 Introduction

A surface plasmon-polariton (SPP) is a transverse-magnetic surface wave that propagates along the interface between a metal and a dielectric at optical wavelengths, as a coupled excitation formed from electromagnetic fields coupled to a charge density wave [1]. SPPs are supported on a variety of metal-dielectric structures, including planar arrangements of dielectric and metal films [2], metal gratings [3], and metal nanoparticles such as spheres, islands, rods [4, 5], or nanostructured resonators (antennas) [6]. SPPs are also involved in optical transmission through one or many sub-wavelength holes in a metal film [7]. SPPs have interesting and useful attributes such as sub-wavelength confinement, energy

P. Berini (✉)
Department of Physics, School of Electrical Engineering and Computer Science,
Centre for Research in Photonics, University of Ottawa, 25 Templeton St., Ottawa
ON K1N6N5, Canada
e-mail: berini@eecs.uottawa.ca

S.I. Bozhevolnyi et al. (eds.), *Quantum Plasmonics*,
Springer Series in Solid-State Sciences 185,
DOI 10.1007/978-3-319-45820-5_9

asymptotes in dispersion curves and high surface and bulk sensitivities, and they can resonate on tiny metallic structures. These attributes and the variety of metallic structures that support SPPs have led to applications in several areas, such as nanophotonics [8], biosensing [9], integrated optics [10], and lasers [11].

An SPP photodetector is a device capable of detecting SPPs or involving SPPs in the photodetection process. Such detectors typically combine photonic detection based on electron-hole pair creation or internal photoemission with a metal structure that supports SPPs. SPP detector architectures are highly varied, following the diversity of detection approaches and detection materials available, and the diversity of the metallic structures that support SPPs [1–8]. The attributes of SPPs are exploited to impart useful properties to photodetectors, such as angular, spectral and/or polarisation selectivity, or to improve their performance in terms of speed, signal-to-noise or photoresponse. The use of SPPs can also lead to small detectors, having dimensions comparable to those of highly integrated electronic elements. Several applications are targeted including high-speed or low-noise detection, near- and mid-infrared imaging, single-plasmon detection, (bio)chemical sensing and photovoltaic solar energy conversion.

Progress on SPP detectors has been rapid, and interest on the topic is vigorous [12]. The objectives of this chapter are to describe the operating principles underpinning SPP detectors, describe examples of detectors from the literature, and highlight avenues that appear promising for future investigation. The properties of propagating SPPs and general photodetection principles are discussed initially, followed by a review of some SPP detector types, with an emphasis on devices exploiting internal photoemission. Avenues that appear promising in terms of performance and application are highlighted and summarised in the last section.

9.2 SPPs and Photodetection Mechanisms

An $e^{+j\omega t}$ time-harmonic dependence is assumed throughout. The relative permittivity is denoted ε_r, and is written for a metal in terms of real and imaginary parts as $\varepsilon_{r,m} = -\varepsilon_R - j\varepsilon_I$. $\boldsymbol{k}$ denotes a generic wavevector and k its magnitude. $k_0 = 2\pi/\lambda_0 = \omega/c_0$ is the wavenumber of plane waves in vacuum, λ_0 the wavelength in vacuum, c_0 the speed of light in vacuum and $\omega = 2\pi\nu$ is the angular frequency.

Metals are dispersive at optical wavelengths. Away from interband transitions, the Drude model for the permittivity captures this character [1]:

$$\varepsilon_{r,m} = -\varepsilon_R - j\varepsilon_I = 1 - \frac{\omega_p^2}{\omega^2 + 1/\tau^2} - j\frac{\omega_p^2/\tau}{\omega(\omega^2 + 1/\tau^2)} \tag{9.1}$$

In the above, ω_p is the plasma frequency and τ the relaxation time. The "Drude region" corresponds to the portion of the electromagnetic spectrum where (9.1) holds, when applied to model the measured relative permittivity of metals. For many metals, this region spans the range from visible wavelengths to the infra-red. The metal approaches a perfect electric conductor as the wavelength increases through the infra-red and beyond.

The simplest planar structure supporting propagating SPPs is an interface between an optically semi-infinite dielectric and a semi-infinite metal (termed the *single-interface*). A good metal for supporting SPPs satisfies $\varepsilon_R \gg \varepsilon_I$. The single-interface supports one purely bound (non-radiative) SPP mode. This mode is transverse magnetic (TM) and may propagate at any angle in the plane. The SPP fields are confined along the perpendicular direction, peaking at the interface and decaying exponentially into both media. The field penetration depth in the metal is much smaller than the field penetration depth in the dielectric. Confinement of the SPP arises because the metal and dielectric have $\mathrm{Re}\{\varepsilon_r\}$ of opposite sign at the wavelength of operation (indeed over a large wavelength range).

The wavenumber of the single-interface SPP is [1]:

$$k_{SPP} = k_0 \left(\frac{\varepsilon_{r,m}\varepsilon_{r,d}}{\varepsilon_{r,m} + \varepsilon_{r,d}} \right)^{1/2} \tag{9.2}$$

For a lossless dielectric cladding ($\mathrm{Im}\{\varepsilon_{r,d}\} = 0$), the above simplifies to the following approximate expressions for the real (phase) and imaginary (attenuation) parts [1]:

$$k'_{SPP} \cong k_0 \left(\frac{\varepsilon_R\varepsilon_{r,d}}{\varepsilon_R - \varepsilon_{r,d}} \right)^{1/2} \quad \text{and} \quad k''_{SPP} \cong k_0 \frac{\varepsilon_I}{2\varepsilon_R^2} \left(\frac{\varepsilon_{r,d}\varepsilon_R}{\varepsilon_R - \varepsilon_{r,d}} \right)^{3/2} \tag{9.3}$$

Figure 9.1a shows a sketch of a Schottky photodiode on n-doped Si (n-Si) [13], used here as an example to describe two generic photon detection mechanisms that can be adapted to the detection of SPPs. A rectifying Schottky contact is formed at the abrupt interface between the top metal and the semiconductor, and an Ohmic (non-rectifying) contact is formed at the interface between the bottom metal and the heavily-doped body. The structure is reverse biased for $V_b < 0$. The complementary structure can also be used, formed by exchanging the dopants (n $\leftrightarrow$ p, $\mathrm{n}^+ \leftrightarrow \mathrm{p}^+$).

Figure 9.1a shows two mechanisms that can be used for photodetection. The first consists of the creation of electron-hole pairs (EHPs) in the semiconductor due to absorption therein of incident radiation of energy $h\nu$ greater than the bandgap energy of the semiconductor ($h\nu > E_g$). This mechanism, sketched in Fig. 9.1a and labeled EHP, involves 3 steps: (i) optical absorption in the semiconductor creating EHPs, (ii) separation of EHPs and transport across the absorption region (with or without gain) under reverse bias, and (iii) collection of EHPs into photocurrent at the device contacts.

The second mechanism consists of the internal photoemission (IPE) of hot carriers created in the metal due to absorption therein of radiation of energy $h\nu$. This mechanism, labeled IPE in Fig. 9.1a, is described in greater detail via the energy band diagram of Fig. 9.1b. IPE is a 3-step process consisting of: the

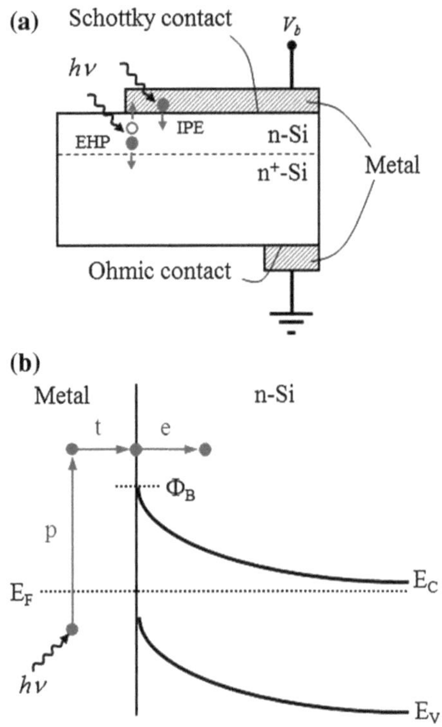

Fig. 9.1 **a** Schottky diode on n-Si illuminated by light of photon energy $h\nu$. Reverse biasing ($V_b < 0$) is assumed. Electrons are depicted by *filled circles* and holes by *unfilled ones*. **b** Energy band diagram of a Schottky contact on n-Si and the 3-step internal photoemission process: p—photoexcitation, t—transport, e—emission. E_C and E_V are the conduction and valence band edges, respectively, E_F is the Fermi level and Φ_B is the Schottky barrier height

photo-excitation of hot (energetic) carriers in the metal by optical absorption (p), the transport and scattering of hot carriers toward the Schottky contact (t), and the emission of hot carriers over the Schottky barrier into the semiconductor (e) where they are collected under a reverse bias as the photocurrent. This mechanism requires that $h\nu$ be greater than the Schottky barrier energy Φ_B ($h\nu > \Phi_B$). IPE is useful for detection at energies below the bandgap of the semiconductor ($\Phi_B < h\nu < E_g$), and is of particular interest to extend the detection bandwidth of technologically important materials such as Si [14]. However, the creation of EHPs in the semiconductor is more efficient and dominates over IPE when $h\nu > E_g$.

The internal quantum efficiency η_i (internal photo-yield) is a useful measure to characterise and optimise the detection mechanism. It is defined as the number of carriers that contribute to the photocurrent I_p per absorbed photon per second:

$$\eta_i = \frac{I_p/q}{S_{abs}/h\nu} \tag{9.4}$$

where S_{abs} is the absorbed optical power leading to the photocurrent and q is the elemental charge. $\eta_i \sim 1$ for detection via the creation of EHPs in high-quality defect-free direct bandgap semiconductors (assuming full absorption in the semiconductor only). In the case of detection via IPE, assuming absorption in the metal only and along the Schottky contact, η_i is given by [15]:

$$\eta_i = \frac{1}{2}\left(1 - \sqrt{\frac{\Phi_B}{h\nu}}\right)^2 \tag{9.5}$$

Typical Schottky barrier heights are, e.g., $\Phi_B = 0.34$, 0.8, 0.58, and 0.72 eV for Au/p-Si, Au/n-Si, Al/p-Si, and Al/n-Si, respectively [13]. Metal silicides can also be used on Si, providing lower Schottky barriers. For detection at, e.g., $\lambda_0 = 1310$ nm, η_i ranges from ~0.3 to 9 %.

The external quantum efficiency η_e (external photo-yield) and the responsivity R describe how well the detector performs when inserted into a system. η_e is defined similarly to η_i except that it depends on the incident optical power S_{inc}:

$$\eta_e = \frac{I_p/q}{S_{inc}/h\nu} \tag{9.6}$$

η_e and η_i are related by:

$$\eta_e = A\eta_i \tag{9.7}$$

where A is the optical absorptance contributing to the photocurrent, defined as:

$$A = \frac{S_{abs}}{S_{inc}} \tag{9.8}$$

The responsivity R is given by the ratio of the photocurrent to the incident optical power, and can be expressed in terms of η_e and η_i:

$$R = \frac{I_p}{S_{inc}} = \frac{\eta_e q}{h\nu} = \frac{A\eta_i q}{h\nu} \tag{9.9}$$

SPP detectors typically combine a metallic structure that supports SPPs with a detector structure such as that shown in Fig. 9.1a. Indeed, the Schottky diode is a convenient structure with which to detect SPPs because SPPs may propagate along the surfaces of the Schottky contact, leading to detection via IPE or the creation of EHPs. IPE is similar to the photo-electric effect, except that the former occurs at the interface between two materials rather than at the metal-vacuum interface. It has long been known that the photo-electric effect can be enhanced by SPPs [16], suggesting that IPE can be similarly enhanced—this has indeed been observed in several structures as discussed in the following sections.

9.3 Grating Detectors

The first types of SPP detectors investigated consisted of prism couplers on a suitable photodetector structure to increase the in-plane momentum of an incident beam in order to match that of SPPs propagating along the detection region [17, 18]. Both the Otto [17] and Kretschmann [18] coupling configurations have been investigated, although the former has been more popular in experimental work because the detection structure can be fabricated independently, then the prism aligned to the detector in the set-up. A grating can be used as an alternative to a prism to increase the in-plane momentum of the incident beam [1], offering manufacturing advantages and compactness over prisms, and ease of use due to simpler alignment.

Corrugated metal surfaces on a metal-semiconductor (Schottky) structure have been investigated, with compelling performance characteristics being reported [19–21]. A corrugated detector combines a semiconductor structure with a corrugated metal grating designed to couple incoming optical waves to SPPs that propagate along the detection region. Two such structures, operating on the basis of IPE, are shown in Figs. 9.2 and 9.3 [19, 21]. Such structures have been used to increase the absorptance in the metal contact of Schottky detectors based on IPE (for sub-bandgap detection) leading to increased responsivity. For example, a 30× increase in responsivity was observed (Fig. 9.2), for a Au/p-InP Schottky detector by exciting SPPs along the air/Au interface at λ_0 = 1150 nm via an integrated corrugated grating (inset of Fig. 9.2), relative to the same structure without the grating [19].

Figure 9.3a shows a cross-sectional sketch of a Au/p-Si SPP photodetector [21]. The structure comprises a Au patch of thickness t on p-Si, with a Au grating consisting of rectangular ridges of width W and thickness H arranged periodically in pitch Λ. The device is illuminated from the top with polarised infrared light

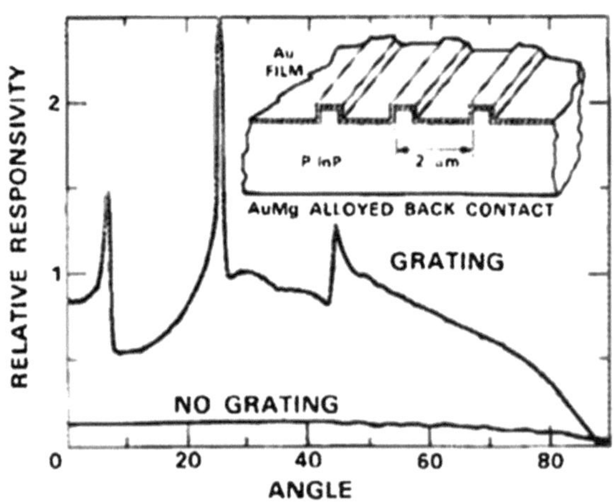

Fig. 9.2 (Adapted from [19]. Copyright (1985) by the American Institute of Physics.) Measured responsivity for a Au/p-InP Schottky detector at λ_0 = 1150 nm with and without a grating coupler. The grating coupler excites SPPs along the air/Au interface at angles that correspond to grating orders

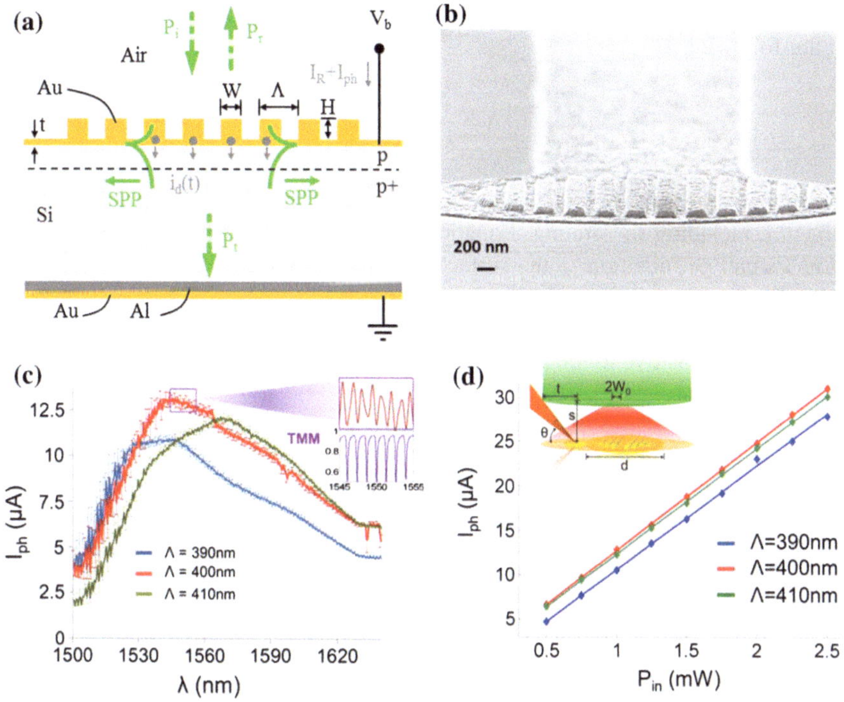

Fig. 9.3 (Adapted from [21]. Copyright (2016) by the Optical Society of America.) **a** Cross-sectional sketch of a Au/p-Si surface plasmon grating photodetector. The structure comprises a Au patch of thickness *t* on p-Si, with a Au grating consisting of *rectangular ridges* of width *W* and thickness *H* arranged periodically in pitch Λ. The device is illuminated from the *top* with polarised infrared light perpendicular to the grating. **b** SEM image of a fabricated structure. **c** Measured photocurrent response of three grating photodetectors; $V_B = -100$ mV, $\Lambda = 390$, 400 and 410 nm, duty cycle of 62 % and patch diameter of 25 μm. A 4-period moving average is plotted on each response as the *bold curve*. The rapid wavelength variations correspond to Fabry-Perot resonances—the *upper right inset* is an enlarged response showing such resonances and the *lower right inset* shows a response calculated using the TMM method. **d** Photocurrent generated by grating detectors versus incident power P_{in} measured for three different pitches: $\Lambda =$ 390 nm ($\lambda_0 = 1537$ nm), $\Lambda = 400$ nm ($\lambda_0 = 1548$ nm) and $\Lambda = 410$ nm ($\lambda_0 = 1570$ nm). A linear fit is applied to the data and the slope corresponds to the responsivity of the devices. *Inset* shows the experimental arrangement

perpendicular to the grating. The grating coupling condition for SPPs on a metal surface is given by:

$$k_{spp} = k_0 \sin(\theta_i) + M \frac{2\pi}{\Lambda} \tag{9.10}$$

where k_{spp} is the wave number of the SPP propagating in the plane of the metal surface, k_0 is the wave number of the incident light, θ_i is the angle of incidence of the light, M is the grating order and Λ is the grating pitch. Setting $\theta_i = 0°$ for broadside excitation along the surface normal and choosing $M = 1$ simplifies (9.10) to the following:

$$\Lambda = \frac{2\pi}{k_{spp}} = \frac{\lambda_0}{n_{eff}} \quad (9.11)$$

where n_{eff} is the average effective refractive index of the SPP propagating along the structure including the grating. Equation (9.10) holds for shallow gratings, where H is a small perturbation to the surface, and thus is used only to obtain an initial value for Λ using n_{eff} for SPPs localised along the Au/Si interface, from which the design of the grating is optimised via numerical modelling. Grating-coupled SPP detectors are sensitive to the angle of incidence, polarisation and wavelength of the incident light through (9.10).

Figure 9.3b shows a SEM image of a fabricated structure. Figure 9.3c shows measured photocurrent responses at $V_B = -100$ mV and $S_{inc} = 1.4$ mW of three grating photodetectors having $\Lambda = 390$, 400 and 410 nm with H = 79 nm, a duty cycle of 62 %, and a patch diameter of 25 μm with a thickness of $t = 29$ nm, due to hot holes emitted into the silicon. A 4-period moving average is plotted on each response as the bold curve. The rapid wavelength variations correspond to Fabry-Perot resonances—the upper right inset is an enlarged response showing such resonances and the lower right inset shows a response calculated using the Transfer Matrix Method (TMM). The Fabry-Perot resonances originate as light transmitted through the grating and metal patch propagates through the semi-transparent substrate (p^+-Si) and is reflected from the backside Ohmic contact. The Fabry-Perot resonances would disappear if a more heavily doped Si substrate was used. The slow wavelength dependency observed in the main panel of Fig. 9.3c is due to the incident light coupling into SPPs due to the grating. The peak of the bold responses red-shifts as the grating pitch increases from 390 to 410 nm, following (9.11).

Figure 9.3d shows the measured photocurrent generated by grating detectors *vs.* the incident power which overlaps with the detector area, for detectors of three different pitches: $\Lambda = 390$ nm ($\lambda_0 = 1537$ nm), $\Lambda = 400$ nm ($\lambda_0 = 1548$ nm) and $\Lambda = 410$ nm ($\lambda_0 = 1570$ nm). A linear fit is applied to the data, and the slope corresponds to the responsivity of the devices (9.9), yielding $R = 11.82$, 12.95 and 12.46 mW/A for the photodetectors having $\Lambda = 390$, 400 and 410 nm, respectively. These are among the highest responsivities measured at telecom wavelengths (~1550 nm) for sub-bandgap detectors on Si.

Grating detectors have many important advantages such as speed, simplicity, compatibility with silicon, and low-cost fabrication. Their small area also allows a high density of integration.

9.4 Nanoparticle and Nanoantenna Detectors

Small metal particles exhibit resonant responses under optical excitation, characteristic of particle size, shape and composition, the dielectric environment in which they find themselves, and the wavelength of illumination [4, 5]. Resonances are excited when the electron charge density oscillates coherently with the illumination. The fundamental resonant mode of, e.g., a spherical metal nanoparticle is dipolar with densities of opposite charge forming at opposite spherical caps of the particle along the polarisation of the illuminating electric field. The wavelength of the dipolar resonance red-shifts as the radius of the particle increases or as the index of the background increases. On resonance, the electric field near the particle (in the dielectric) is strongly enhanced compared to the illuminating field. Resonant features appear in the measured extinction and scattering spectra of metal nanoparticles, in correlation with the plasmon resonances that are excited thereon [22].

In the quasi-static limit, the absorption and scattering cross-sections (C_{abs}, C_{sca}) of a small metal nanosphere are given by [1]:

$$C_{abs} = \frac{2\pi}{\lambda_0} \mathrm{Im}\{\alpha\} \tag{9.12}$$

and

$$C_{sca} = \frac{1}{6\pi} \left(\frac{2\pi}{\lambda_0}\right)^4 |\alpha|^2 \tag{9.13}$$

where α is the sphere's polarisability, given by the following expression for the case of the dipolar resonant mode:

$$\alpha = 4\pi r^3 \left(\frac{\varepsilon_m - \varepsilon_d}{\varepsilon_m + 2\varepsilon_d}\right) \tag{9.14}$$

In the above, ε_m and ε_d are the permittivities of the metal and surrounding dielectric, respectively, and r is the radius of the nanosphere. The extinction cross-section C_{ext} is then given by $C_{ext} = C_{sca} + C_{abs}$ and the scattering efficiency Q_{sca} by:

$$Q_{sca} = \frac{C_{sca}}{C_{sca} + C_{abs}} \tag{9.15}$$

Given the typical dispersion of metals (9.1), it is possible for the polarisability of a nanosphere to become large. For the dipolar mode, this occurs at the frequency where the denominator of (9.15) becomes smallest in magnitude—termed the dipolar surface plasmon resonance frequency. On resonance, the absorption, scattering and extinction cross-sections become large, as do the electric fields in the dielectric region near the particle. Substituting (9.15) into (9.13) and (9.14) reveals that C_{abs} scales as r^3 and C_{sca} scales as r^6. Thus, for very small spheres, absorption dominates, whereas for larger spheres, scattering dominates and the scattering efficiency (9.15) approaches unity.

These properties render metal nanoparticles interesting for photodetector applications. As light strikes the nanoparticles, resonances are excited thereon, large near-field enhancement factors can be generated, and strong scattering can occur. These phenomena can lead to improved detector performance, or to the maintenance of performance using less material for optical absorption [23]. Metal nanoparticles can be integrated with a photodetector by depositing them on top of the detection medium, integrating them below, or embedding them within the detection medium.

Detectors involving metal nanoparticles embedded, e.g., into Si [24–27] or GaAs [28, 29] were among the first investigated. Figure 9.4a shows a sketch of a substrate-illuminated (infra-red) Co/p-Si Schottky diode with $CoSi_2$ nanoparticles embedded into single-crystalline Si [27]. Figures 9.4b and c give TEM images of such nanoparticles in plan and cross-sectional views, respectively [25]. The nanoparticles, 10–50 nm in dimensions and lattice-matched to Si, were formed via columnar molecular beam epitaxy and their size can be tailored to support SPP resonances at wavelengths of interest [25]. Embedding nanoparticles in this manner allows detection of infrared radiation (λ_0 = 1–2 μm) via IPE [24, 27] because the $CoSi_2$ nanoparticles form Schottky barriers at all interfaces with the Si, and as an SPP resonance on a nanoparticle decays, it generates hot carriers therein, leading to photocurrent [26]. The nanoparticles may become charged during illumination because they are not contacted to an electrode (i.e., they are "floating"). Discharging occurs via compensating charge injection from the host medium (Si) but this can be a slow process. The GaAs devices [28, 29] operate in the same manner, with As precipitates forming the embedded metal nanoparticles.

More recently, detectors involving structured nanoparticles or nanoantennas have been reported. [6, 30] are passive structures that, on reception, capture part of an incident optical wave to produce oscillating near-fields at defined locations along its geometry. In transmission, the antenna converts oscillating near-fields into radiated fields. Combining the antenna reception function with photodetection leads to integrated nanoscale detectors having compelling characteristics [31–37].

Resonant optical antennas based on metal nanowires, inspired by radio-frequency or microwave wire antennas such as monopole, dipole, bow-tie, Yagi-Uda and log-periodic antennas, were among the first investigated [6, 30]. Such antennas operate qualitatively in the same way as their microwave counterparts, and their performance depends on the geometry in a similar manner: e.g., the resonant wavelength of an optical dipole antenna depends strongly on its length and on the capacitance of the gap between the nanowires.

However, contrary to microwave antennas, propagating SPPs are excited on metal nanowires at optical wavelengths, implying that the antenna performance also depends on the propagation characteristics of the SPPs excited thereon [32]. This means that the nanowire cross-section and the dielectric environment affect the performance of optical antennas, in addition to the nanowire length: e.g., the length of a "half-wavelength" monopole antenna is less than $\lambda_0/2$ and depends on the

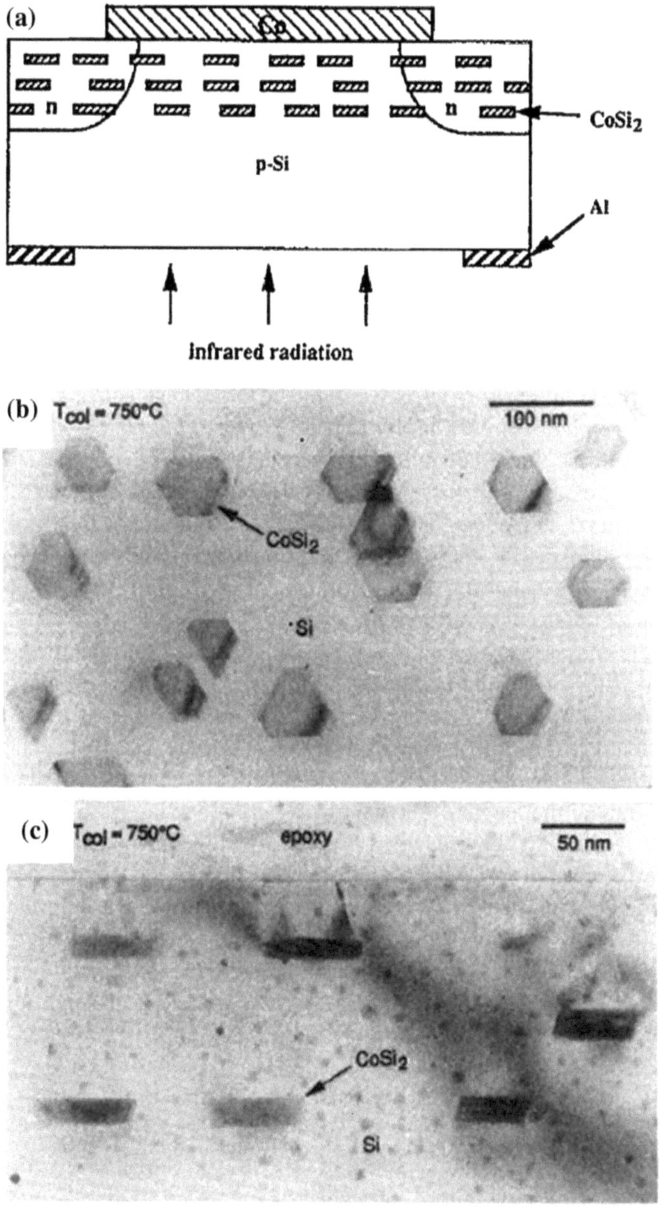

Fig. 9.4 **a** (Adapted from [27]. Copyright (1993) by the American Institute of Physics.) Sketch of a Co/p-Si Schottky diode with epitaxial $CoSi_2$ nano-particles embedded into the single-crystal Si substrate. **b** and **c** (Adapted from [25]. Copyright (1991) by the American Physical Society.) TEM images of $CoSi_2$ nano-particles in Si, in (**b**) plan and (**c**) cross-sectional views

penetration depth of the SPP into the bounding dielectric media, as well as on its n_{eff} [32]. It is also noteworthy that the electric near-fields produced by antennas can be of much greater intensity than the incident electric fields [30, 32]—this property, or equivalently the large antenna effective aperture, is interesting for photodetector applications as it allows the detection volume to shrink, leading potentially to fast devices with compelling signal-to-noise characteristics [36]. The integration of photodetection capabilities with an antenna must be carried out while considering several factors including the location and strength of the near-fields, means for collecting the photocurrent, and the effects that the detection medium have on the antenna performance.

Arrays of uncoupled monopole nanoantennas, each 30 nm thick, 50 nm wide and ~110–160 nm long, formed as Au/Ti on n-Si, isolated by SiO_2, and covered by indium tin oxide to establish electrical contact to the nanoantennas were investigated [31]. The photocurrent was generated via IPE by hot electrons created within the nanoantennas under perpendicularly incident light polarised along the nanoantenna axes. The spectral response of the photocurrent followed the absorptance spectrum of the nanoantennas, exhibiting a maximum at the resonant wavelength of the dipolar SPP resonance excited thereon. Responsivities of ~10 μA/W were measured at $\lambda_0 = 1250$ nm. Application as a miniaturised polarisation-sensitive spectrometer was suggested.

Monopole and dipole antenna arrays were investigated for photodetection based on IPE, targeting biosensing applications [33]. The systems investigated consisted of Au nanowires on Si covered by H_2O and a thin adlayer representing a biochemical coating. The resonance excited on such antennas by perpendicularly incident light polarised along their axis is dipolar, and comprised of an SPP mode propagating along the length of the nanowires and reflecting from the ends to form standing waves along the antenna [32]. Changes in antenna performance with geometry were related to changes in the propagation characteristics of this mode, and an equivalent circuit for dipole antennas was proposed as capacitor-coupled open-circuited transmission lines, where the capacitor models the gap between the dipole arms [32]. Using a thin-film IPE model [15], responsivities of up to 100 mA/W are predicted at 1310 nm, along with spectral biosensing sensitivities of ~250 nm/RIU (bulk) and ~8 nm/nm (surface) [33].

9.5 Waveguide Detectors

The single-interface is one example of an SPP waveguide. Other popular waveguides include 1-D structures such as the thin metal slab bounded by dielectrics and the thin dielectric slab cladded by metals [2]. These structures support super-modes formed from the coupling of single-interface SPPs through the thin intervening layer. The structures can be designed such that they support super-modes that are low loss but weakly confined (metal slab) or strongly confined but high loss (metal clads), and thus that are at opposite ends of the confinement-attenuation trade-off [2].

Metal stripes in symmetric and asymmetric dielectric environment are also useful as SPP waveguides [2].

SPP waveguides can be integrated with photodetector structures to form SPP detectors. Approaches consist of replacing a dielectric region with a semiconductor medium. Such detectors are attractive because they can be integrated with other plasmonic or photonic waveguide structures. Other motivating factors include a desire to increase the absorptance of photodetectors, shrink their dimensions, or enhance the detection process by exploiting enhanced or localised guided fields. SPP waveguide detectors based on Si in particular are of current interest, strongly motivated by the technological importance of this material and by applications such as high-speed photodetection at telecom wavelengths or clock distribution in CMOS electronics. Several approaches involving SPP waveguides on Si have been proposed. A popular approach consists of integrating a Schottky detector with a waveguide such that SPPs may be excited thereon, and exploiting IPE as the detection mechanism to enable infrared detection below the bandgap of Si. Several SPP waveguide detectors based on absorption in metals and IPE have been reported [38–52].

Detectors where the Schottky metal contact is also a metal stripe waveguide [2], have been proposed on n-Si and on p-Si in symmetric [41, 46] and asymmetric [40, 42, 43, 50, 52] cladding configurations. Figure 9.5a gives a sketch of an asymmetric SPP waveguide detector consisting of a thin narrow metal stripe on Si with air on top [42]. Structures were fabricated as Au or Al stripes on n-Si forming Schottky contacts thereon. The sa_b^0 mode, localised to the metal-Si interface, was excited at wavelengths below the bandgap of Si via butt-coupling to a tapered polarisation-maintaining single mode fiber (PM-SMF), as shown schematically in Fig. 9.5a. This mode propagates along the metal-Si interface with strong absorption, creating hot carriers in the metal along the Schottky contact, some of which cross the Schottky barrier and are collected as the photocurrent (IPE). Figure 9.5b gives the spectral response in Fowler form of a Au stripe of width 2.5 μm and thickness 135 nm on n-Si, for a reverse bias of 200 mV and an incident optical power of 2 mW [42]. The intercept with the abscissa yields the cut-off photon energy, ~0.765 eV, corresponding to a cut-off wavelength of $\lambda_0 \sim 1620$ nm. Responsivities up to 1 mA/W were reported with this detector scheme, below what can be obtained using a grating detector on p-Si [21], due to the limited coupling efficiency of the fibre coupling scheme in this configuration and the higher barrier height of a Au contact to n-Si (compared to p-Si).

Figure 9.5c shows a measured photocurrent map of a Au on p-Si detector, generated by scanning a tapered PM-SMF over the end facet (following Fig. 9.5a) using a piezoelectric nanopositioner [50]. The Au stripe width and thickness in this case were 1.5 μm and 40 nm, the reverse bias was 100 mV, the incident optical power was 1 mW, and the excitation wavelength was 1310 nm. The photocurrent was largest when the PM-SMF was perfectly aligned with the stripe maximising the overlap between the incident beam and the sa_b^0 mode of the stripe.

Other studies involving asymmetric metal stripe Schottky detectors include operation under a strong reverse bias approaching breakdown where a 10× increase in responsivity relative to low bias was reported [43], and a performance comparison

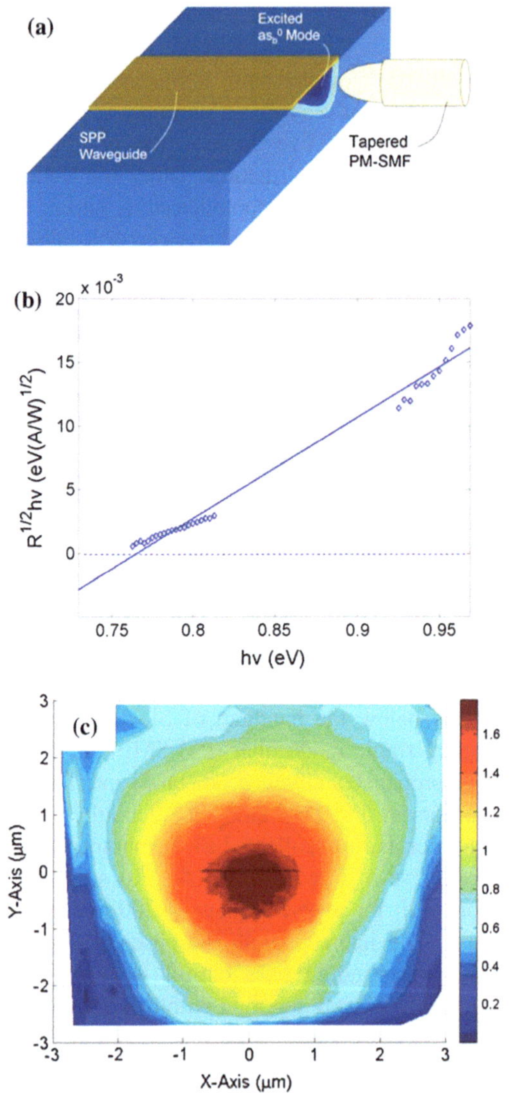

Fig. 9.5 **a** (Adapted from [42]. Copyright (2010) by the Optical Society of America.) Sketch of an SPP waveguide detector consisting of a metal stripe forming a Schottky contact on Si, excited in the sa_b^0 mode via butt-coupling to a tapered polarisation-maintaining single mode fibre (PM-SMF). **b** (Adapted from [42]. Copyright (2010) by the Optical Society of America.) Spectral response of a Au on n-Si detector as sketched in (**a**), plotted in Fowler form. The Au stripe width and thickness were 2.5 μm and 135 nm, the reverse bias was 200 mV and the incident optical power was 2 mW. **c** (Adapted from [50]. Copyright (2012) by the Institute of Physics.) Photocurrent map (μA) of a Au on p-Si detector as sketched in (**a**), generated by scanning a tapered PM-SMF over the end facet using a piezoelectric nanopositioner. The Au stripe width and thickness were 1.5 μm and 40 nm (sketched as the *black horizontal line*), the reverse bias was 100 mV, the incident optical power was 1 mW, and the excitation wavelength was 1310 nm

of such detectors for two metals on p-Si and n-Si, two SPP modes of operation, and several stripe geometries [52].

Photodetector studies have also been carried out using metal stripes in symmetric claddings [41, 46] along which SPPs are guided. A particularly promising configuration consists of a thin $CoSi_2$ stripe buried in p-Si forming Schottky contacts around its periphery, for which a responsivity of ~100 mA/W and a minimum detectable power of −20 dBm were predicted at $\lambda_0 = 1550$ nm under end-fire coupling. The predicted performance of this detector is due to efficient coupling to SPPs guided by the stripe (the ss_b^0 mode [2]), full absorption of SPPs along the stripe length generating hot carriers therein close to the interfaces, and the improved probability for carrier escape from the stripe (IPE) due to multiple internal carrier reflections and the increased number of barriers over which carrier escape can occur [15]. High-speed operation is possible, with receiver sensitivities of about -15 dBm predicted for error-free digital transmission (BER $< 10^{-12}$) [46]. The higher attenuation of SPPs in asymmetric structures yields a shorter device with a lower dark current [40, 42] compared to operating with SPPs in symmetric claddings [41, 46], but the responsivity of asymmetric detectors is lower.

The integration of SPP detectors with SOI waveguides has also been proposed [38, 39, 44, 45, 48, 51], and the LOCOS (local oxidation of silicon) process used to fabricate some of the structures [44, 45, 51]. Figure 9.6a shows a SEM image of an SOI waveguide, with a Si core that is 310 nm wide by 340 nm thick, fabricated with this process [45], and Fig. 9.6b shows a 50 nm thick Au film forming a Schottky contact to the p-Si core of an underlying SOI waveguide, along which SPPs are excited. The photoresponse of such detectors was measured at several infra-red wavelengths under a reverse bias of 100 mV by butt-coupling to a lensed PM-SMF, yielding an internal quantum efficiency of, e.g., ~0.02 % at $\lambda_0 =$ 1550 nm. An improved detector was reported [51], also fabricated via the LOCOS process, but followed by an etch step to expose more of the Si core. A 100 nm thick, 1 μm long Al film was then deposited on the top and sides of the p-Si core, forming a Schottky contact thereon, as shown in the SEM image of Fig. 9.6c. The photoresponse of such detectors was measured at $\lambda_0 = 1550$ nm by butt-coupling to a lensed PM-SMF, yielding an internal quantum efficiency of ~1 %. The improvement over [45] is due to Al deposited on three sides of the core and roughness along the sidewalls (leading to a roughened Schottky contact).

Metal silicide Schottky detectors were also integrated with SOI waveguides. A detector fabricated as a ~10 nm thick ~20 μm long $NiSi_2$ film on p-Si produced ~5 mA/W of responsivity (internal to the SOI waveguide) near 1550 nm with an electrical bandwidth of 3 GHz [38]. A similar detector was integrated into the collector of a bipolar transistor in an SOI waveguide, producing a responsivity of 150 mA/W at 1550 nm, due partly to the current gain provided by the transistor [39].

A Si pn junction with NiSi nanodisks embedded therein and forming Schottky contacts at all interfaces, was integrated with an SOI waveguide [48]. They remained electrically floating within the pn junction. The size of the nanodisks was selected such that absorption dominates over scattering and that SPPs would be excited thereon. A peak responsivity of 30 mA/W (internal to the SOI waveguide)

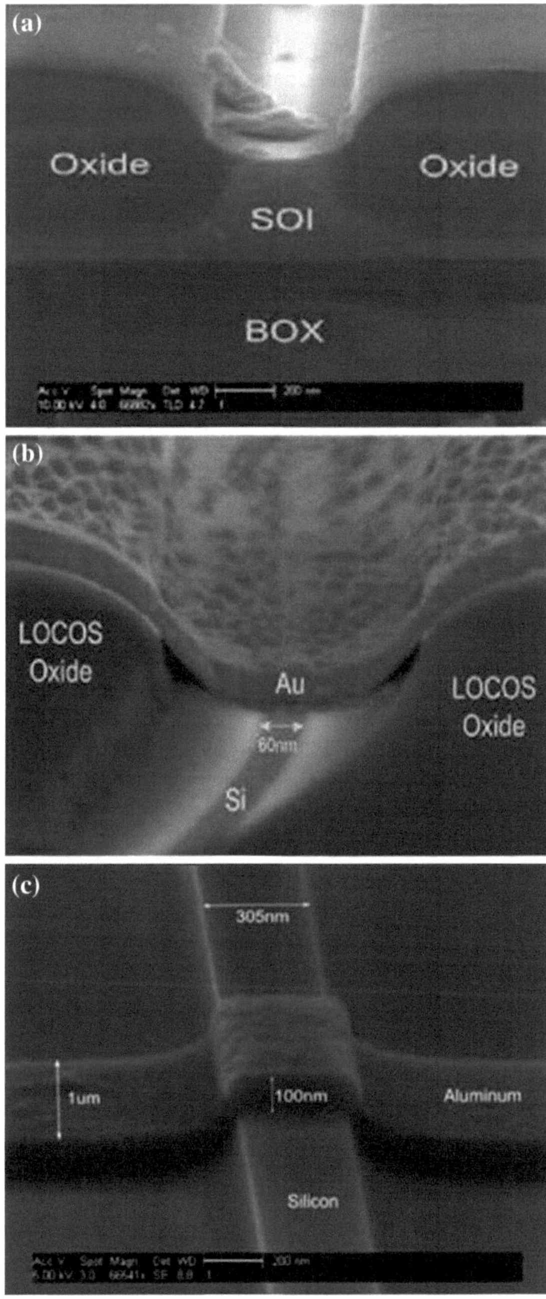

Fig. 9.6 **a** (Adapted from [45]. Copyright (2011) by the American Chemical Society.) SEM image of an SOI waveguide fabricated via local oxidation of silicon (LOCOS). **b** (Adapted from [45]. Copyright (2011) by the American Chemical Society.) SEM image of a Au/Si SPP Schottky detector integrated with a an SOI waveguide. **c** (Adapted from [51]. Copyright (2012) by the Optical Society of America.) SEM image of a Al/Si SPP detector integrated with an SOI waveguide with etched sidewalls

was measured at 1550 nm, over an electrical bandwidth of 6 GHz. IPE of both electrons and holes from the nanodisks into the Si generated the photocurrent.

A metal-cladded detector structure incorporating thin metal silicide to Si Schottky detectors, along which SPPs are excited, was proposed and modelled [47]. $TaSi_2$ was selected among several silicides due to its large absorption. A responsivity of 70 mA/W, an electrical bandwidth of 66 GHz and a minimum detectable power of −29 dBm were predicted at 1550 nm for an optimised structure, with detection occurring via IPE.

9.6 Summary and Prospects

SPP detectors typically combine a metallic structure that supports SPPs, such as a waveguide or grating structure, nanoparticles or nanoantennas, with a semiconductor photodetector structure such as a Schottky junction. Properties inherent to SPPs are exploited to convey additional characteristics to a detector such as polarisation, angular or spectral selectivity, or to enhance the absorptance of the detector. SPP detector architectures are highly varied, due to the diversity of metallic structures that support SPPs, and the diversity of detection schemes and materials that are available. While studies on SPP detectors have been carried out on a broad range of detection materials, a major emphasis has been on Si given its technological importance.

Interest in IPE as the detection mechanism is motivated by the fact that it enables sub-bandgap detection so the optical bandwidth of a detector can be extended to photon energies below the bandgap of the semiconductor. Several investigations have been reported aiming at improving the efficiency of detectors based on IPE by exciting SPPs along a metal contact(s) forming the device. Target applications of IPE are very broad, but improving the efficiency of solar cells and providing infrared detection at telecom wavelengths in Si are frequently evoked.

Early investigations exploited prism- or grating-coupled SPP detector structures. Prism coupling in the Otto configuration was used primarily for experimental convenience and to carry out fundamental studies. Corrugated grating couplers were used initially to create integrated and compact structures—features that remain useful today. Both types of coupling provided early evidence of SPP-enhanced IPE. Grating-coupled detectors remain among the best performing structures to date.

Metal nanoparticles supporting SPP resonances are useful to increase the absorptance of detectors, particularly those based on a thin absorption layer, or at wavelengths where the absorption is low. Metal nanoparticles have been integrated with detection media by depositing them on top of the detection medium, integrating them below, or embedding them within the detection medium—the latter, although technologically more challenging, seems to hold promise, especially for the enhancement of IPE.

Metallic nanoantennas capture part of an incident optical wave to produce oscillating SPPs with near-fields at defined locations along its geometry. Combining this function with photodetection leads to integrated nanoscale detectors having compelling characteristics, as the antennas allow the detection volume to shrink, leading to fast devices with good signal-to-noise characteristics.

SPP waveguides can be formed into SPP detectors by replacing a dielectric region with a semiconductor medium. Such detectors are attractive because they can be integrated with other plasmonic or photonic waveguides, they can increase the absorptance, shrink dimensions, enhance the detection process by exploiting enhanced or localised guided fields, and provide broad electrical and optical bandwidths.

SPP photodetectors are of vigorous current interest. Detector architectures are highly varied, reflecting broad diversity in materials and architectures. Properties inherent to SPPs are exploited to convey useful characteristics to detectors in addressing applications. The prospects for further enhancement of IPE-based detectors are very strong.

References

1. S.A. Maier, *Plasmonics: Fundamentals and Applications* (Springer Science+Business Media, New York, 2007)
2. P. Berini, Adv. Opt. Phot. **1**, 484–588 (2009)
3. W.L. Barnes, T.W. Preist, S.C. Kitson, J.R. Sambles, Phys. Rev. B **54**, 6227–6244 (1996)
4. K.L. Kelly, E. Coronado, L.L. Zhao, G.C. Schatz, J. Phys. Chem. B **107**, 668–677 (2003)
5. M. Pelton, J. Aizpurua, G. Bryant, Laser Phot. Rev. **2**, 136–159 (2008)
6. P. Bharadwaj, B. Deutsch, L. Novotny, Adv. Opt. Phot. **1**, 438–483 (2009)
7. G. Genet, T.W. Ebbesen, Nature **445**, 39–46 (2007)
8. W.L. Barnes, A. Dereux, T.W. Ebbesen, Nature **424**, 824–830 (2003)
9. J. Homola, Chem. Rev. **108**, 462–493 (2008)
10. T.W. Ebbesen, C. Genet, S.I. Bozhevolnyi, Phys. Today **61**, 44–50 (2008)
11. P. Berini, I. De Leon, Nat. Photonics **6**, 16–24 (2012)
12. P. Berini, Laser Phot. Rev. **8**, 197–220 (2014)
13. S.M. Sze, *Physics of Semiconductor Devices* (Wiley, New York, 1981)
14. M. Casalino, G. Coppola, M. Iodice, I. Rendina, L. Sirleto, Sensors **10**, 10571 (2012)
15. C. Scales, P. Berini, IEEE J. Quantum Electron. **46**, 633–643 (2010)
16. J. Sipe, J. Becher, J. Opt. Soc. Am. **71**, 1286–1288 (1981)
17. J. Derov, Y.Y. Teng, A.S. Karakashian, Phys. Lett. A **95**, 197–200 (1983)
18. C. Daboo, M.J. Baird, H.P. Hughes, N. Apsley, G.A.C. Jones, J.E.F. Frost, D.C. Peacock, D.A. Ritchie, Thin Solid Films **189**, 27–38 (1990)
19. S.R.J. Brueck, V. Diadiuk, T. Jones, W. Lenth, Appl. Phys. Lett. **46**, 915–917 (1985)
20. A. Sobhani, M. Knight, Y. Wang, B. Zheng, N.S. King, L.V. Brown, Z. Fang, P. Nordlander, N.J. Halas, Nat. Commun. **4**, 1643 (2013)
21. M. Alavirad, A. Olivieri, L. Roy, P. Berini, Opt. Express **24**, 22544 (2016)
22. A. Calà Lesina, A. Vaccari, P. Berini, L. Ramunno, Opt. Express **23**, 10481–10497 (2015)
23. H. Zhang, A.O. Govorov, J. Phys. Chem. C **118**, 7606–7614 (2014)
24. R.W. Fathauer, J.M. Iannelli, C.W. Nieh, S. Hashimoto, Appl. Phys. Lett. **57**, 1419–1421 (1990)
25. R.W. Fathauer, A. Ksendzov, J.M. Iannelli, T. George, Phys. Rev. B. **44**, 1345–1348 (1991)
26. J.R. Jimenez, L.J. Schowalter, R.W. Fathauer, Mat. Res. Soc. Symp. Proc. **220**, 513–518 (1991)

27. R.W. Fathauer, S.M. Dejewski, T. George, E.W. Jones, T.N. Krabach, A. Ksendzov, Appl. Phys. Lett. **62**, 1774–1776 (1993)
28. A.C. Warren, J.H. Burroughes, J.M. Woodall, D.T. McInturff, R.T. Hodgson, M.R. Melloch, I.E.E.E. Electr, Dev. Lett. **12**, 527–529 (1991)
29. D.T. McInturff, J.M. Woodall, A.C. Warren, N. Braslau, G.D. Pettit, P.D. Kirshner, M.R. Melloch, Appl. Phys. Lett. **60**, 448–450 (1992)
30. L. Novotny, N. van Hulst, Nat. Photonics **5**, 83–90 (2011)
31. M.W. Knight, H. Sobhani, P. Nordlander, N.J. Halas, Science **332**, 702–704 (2011)
32. S. Siadat Mousavi, P. Berini, D. McNamara, Opt. Express **20**, 18044–18065 (2012)
33. M. Alavirad, S. Siadat Mousavi, L. Roy, P. Berini, Opt. Express **21**, 4328–4347 (2013)
34. M. Alavirad, L. Roy, P. Berini, IEEE J. Sel. Top. Quant. **20**, 7–14 (2014)
35. M.W. Knight, Y. Wang, A.S. Urban, A. Sobhani, B.Y. Zheng, P. Nordlander, N.J. Halas, Nano Lett. **13**, 1687–1692 (2014)
36. S. Siadat Mousavi, A. Stöhr, P. Berini, Appl. Phys. Lett. **104**, 143112 (2014)
37. B. Desiatov, I. Goykhman, N. Mazurski, J.O. Shappir, J.B. Khurgan, U. Levy, Optica **2**, 335–338 (2015)
38. S. Zhu, M.B. Yu, G.Q. Lo, D.L. Kwong, Appl. Phys. Lett. **92**, 081103 (2008)
39. S. Zhu, G.Q. Lo, M.B. Yu, D.L. Kwong, Appl. Phys. Lett. **93**, 071108 (2008)
40. A. Akbari, P. Berini, Appl. Phys. Lett. **95**, 021104 (2009)
41. C. Scales, I. Breukelaar, P. Berini, Opt. Lett. **35**, 529–531 (2010)
42. A. Akbari, R.N. Tait, P. Berini, Opt. Express **18**, 8505–8514 (2010)
43. A. Olivieri, A. Akbari, P. Berini, Phys. Stat. Sol. RRL **4**, 283–285 (2010)
44. S. Li, N.G. Tarr, P. Berini, Proc. SPIE **7750**, 77501M (2010)
45. I. Goykhman, B. Desiatov, J. Khurgin, J. Shappir, U. Levy, Nano Lett. **11**, 2219–2224 (2011)
46. C. Scales, I. Breukelaar, R. Charbonneau, P. Berini, J. Lightwave Technol. **29**, 1852–1860 (2011)
47. S. Zhu, G.Q. Lo, D.L. Kwong, Opt. Express **19**, 15843–15854 (2011)
48. S. Zhu, H.S. Chu, G.Q. Lo, P. Bai, D.L. Kwong, Appl. Phys. Lett. **100**, 061109 (2012)
49. D.-S. Ly-Gagnon, K.C. Balram, J.S. White, P. Wahl, M.L. Brongersma, D.A.B. Miller, Nanophotonics **1**, 9–16 (2012)
50. P. Berini, A. Olivieri, C. Chen, Nanotechnology **23**, 444011 (2012)
51. I. Goykhman, B. Desiatov, J. Khurgin, J. Shappir, U. Levy, Opt. Express **20**, 28594–28602 (2012)
52. A. Akbari, A. Olivieri, P. Berini, IEEE J. Sel. Top. Quantum Electron. **19**, 4600209 (2013)

Chapter 10
Antenna-Coupled Tunnel Junctions

Markus Parzefall, Palash Bharadwaj and Lukas Novotny

Abstract Forty years after the discovery of light emission from inelastic electron tunneling (LEIT), we review its historical development in relation to the recent demonstrations of light emitting optical antennas driven by quantum tunneling. A general outlook on promising research directions towards efficient, electrically driven, nanoscale light emitters is given.

10.1 Introduction

An optical antenna in the most general sense is a transducer between optical and electronic signals at the nanoscale. More specifically, metallic nanostructures—depending on their geometry and dielectric environment—are able to resonantly enhance the localization of electromagnetic energy in ultrasmall volumes [1–3]. This local field enhancement has made them very attractive for potential applications in microscopy and sensing [4].

Optical antennas can be likened to their macroscopic radio- and microwave counterparts in many aspects. Despite some differences in the underlying physics, the same design principles have been successfully applied to optical antennas which had been originally developed for antennas in other frequency domains. Engineering geometry and arrangement of antenna constituents allows for the tuning of their operating frequencies as well as the determination of their spatial radiation characteristics—ranging between omni- and unidirectionality.

The main characteristic in which they differ thus far is the mode of operation. Radio- and microwave antennas are generally used to convert electrical to optical signals simply by driving the antenna with an AC voltage at a frequency corresponding

M. Parzefall · P. Bharadwaj · L. Novotny (✉)
Photonics Laboratory, ETH Zürich, 8093 Zürich, Switzerland
e-mail: lnovotny@ethz.ch

S.I. Bozhevolnyi et al. (eds.), *Quantum Plasmonics*,
Springer Series in Solid-State Sciences 185,
DOI 10.1007/978-3-319-45820-5_10

to the frequency of the emitted radiation. On the other hand, optical antennas have been historically operated on a 'light-in/light-out' basis, i.e. the antenna receives radiation from the far-field as well as emits radiation back into the far-field [4]. The obvious question to ask is whether it is feasible to use optical antennas as an interface between electrical and optical signals. This question is interesting not only from a purely scientific point of view. Electrically driven optical antennas could also serve as a highly integrable, on-chip, nanoscopic transducer, facilitating the conversion of electrical to optical signals. Furthermore, they could serve as nanoscopic sources of light for spectroscopy and sensing.

A mechanism which could potentially facilitate the realization of these goals was discovered by Lambe and McCarthy in 1976 [5], who observed light emission from a metal-insulator-metal (MIM) tunnel junction (c.f. Sect. 10.3.1). Applying a bias voltage V_b to a tunnel junction supplies electrons in one electrode with excess potential energy $|eV_b|$. In general, as will be discussed in more detail in the following section, light emission is the result of a two-step process. First, electrons tunnel inelastically, exciting a mode of energy $\hbar\omega$ provided that $|eV_b| \geq \hbar\omega$ (process **1**). Second, the localized mode decays which results in the emission of a photon with a probability given by the antenna efficiency $\eta_{antenna}$. This mechanism was termed light emission from inelastic electron tunneling (LEIT) and suggested as the cause of light emission by Lambe and McCarthy [5]. Kirtley et al. later on suggested that under certain experimental conditions the excitation of optical modes is primarily caused by an alternative mechanism shown as process **2** in Fig. 10.1 [6]. Here the mode is excited by elastic tunneling with subsequent hot electron decay. We will elaborate on the importance of the two mechanisms in the following section.

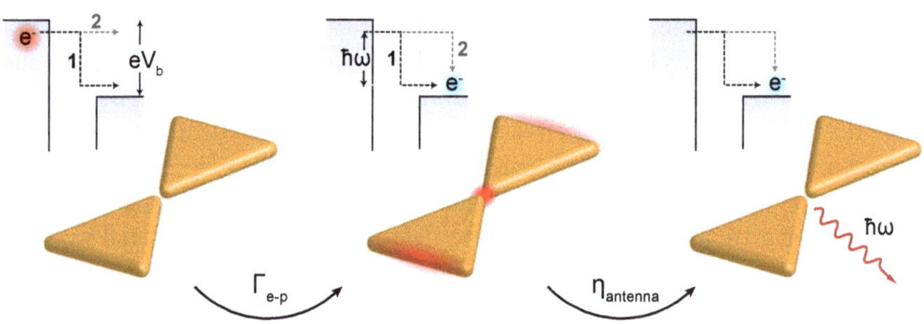

Fig. 10.1 Illustration of light emission from an antenna-coupled tunnel junction, described as a two-step process: Applying a voltage V_b to a MIM tunnel junction supplies electrons with excess energy of $\Delta E \leq |eV_b|$ in one of the electrodes. Tunneling electrons may excite localized modes of energy $\hbar\omega$ supported by an optical antenna by either (**1**) inelastic electron tunneling or (**2**) elastic tunneling with subsequent hot electron decay. The rate of electron-to-plasmon conversion is is given by Γ_{e-p}. Finally, the antenna mode decays and a photon is emitted with a probability determined by the antenna efficiency $\eta_{antenna}$

10.2 Theoretical Framework

10.2.1 Historical Survey

Several theoretical works aiming to explain experimental findings have been published since the discovery of LEIT. The first of these works goes back to Davis [7]. His model describes the light emission process similar to the energy-loss of electron beams, employing a quantum-mechanical transition current density as the source term. From a different point of view, D. Hone et al. argue that fluctuations in the tunneling current act as a generator for a radiating dipolar mode [8]. Such fluctuations can be likened to shot noise of the tunneling electrons [9]. This approach leads to a fairly simple source term for the excitation of optical modes which was subsequently used to devise theoretical models for different sample geometries and outcoupling modalities [9–15].

Renewed interest in a theoretical description was sparked after the discovery of light emission from the scanning tunneling microscope (STM, c.f. Sect. 10.3.2) and resulted in a series of works in which LEIT is modeled by calculating the radiated power of a localized, quantum mechanical current source [16–19], following the basic idea of Davis [7]. A theory for STM light emission based on the power spectrum of tunneling current fluctuations has also been developed [20].

Alternatively, inelastic electron tunneling processes, in addition to elastic tunneling, can be treated perturbatively in the framework of the transfer-Hamiltonian model [21–23]. Several works have followed this approach to describe the interaction of electronic states and optical modes [24–27]. In one of these, Persson and Baratoff estimate the relative probabilities for the two processes mentioned in the previous section [24]. Their estimates suggest that the probability for mode excitation from inelastic tunneling is two orders of magnitude higher than the corresponding probability for hot electron decay. This estimate is also corroborated by the fact that most measurements, for which the radiative optical mode is localized to the tunnel junction, can be well explained by inelastic electron tunneling. However, we do emphasize that the two processes are difficult to distinguish experimentally.

The goal of this section is to develop a model which describes the fundamental physics involved in LEIT. This will serve as a basis for the discussion of experimental results in Sect. 10.3. The model described in the following treats the interaction of electrons and optical modes in a perturbative manner. We choose this method because it allows us to relate the probability of inelastic electron tunneling to the local density of electromagnetic states (LDOS), an important quantity in optical antenna theory [2, 3].

10.2.2 Photon Emission: A Two-Step Process

As illustrated in Fig. 10.1, LEIT can be described as a two-step process. Initially, electrons which tunnel inelastically excite electromagnetic modes of the particular junction geometry. This process takes place at a rate $\Gamma_{\mathrm{e-p}}$, the electron to plasmon conversion rate. Secondly, these modes may either decay radiatively or non-radiatively. The splitting ratio between these two processes determines the antenna radiation efficiency [1]

$$\eta_{\mathrm{antenna}} = \frac{P_{\mathrm{r}}}{P_{\mathrm{r}} + P_{\mathrm{nr}}} \tag{10.1}$$

with P_{r} and P_{nr} denoting the radiative and non-radiatively dissipated power, respectively.

We express the emitted power spectrum of the antenna as:

$$p_{\mathrm{em}}(\omega, V_{\mathrm{b}}) = \hbar\omega\, \Gamma_{\mathrm{e-p}}(\omega, V_{\mathrm{b}})\, \eta_{\mathrm{antenna}}(\omega). \tag{10.2}$$

In analogy to classical antennas we can view $\Gamma_{\mathrm{e-p}}$ as a quantized rate of power dissipation $(\hbar\omega)^{-1} dW/dt$. We will parametrize $\Gamma_{\mathrm{e-p}}$ in the next section in the framework of the transfer-Hamiltonian formalism.

10.2.3 *Tunneling Rates*

The rate of electron to plasmon conversion $\Gamma_{\mathrm{e-p}}$ is given by the rate of inelastic electron tunneling caused by the perturbation of the electronic system by the presence of electromagnetic modes in the tunnel junction. To derive an expression for $\Gamma_{\mathrm{e-p}}$, we will first introduce the fundamentals of Bardeen's transfer-Hamiltonian approach to elastic tunneling [21] based on the derivation of Reittu [28]. For an elaborate discussion of this formalism, see [23, 29, 30].

10.2.3.1 Transfer-Hamiltonian Formalism

The fundamental idea behind this formalism is to treat the two electrodes constituting the junction separately, i.e. in the absence of the other electrode. The transfer of electrons between the electrodes is introduced via time-dependent perturbation theory.

The two (initially) independent electrodes depicted in Fig. 10.2 are described by the single-particle Hamiltonians $\hat{H}_{\mathrm{L}}$ and $\hat{H}_{\mathrm{R}}$ determining the wavefunctions ψ_μ and ψ_ν of left and right electrode, respectively:

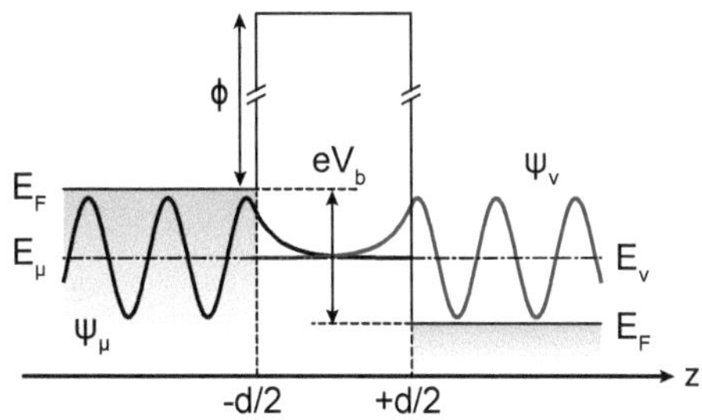

Fig. 10.2 Elastic tunneling in the framework of the transfer-Hamiltonian approach. For a given bias V_b, the tunneling rate is determined by the overlap of occupied states in the *left* electrode ψ_μ and unoccupied states in the *right* electrode ψ_ν with $E_\mu = E_\nu$

$$\hat{H}_L \left|\psi_\mu\right\rangle = E_\mu \left|\psi_\mu\right\rangle \tag{10.3a}$$

$$\hat{H}_R \left|\psi_\nu\right\rangle = E_\nu \left|\psi_\nu\right\rangle . \tag{10.3b}$$

The wave functions ψ_μ are solutions to (10.3a) in the absence of the right electrode and ψ_ν are solutions to (10.3b) in the absence of the left electrode. The Hamiltonians for a one-dimensional free electron model system are:

$$\hat{H}_L = -\frac{\hbar^2}{2m}\frac{d^2}{dz^2} + U_L(z) \tag{10.4a}$$

$$\hat{H}_R = -\frac{\hbar^2}{2m}\frac{d^2}{dz^2} + U_R(z) \tag{10.4b}$$

with m being the electron mass and $U_{L/R}(z)$ being the potential energies of the electrodes. Once the electrodes are brought to a distance where the exponentially decaying wavefunctions start to overlap, the presence of the right electrode can be treated as a perturbation. The effective system Hamiltonian reads as:

$$\hat{H} = \hat{H}_L + \hat{H}_T. \tag{10.5}$$

Electron tunneling between the two electrodes is described via the transfer-Hamiltonian $\hat{H}_T$. This transfer may take place either elastically or inelastically, hence we can write

$$\hat{H}_T = \hat{H}_{el} + \hat{H}_{inel}. \tag{10.6}$$

Both, elastic and inelastic tunneling can give rise to photoemission. In the elastic case—process **2** in Fig. 10.1—an electron first tunnels without losing energy and then relaxes to the Fermi level either by coupling to phonons or to photons. In the inelastic case—process **1** in Fig. 10.1—the electron loses energy during the tunneling process and the energy lost is coupled to plasmonic modes.

In the following we will assume that photoemission is dominated by inelastic tunneling and that an elastic tunneling event is followed by coupling to phonons and the generation of heat.

10.2.3.2 Elastic Tunneling

In first order perturbation theory, the rate of elastic tunneling between a state ψ_μ in the left electrode and a state ψ_ν in the right electrode is given by [30]

$$\Gamma_{\mu\rightarrow\nu} = \frac{2\pi}{\hbar} \left|t_{\mu\nu}\right|^2 \delta\left(E_\mu - E_\nu\right) \tag{10.7}$$

with $t_{\mu\nu} = \left\langle\psi_\nu|\hat{H}_{\rm el}|\psi_\mu\right\rangle$ being the transfer matrix element for elastic electron tunneling. The Hamiltonian for elastic tunneling is given by

$$\hat{H}_{\rm el} = \Theta(z - z_0)\left[U_{\rm R}(z) - U_{\rm L}(z)\right] \tag{10.8}$$

with z_0 defining an arbitrary point within the barrier region effectively separating the two electrodes. After insertion, the matrix element reads as

$$t_{\mu\nu} = \int_{z_0}^{\infty} \psi_\nu^* \left[U_{\rm R}(z) - U_{\rm L}(z)\right]\psi_\mu dz. \tag{10.9}$$

By inserting (10.4a) and applying (10.3a), we arrive at

$$t_{\mu\nu} = \int_{z_0}^{\infty} \psi_\nu^* \left[U_{\rm R}(z) - E_{\rm L} - \frac{\hbar^2}{2m}\frac{d^2}{dz^2}\right]\psi_\mu dz. \tag{10.10}$$

Integrating the last term twice by parts and making use of (10.4b) leads to

$$t_{\mu\nu} = \frac{\hbar^2}{2m}\left[\psi_\mu \frac{d\psi_\nu^*}{dz} - \psi_\nu^* \frac{d\psi_\mu}{dz}\right]_{z=z_0} + \int_{z_0}^{\infty} \psi_\mu(\hat{H}_{\rm R} - E_{\rm L})\psi_\nu^* dz. \tag{10.11}$$

The second term of (10.11) vanishes if we apply (10.3b) and take into account that $E_\mu = E_\nu$. The final expression for the matrix element is

$$t_{\mu\nu} = \frac{\hbar^2}{2m}\left[\psi_\mu \frac{d\psi_\nu^*}{dz} - \psi_\nu^* \frac{d\psi_\mu}{dz}\right]_{z=z_0}. \tag{10.12}$$

This expression, which was first derived by Bardeen [21], describes the rate of elastic electron transfer between the electrodes. It reproduces the results of exact, non-perturbative models in the limit of weak tunneling [23]. The total rate of elastic electron tunneling Γ is determined by summing over all available initial and final states. Furthermore we need to take into account the probability for the initial state to be occupied and the final state to be unoccupied by introducing the Fermi factors $f\left(E_\mu\right)$ and $1 - f\left(E_\nu\right)$, respectively

$$\Gamma = \frac{2\pi}{\hbar} \sum_{\mu,\nu} \left|t_{\mu\nu}\right|^2 f\left(E_\mu\right)\left(1 - f\left(E_\nu\right)\right)\delta\left(E_\mu - E_\nu\right). \tag{10.13}$$

After the elastic tunneling event the electron relaxes from E_ν to E_F (c.f. Fig. 10.2). If this relaxation happens radiationless, e.g. by coupling to phonons, there is no photon emission associated with elastic tunneling.

10.2.3.3 Inelastic Tunneling

The driving force for light emission is the inelastic transfer of electrons given by $\hat{H}_{\text{inel}}$. Just like elastic tunneling—in the limit of weak coupling—we can treat inelastic tunneling between the electrodes perturbatively. $\hat{H}_{\text{inel}}$ is given by the light-matter interaction Hamiltonian, which reads as $\hat{H}_{\text{inel}} = -e/m\, \hat{\mathbf{A}} \cdot \hat{\mathbf{p}}$ with $\hat{\mathbf{A}}$ being the vector potential operator and $\hat{\mathbf{p}} = -i\hbar\nabla$ being the momentum operator.

Let us now consider an arbitrary state ψ_μ of energy E_μ in the left electrode and a corresponding state ψ_ν of energy E_ν in the right electrode such that $E_\mu \neq E_\nu$ and $E_\mu > E_\nu$. The rate of spontaneous transitions, i.e. the inelastic tunneling rate, from ψ_μ to ψ_ν can be expressed in terms of Fermi's golden rule

$$\Gamma'_{\mu\to\nu} = \frac{2\pi}{\hbar} \sum_l \left|\left\langle \psi_\nu, \{1_{\omega_l}\} \left|\hat{H}_{\text{inel}}\right| \psi_\mu, \{0\}\right\rangle\right|^2 \delta\left(E_\mu - E_\nu - \hbar\omega_l\right). \tag{10.14}$$

The transition between the two electronic states is accompanied by an optical transition from the zero-photon state $|\{0\}\rangle$ to the one-photon state $\left|\{1_{\omega_l}\}\right\rangle$. The summation over l accounts for the total number of available optical modes of energy $\hbar\omega_l$. The energy difference between the electronic states $E_\mu - E_\nu$ is given by $\hbar\omega$. Since our system exhibits a continuum of final optical states as well as a continuum of final electronic states, we rewrite (10.14) as

$$\Gamma'_{\mu\to\nu} = \frac{2\pi}{\hbar} \int \sum_l \left|\left\langle \psi_\nu, \{1_{\omega_l}\} \left|\hat{H}_{\text{inel}}\right| \psi_\mu, \{0\}\right\rangle\right|^2 \delta\left(\omega - \omega_l\right)\delta\left(E_\mu - E_\nu - \hbar\omega\right) d\omega. \tag{10.15}$$

In the limit of a single final electronic or optical state energy, (10.15) reduces back to (10.14). From this we extract the inelastic tunneling rate per unit of energy (inelastic tunneling rate density)

$$\frac{1}{\hbar}\frac{d\Gamma'_{\mu\to\nu}}{d\omega}=\frac{2\pi}{\hbar^2}\sum_l\left|\left\langle\psi_\nu,\{1_{\omega_l}\}\left|\hat{H}_{\text{inel}}\right|\psi_\mu,\{0\}\right\rangle\right|^2\delta\left(\omega-\omega_l\right)\delta\left(E_\mu-E_\nu-\hbar\omega\right). \tag{10.16}$$

We now rewrite the initial and final states as the products of electronic and optical states

$$\left|\psi_\mu,\{0\}\right\rangle=\left|\psi_\mu\right\rangle\left|\{0\}\right\rangle \tag{10.17a}$$

$$\left|\psi_\nu,\{1_{\omega_l}\}\right\rangle=\left|\psi_\nu\right\rangle\left|\{1_{\omega_l}\}\right\rangle. \tag{10.17b}$$

Furthermore, for the one-dimensional system depicted in Fig. 10.3a, we find that $\hat{\mathbf{A}}\cdot\hat{\mathbf{p}}\left|\psi_\mu\right\rangle=\hat{A}_z\hat{p}_z\left|\psi_\mu\right\rangle$. Assuming the vector potential to spatially vary much slower than the electronic wavefunction across the tunneling gap, we rewrite (10.16) as [1, 31]

$$\frac{1}{\hbar}\frac{d\Gamma'_{\mu\to\nu}}{d\omega}=\frac{2\pi}{\hbar^2}\frac{e^2}{m^2}\sum_l\left|\left\langle\{1_{\omega_l}\}\left|\hat{A}_z\right|\{0\}\right\rangle\right|^2\left|p_{\mu\nu}\right|^2\delta\left(\omega-\omega_l\right)\delta\left(E_\mu-E_\nu-\hbar\omega\right) \tag{10.18}$$

with $p_{\mu\nu}=\left\langle\psi_\nu\left|\hat{p}_z\right|\psi_\mu\right\rangle$ being the momentum matrix element. The vector potential operator component $\hat{A}_z$, expressed in terms of annihilation and creation operators $\hat{a}_l$ and $\hat{a}_l^\dagger$ and normal modes $\mathbf{u}_l$, reads as

$$\hat{A}_z=\sqrt{\frac{\hbar}{2\varepsilon_0\omega_l}}\sum_l\left[\mathbf{u}_l\hat{a}_l e^{-i\omega_l t}+\mathbf{u}_l^*\hat{a}_l^\dagger e^{i\omega_l t}\right]\hat{z}. \tag{10.19}$$

By inserting (10.19) into (10.18) and applying commutation rules for annihilation and creation operators we obtain

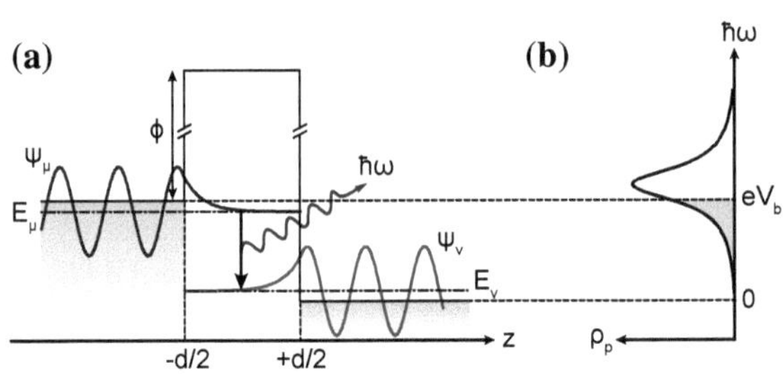

Fig. 10.3 Inelastic tunneling via the excitation of an optical mode. **a** Occupied states in the *left* electrode ψ_μ and unoccupied states in the *right* electrode ψ_ν with $E_\mu=E_\nu+\hbar\omega$ are perturbatively coupled to optical modes of energy $\hbar\omega$. **b** The rate of coupling to any given mode $\hbar\omega$ depends on the density of electromagnetic states $\rho_\text{p}(\hbar\omega)$. The range of mode energies which can be excited is determined by the applied bias V_b via $\hbar\omega\le|eV_\text{b}|$

$$\frac{1}{\hbar}\frac{d\Gamma'_{\mu\to\nu}}{d\omega} = \frac{\pi e^2}{3\hbar m^2 \varepsilon_0 \omega}\rho_{\mathrm{p}}(\omega)\left|p_{\mu\nu}\right|^2 \delta\left(E_\mu - E_\nu - \hbar\omega\right) \tag{10.20}$$

with $\rho_{\mathrm{p}}(\omega) = 3\sum_l [\hat{z}(\mathbf{u}_l\mathbf{u}_l^*)\hat{z}]\delta(\omega_l - \omega)$ being the partial local density of *optical* states (LDOS) in the direction of electron flow. To obtain the total inelastic tunneling rate density we have to sum over all initial and final *electronic* states

$$\frac{1}{\hbar}\frac{d\Gamma_{\mathrm{e-p}}}{d\omega} = \frac{\pi e^2}{3\hbar m^2 \varepsilon_0 \omega}\rho_{\mathrm{p}}(\omega)\sum_{\mu,\nu}\left|p_{\mu\nu}\right|^2 f\left(E_\mu\right)\left(1 - f\left(E_\nu\right)\right)\delta\left(E_\mu - E_\nu - \hbar\omega\right). \tag{10.21}$$

Hence we find that the rate of mode excitation for any given optical mode energy $\hbar\omega$ is directly proportional to the LDOS at this energy in the tunnel gap. The total rate of inelastic tunneling $\Gamma_{\mathrm{e-p}}$ is obtained by integrating (10.21) over all frequencies ω:

$$\Gamma_{\mathrm{e-p}} = \int_0^\infty \frac{\pi e^2}{3m^2 \varepsilon_0 \omega}\rho_{\mathrm{p}}(\omega)\sum_{\mu,\nu}\left|p_{\mu\nu}\right|^2 f\left(E_\mu\right)\left(1 - f\left(E_\nu\right)\right)\delta\left(E_\mu - E_\nu - \hbar\omega\right)d\omega. \tag{10.22}$$

10.2.3.4 LDOS and Device Efficiency

The density of states is a powerful quantity in that it is capable of describing the decay rate of a quantum mechanical system as well as the power dissipation of a classical dipole, hence bridging a quantum and classical description. This connection is summarized by the relation [1]

$$\frac{\rho_{\mathrm{p}}}{\rho_0} = \frac{P}{P_0} = \frac{P_{\mathrm{r}} + P_{\mathrm{nr}}}{P_0} \tag{10.23}$$

with $\rho_0 = \pi^2\omega^2c^{-3}$ being the vacuum LDOS, P being the total dissipated power of a dipole and P_0 being the power radiated by a dipole in vacuum. This relationship allows for the calculation of the LDOS of any given geometry using classical electrodynamics.

Finally we express the external quantum efficiency of antenna-coupled, light emitting tunneling devices in terms of inelastic and elastic tunneling rates as well as the antenna efficiency

$$\eta_{\mathrm{ext}} = \frac{\Gamma_{\mathrm{e-p}}}{\Gamma + \Gamma_{\mathrm{e-p}}}\eta_{\mathrm{antenna}} \approx \frac{\Gamma_{\mathrm{e-p}}}{\Gamma}\eta_{\mathrm{antenna}}. \tag{10.24}$$

The framework laid out in this section constitutes a toolbox for the interpretation of experimental results presented in Sect. 10.3. In order to arrive at (10.21) we had to assume the separability of the optical and electronic subsystems. This step implicitly assumes that the field operator $\hat{\mathbf{A}}$ does not vary appreciably over the relevant length scales of the electronic wave functions, which is generally a good approximation in free space. However, as we will see further on, the wave vectors of optical modes in (antenna-coupled) tunnel junctions can be orders of magnitude higher than in free space. This sets limits to the theory outlined above and necessitates more refined models.

10.3 Coupling Tunnel Junctions to Free Space

Light emission from tunnel junctions has been studied experimentally for four decades. These studies can historically be separated into three stages which we will discuss in the following three subsections. Furthermore, these stages can be assigned to different research communities. The first experiments emerged from the solid state research community investigating lateral tunnel junctions primarily based on the insulator formed by oxidizing evaporated aluminum electrodes in the mid 1970s [32]. Approximately ten years later, LEIT was rediscovered in the scanning tunneling microscopy (STM) community and investigated in terms of its potential as a new spectroscopic tool [33]. Recently, renewed interest has emerged within the optics community based on the goal to realize electrically driven optical antennas by coupling them to tunnel junctions in an integrated device. Such devices constitute nanoscopic light sources, i.e. highly integrable links between electrical and optical functional units.

10.3.1 Macroscopic Solid State Tunnel Devices

10.3.1.1 Optical Modes of MIM Tunnel Junctions

A typical metal-insulator-metal (MIM) sample configuration—as investigated in the first experiments of Lambe and McCarthy [5]—is shown in Fig. 10.4. First, a aluminum electrode is fabricated on a substrate via evaporation. Subsequently, the natural oxide is allowed to form on the electrode either by oxygen plasma anodization [5, 34–37], thermal oxidation [6, 38–44] or exposure to oxygen-rich environments [45–47]. Similar experiments have also been carried out with oxidized magnesium electrodes [48]. The tunnel junction is finalized by crossing the oxidized aluminum electrode with a second electrode, in most cases consisting of metals with good plasmonic properties.

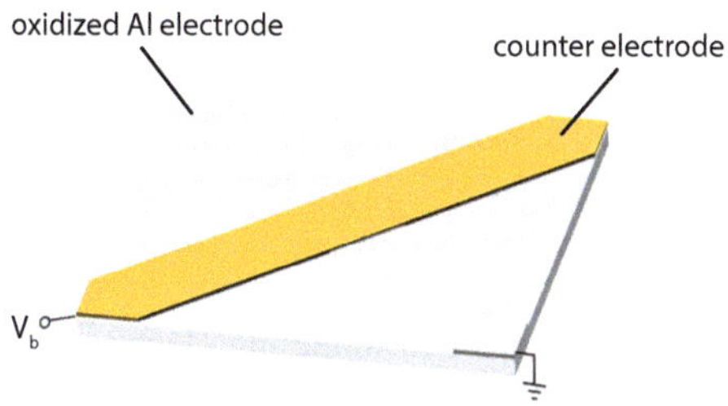

Fig. 10.4 Illustration of the sample configuration employed in the first experiments investigating LEIT. A bottom aluminum electrode is evaporated, oxidized to form the tunnel barrier and subsequently crossed with a counter electrode to form a tunnel junction

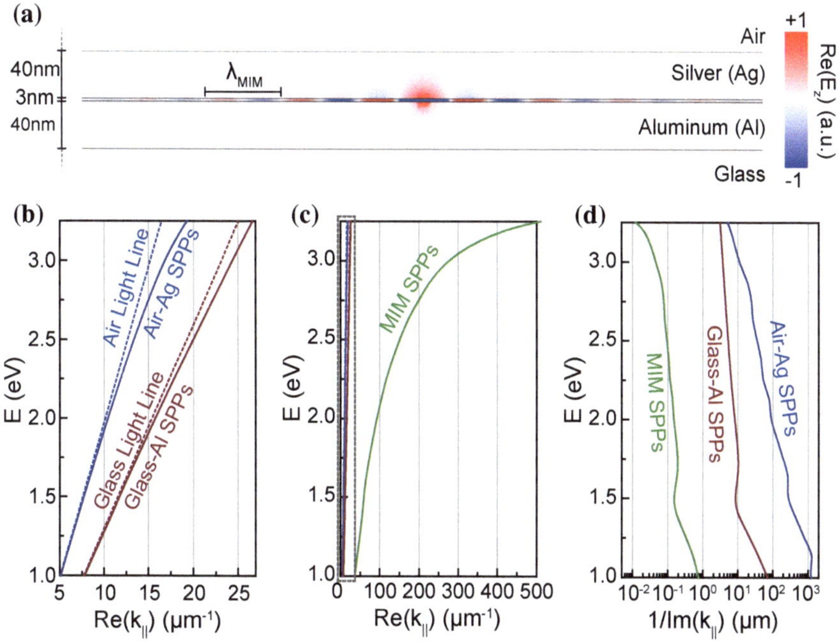

Fig. 10.5 Optical modes supported by MIM tunnel junctions. **a** Electric field distribution of a point dipole of frequency $\omega = 3 \times 10^{15}$Hz ($E_{ph} \sim 2$eV) located in the center of the insulator of a Al-Al_2O_3-Ag stack. The dipole primarily couples to MIM SPP modes localized to the insulator region. **b** Dispersion relation of the SPP modes supported by the air-silver and glass-aluminum interfaces supported by the geometry depicted in (**a**). **c** Dispersion relation of the MIM SPP mode, the dashed square indicates the region shown in (**b**). **d** Propagation lengths as a function of mode energy for the three modes

As shown in Sect. 10.2.3.3, the rate of inelastic electron tunneling depends on the density of optical modes ρ_p in the tunnel barrier. Figure 10.5 summarizes the optical properties of planar MIM tunnel barriers exemplarily considering the case of a Al-Al_2O_3-Ag tunneling device fabricated on top of a glass substrate. Such geometries are well known to support electromagnetic modes at the interfaces between the different materials [49]. The first two modes of interest are surface plasmon polariton (SPP) modes supported by the bottom and top metal-dielectric interfaces. The third mode, the MIM SPP mode, is localized to the insulator region between the two metal electrodes. The dispersion relations for these three modes are shown in panels (b) and (c) of Fig. 10.5. Considering the case of semi-infinitely thick metal electrodes it is obvious that tunneling electrons can only couple to the MIM SPP mode. However, for finite metal electrode thicknesses we have to keep in mind that the electric fields of the interface modes at top and bottom decay exponentially away from the interface. The penetration depth of these modes is approximately given by [1]

$$\delta_{\mathrm{m}} \approx \frac{c}{\omega}\sqrt{\frac{\varepsilon_{\mathrm{m}}+\varepsilon_{\mathrm{d}}}{\varepsilon_{\mathrm{m}}^{2}}} \tag{10.25}$$

with $\varepsilon_{\mathrm{d/m}}$ being the dielectric constants of the dielectric and metal, respectively. This quantity is usually of the order of a few tens of nanometers. Hence, for thin metal electrodes, the coupling of electrons to the evanescent tail of the top and bottom SPP modes has to be taken into account. Figure 10.5a shows the electric field distribution of a point dipole in the center of the insulator inside an MIM structure, emulating the behavior of a tunneling electron (c.f. (10.23)). Evidently, the dipole primarily couples to the MIM SPP mode.

10.3.1.2 Coupling Bound Modes to Free Space

It has already been shown in early theoretical calculations that the coupling to this mode can be fairly efficient with inelastic to elastic tunnel rate ratios of the order of a few percent [7]. However, in order to realize an efficient light source one has to optimize the conversion of these SPPs to photons, a process which is characterized by the parameter η_{antenna} in (10.2). The main challenge here is that by definition SPPs are *bound* interface modes. The non-radiative nature of these modes originates from the momentum mismatch between the mode and free space photons as seen in Fig. 10.5b, c. For each energy, the light line given by $\omega = ck_{\parallel}/\sqrt{\varepsilon_d}$ determines the maximum momentum available for photon emission into the two media. Hence, in order to observe any light emission at all, it is necessary to develop sample geometries which allow these modes to radiate. The first series of experiments related to LEIT was devoted to the exploration of different schemes to overcome this mismatch.

A broad range of momenta can be made available by introducing roughness in the electrodes, which may be present as a result of the fabrication procedure or introduced deliberately. This scheme was employed in the first demonstration of LEIT [5] and was studied in more detail later on [34, 36, 37, 47, 48, 50, 51]. Even though this approach proved reasonably effective in achieving light output, the random nature of surface roughness complicates comparison between theory and experiment [11] and led to some disagreements about whether the MIM SPP mode [5, 34, 47, 48] or the surface SPP mode [37, 42, 50, 51] is responsible for the observed light emission. Another method to realize the conversion of non-radiative evanescent fields into photons is to introduce scatterers in the form of metallic nanoparticles. The feasibility of this approach was demonstrated for nanoparticles placed on top of the junction [34, 52] as well as for top electrodes consisting of nanoparticles interconnected by a thin metallic film [45–47].

A different approach was taken by Kirtley et al. and others who fabricated tunnel junctions on top of holographic diffraction gratings with sinusoidal profiles [35, 38, 39]. This results in the periodic modulation of the three interfaces of the junction which introduces well-defined Fourier components/momenta $\Delta k_n = n(2\pi/a)$ with a being the grating period and n being a positive integer. This momentum exchange between SPP modes and the grating enables the coupling of the SPP modes to free space. Due to the discreteness of the Fourier components introduced this results in narrow spectral emission lines when analyzing the light emitted at certain angles with respect to the sample surface, essentially mapping out the (folded) dispersion relation of the radiating mode. Since the period was of the order of the free space wavelength of the emitted photons, the grating is most efficient at radiating the interface SPP as opposed to the MIM-SPP. Deviations between their experiments and theoretical calculations based on a model by Laks and Mills [12] led to the suggestion that hot electron decay may play an important role in the light emission process from the interface SPP modes [6]. The coupling of the MIM SPP mode via short period gratings has also been demonstrated [44, 53].

Most experiments analyzed light emission from the top (air) surface of planar tunnel junctions. In this situation, for nominally smooth films, indeed none of the modes can radiate without further modifications. However, considering the geometry depicted in Fig. 10.5a, when observing light emission from the bottom surface, the dispersion relation of the Air-Ag SPP mode lies to the left of the glass light line. Consequently, if the total sample thickness is of the order of the penetration depth, the top SPP mode may radiate into the substrate. This mode of operation was investigated by Ushioda et al. [40, 41, 43] who studied tunnel junctions fabricated on prisms. Similar to Kirtley et al. they found some discrepancies between model and experiments [14], which they argued was primarily caused by the conversion of the MIM SPP mode to the top SPP mode by surface roughness [15, 43].

Perhaps it is interesting to note that Laks and Mills theoretically predicted the existence of direct emission of light without an intermediate SPP excitation [11]. This mechanism was assumed to be responsible for the experimentally observed light emission in Al-Al_2O_3-Au tunnel junctions above a 'cut-off' energy of around 2.4eV for the MIM-SPP mode. However, as shown later [42, 54], the dispersion relation of this mode does not actually cut off despite being strongly damped beyond this energy. Further experiments did not find any experimental evidence for this process to play a major role [36, 50] with the exception of one report [55].

10.3.2 Scanning Tunneling Microscope

Photon emission from the scanning tunneling microscope (STM) was first observed in 1988 by Gimzewski et al. [56–58]. Since then, light emission with molecular [59, 60] and even atomic resolution [61–64], spectroscopic analysis of molecular vibrations [65–68], imaging of electronic wavefunctions [69, 70] and visualization of interactions between molecules [71] have been demonstrated, rendering STM light emission an invaluable spectroscopic tool within the STM community. In the words of R. Berndt, 'spatial mapping of these physical quantities permits the addition of true color to STM images' [33]. In the context of this chapter, we will focus on a limited number of publications that are important in the context of antenna-coupled tunnel junctions (c.f. Sect. 10.3.3). For a comprehensive review and overview over the topic, see [33, 72, 73].

In a configuration as depicted in Fig. 10.6, the combination of STM tip and metallic film leads to the formation of (dipolar) localized plasmonic modes with resonances in the visible and near-infrared part of the electromagnetic spectrum [16]. These modes can be understood when approximating the geometry as a nanoscopic sphere located over a metallic substrate [16]. The sphere effectively behaves as a dipole which induces a mirror dipole in the metallic substrate, forming a coupled system with amplified LDOS in the region between sphere and substrate [1]. STM experiments with tungsten tips on noble metal surfaces by Berndt et al. showed that these modes can be excited by inelastic electron tunneling from tip to sample, detectable as light emission in the far field [74]. Resonance position and spectral shape were shown to depend on the substrate material, in good agreement with theoretical calculations by Johansson et al. [16, 17]. While modes in MIM tunnel junctions are non-radiative in ideal geometries, the localized plasmon mode formed by tip and sample is inherently radiative and can be observed directly. This localized mode however also couples to SPP modes of the surface of the metallic film, turning the STM tip into a local excitation source for surface plasmons. For thin metal films these SPPs become radiative due to the higher refractive index of the substrate as described earlier. Radiation from these modes can be observed by prism-coupling [75, 76] or by high NA oil-immersion objectives [77, 78]. Consequently, the STM

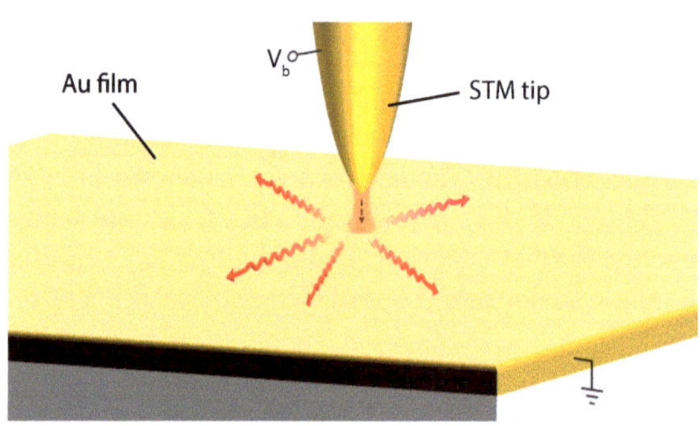

Fig. 10.6 Light emission from a scanning tunneling microscope (STM). Inelastic electron tunneling excites propagating surface plasmons as well as strongly localized (dipolar) plasmonic mode which are formed by the STM tip and the metallic film

was used to excite propagating SPPs confined to nanowires [79] and metallic stripes [80]. Furthermore, STM excitation of localized SPP resonances of arrays of as well as isolated nanostructures has been demonstrated [81–84].

The main advantage of the STM is its ability to analyze light emission properties of samples, e.g. molecular layers, as a function of position with sub-nanometer resolution [73]. STM light emission experiments have greatly contributed to the understanding of the physics behind LEIT and have led the way to the development of antenna-coupled tunnel junctions (ACTJs)—the main focus of this chapter—which will be discussed in the following.

10.3.3 Antenna-Coupled Tunnel Junctions

10.3.3.1 LEIT—An Antenna Problem

The problem at hand from an electrodynamics point of view is summarized in (10.24). In order to realize an efficient light emitter based on inelastic electron tunneling one has to optimize two parameters, namely the LDOS ρ_{p} as well as the radiation efficiency η_{antenna}. We have introduced optical antennas as elements that are able to resonantly enhance the localization of electromagnetic fields (c.f. Sect. 10.1). By reciprocity this corresponds to a high LDOS in the active region of the antenna. Furthermore, impedance matching of optical antennas allows for the optimization of their radiation efficiency [85, 86]. In fact, since the very early days of LEIT the conversion of localized optical modes to free space radiation has been referred to as an 'antenna problem' [8].

Two groups have recently developed antenna-coupled tunnel junctions (ACTJs). J. Kern et al. after initially demonstrating the feasibility of attaching electrical leads to a linear gap antenna [87], realized a light emitting device by placing a gold nanoparticle into the feedgap of such an antenna [88]. The tunnel junction is formed by the ligand shell surrounding the nanoparticle and separating it from the antenna arms. On the other hand, our group simultaneously demonstrated antenna-coupled light emission from vertical MIM tunnel devices. Arrays of resonant slot antennas were fabricated into one of the electrodes to efficiently outcouple the energy associated with MIM gap modes [89]. In the following we will discuss the operating principle of these antenna-coupled MIM devices.

Another mode of operation has been demonstrated by the group of A. Bouhelier. Blackbody-like light emission from electromigrated tunnel junctions was observed and shown to originate from the spontaneous emission of hot electrons [90, 91].

10.3.3.2 Antenna Geometry

The ACTJ geometry employed by us is depicted in Fig. 10.7. The optical properties of the device are determined by the layout of the gold bottom electrode shown in

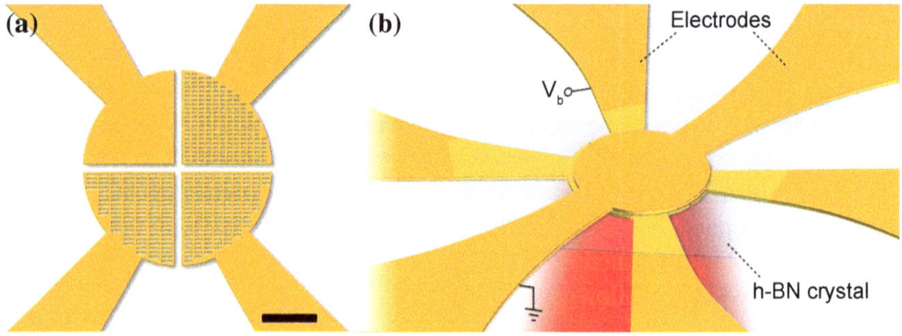

Fig. 10.7 Design of a four-quadrant ACTJ geometry. **a** Sketch of the bottom electrode design. A gold electrode is sectioned into electrically isolated quadrants. While one electrode remains unstructured to serve as a reference, the remaining three electrodes are patterned into arrays of slot antennas with varying aspect ratios. Scale bar 2 μm. **b** Illustration of the final ACTJ device including few-layer h-BN and top electrode. The geometry favors photon emission into the substrate where light is collected. The emission spectrum is tailored by the optical properties of the slot antennas

panel (a). A gold electrode is patterned into four electrically isolated quadrants. One of these electrodes remains unstructured and finally constitutes a reference device similar in terms of its optical mode properties to the MIM devices discussed in Sect. 10.3.1. The remaining three electrodes are patterned into arrays of slot antennas which form the basis for the final ACTJs. The physical dimensions of the slot antennas determine their optical properties, i.e. resonance wavelength. The insulator of the vertical tunnel junction is formed by few-layer hexagonal boron nitride (h-BN) which is exfoliated from bulk crystals [92]. h-BN provides a stable tunnel barrier at ambient conditions and is characterized by an electronic band gap of ~6 eV [93–95]. The ACTJs are finalized by the formation of a common Au top electrode.

10.3.3.3 Optical Modes of Slot ACTJs

According to (10.21), the electron-to-plasmon conversion rate is directly linked to the partial LDOS ρ_p along the direction of electron flow. The fraction of plasmons which are emitted as photons is determined by the antenna efficiency $\eta_{antenna}$. For simplicity we may combine these two parameters into a single parameter, namely the radiative LDOS $\rho_{rad} = \rho_p \times \eta_{antenna}$. The relation between the normalized LDOS and the power dissipation of a dipole (c.f. (10.23)) allows us to numerically determine the optical mode density of the tunnel gap in our geometry, shown in Fig. 10.8 exemplarily for a $250 \times 50 \times 50\,nm^3$ slot antenna integrated in a 50 nm–3 nm–15 nm Au–h-BN–Au MIM tunnel junction.

Slot ACTJs support a magnetic dipole mode along the long axis of the slot [89]. The resonance position can be tuned by changing the dimensions of the antenna. In particular, increasing the aspect ratio of the slot leads ot a shift of the resonance towards lower energies. Interestingly, slot ACTJs not only support this magnetic

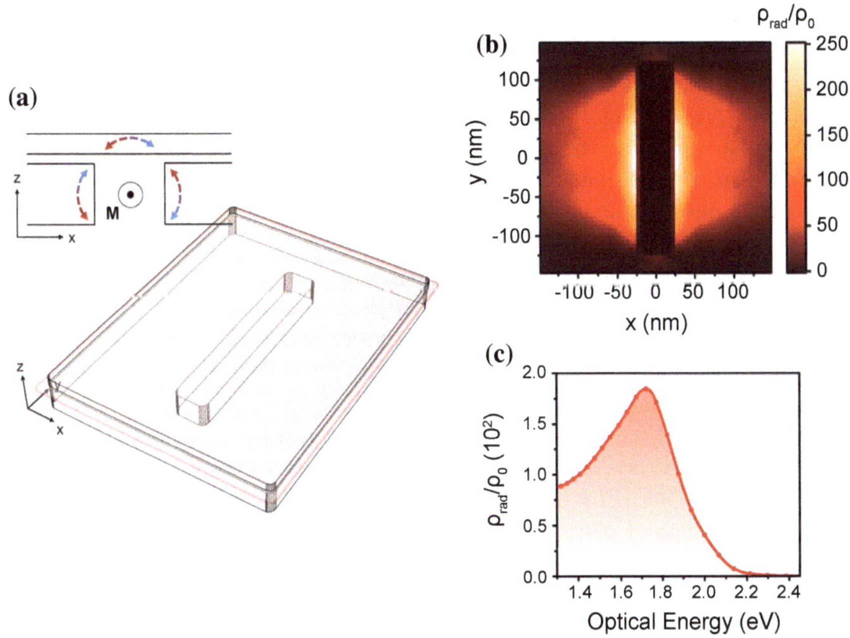

Fig. 10.8 **a** Sketch of an isolated slot antenna integrated in a MIM geometry. The plane indicates the region over which we evaluate the LDOS in the tunnel gap (c.f. panel (**b**)). Electron to far-field radiation coupling is facilitated by a magnetic dipole mode along the long axis of the mode as indicated in the cross-section. Arrows represent the current distribution associated with this mode. **b** Normalized radiative LDOS $\rho_{rad} = \rho_p \times \eta_{antenna}$ calculated as a function of position in the tunneling gap at an energy of 1.7 eV. The LDOS is maximal in the vicinity of the long axis of the $250 \times 50 \times 50\,nm^3$ slot. **c** Radiative LDOS spectrum at a distance of 5 nm from the antenna edge

mode but also an electric dipole mode along the short axis of the slot. However, due to the symmetry of the mode tunneling electrons primarily couple to the magnetic mode. Simplistically we illustrate this symmetry by the arrows—representing current oscillations—in the cross-section through the antenna shown in Fig. 10.8a. This current distribution facilitates strong electric fields along z in the tunnel gap which corresponds to an enhancement of the partial LDOS along the direction of electron flow. Numerical simulations allow us to map out the LDOS of a resonant slot ACTJ as a function of position in the tunnel gap as shown in Fig. 10.8b. The LDOS enhancement is strongest in the vicinity of the antenna and falls of gradually with increasing distance. The resonant character of the LDOS enhancement becomes apparent when analyzing ρ_{rad} as a function of energy as shown in Fig. 10.8c. The spectrum exhibits a resonance at an energy of 1.7 eV. The asymmetry of the spectrum is a result of the coupling of the localized magnetic dipole mode and the MIM-SPP mode, the damping of which strongly decreases with increasing mode energy [89].

10.3.3.4 Light Emission Characteristics

The resonant nature of the slot antennas becomes apparent from the optical transmission image shown in Fig. 10.9a. While the reference device simply appears dark due to the broad-band absorption of the bottom and top electrode, the antenna arrays integrated in the remaining three quadrants resonantly enhance the transmission through the device. The change in resonance wavelength towards lower energies (longer wavelengths) with increasing antenna aspect ratio (in clockwise order) of the quadrants results in a change of observed color.

We now turn to the light emission properties of slot ACTJs. Figure 10.9c shows a real space image of the light emitted by an unstructured reference device. As seen from the image, emission is limited to the edges of the device. We can understand this result by recalling the discussion of planar MIM tunnel junctions in Sect. 10.3.1. Since no surface roughness was deliberately introduced we would naturally expect very little light emission from the reference device. However, device edges act as abrupt perturbations and provide a broad range of momenta needed to couple SPP modes to free space. Contrarily, ACTJs emit from the entire device area as seen in Fig. 10.9d. Local variations of light emission intensity are caused by fabrication-related device inhomogeneities.

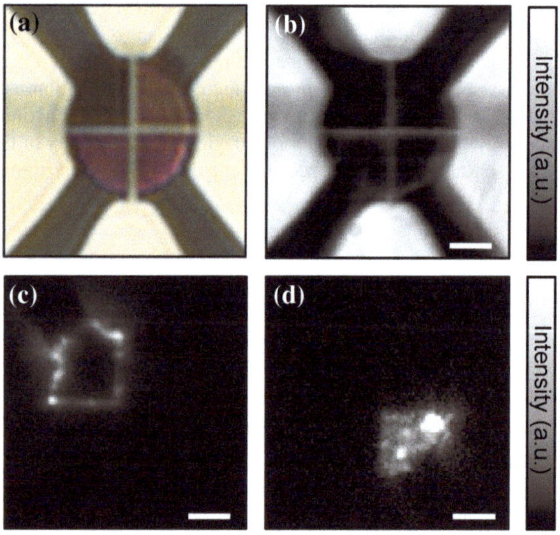

Fig. 10.9 **a** Optical transmission image of a four-quadrant ACTJ device. Slot antennas cause polarization-sensitive, resonant transmission enhancement as evident from the color variations amongst the structured quadrants. **b–d** EMCCD image of light transmitted through the ACTJ device (**c**) and of light emitted by the reference device (**c**) as well as one of the structured devices (**d**). While light emission of the reference device is limited to the device edge, the slot ACTJ quadrant emits from the entire device area. Scale bar 2 μm

A characteristic feature of LEIT is the restriction on photon energies given by the applied bias as $\hbar\omega \leq |eV_b|$. This allows for the color of the emitted light to be tuned by the bias voltage. It also serves as an electrical control to turn photoemission 'on' or 'off'. This is demonstrated in the light emission spectra shown in Fig. 10.10. The spectra exhibit a cut-off which is controlled by the applied bias and may be tuned continuously. The appearance of a resonance peak shows that slot antennas indeed control and shape the emission spectrum of the device as opposed to acting as mere perturbations similar to the edges of the reference device. The same figure shows calculated device efficiency spectra at the corresponding bias values based on the radiative LDOS shown in Fig. 10.8c and the theory presented in Sect. 10.2. We find good agreement between theory and experiment. The apparent discrepancy at $V_b = 1.75\ V$ is primarily caused by the slight difference in resonance position. Our theoretical calculations slightly overestimates the external device efficiency, primarily because we have neglected the position-dependence of the LDOS spectrum.

The slot antennas used in our devices also define the radiation pattern and the polarization of the emitted light. We find an emission pattern and polarization properties that agree with a magnetic dipole [89]. On the other hand, in the case of linear gap antennas, the emission was found to agree with an electric dipole mode [88].

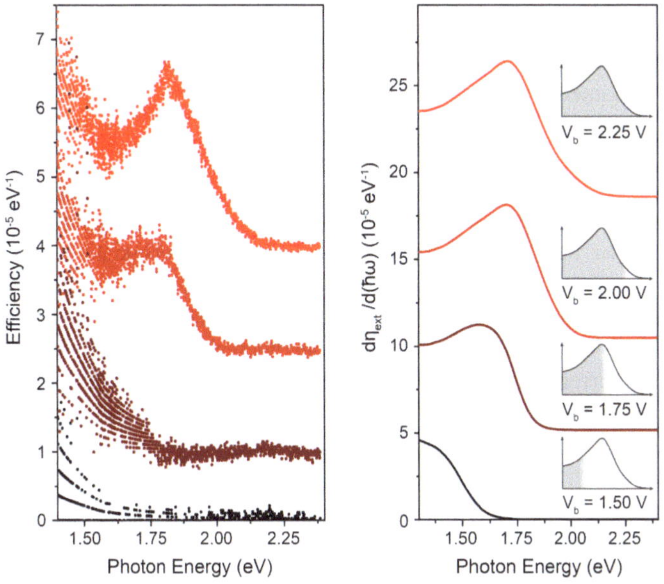

Fig. 10.10 *Left* Experimentally measured light emission spectra of a slot ACTJ device as a function of bias. The integrated slot antennas are $250 \times 50 \times 50\,\text{nm}^3$ in size. The applied bias is increased from 1.50 V to 2.25 V in steps of 0.25 V from *bottom* to *top*. *Right* Corresponding theoretically calculated external device efficiency spectra for a one-dimensional model assuming free-electron-like dispersion relations, a barrier height of 1 eV and width of 3 nm. The inset indicates the range of photon energies accessible for each bias value in relation to the LDOS spectrum (c.f. Fig. 10.8c)

10.3.4 Conclusion

While modern nanofabrication techniques as well as the development of new materials have enabled the realization of nanoscale light emitting devices based on inelastic electron tunneling, these devices are still far from reaching the application stage. Device efficiencies are still lower than 10^{-3} photons per electron. However, we are convinced that many concepts which promise interesting physical insights as well as the prospect to further increase efficiencies remain to be explored.

10.4 Outlook

We conclude this chapter by a discussion of future research directions and potential applications.

10.4.1 Ultrafast Photon/SPP Sources

Despite of their low efficiencies, ACTJs are of interest for ultrafast electronic-optical signal transduction from the electronic into the optical domain by LEIT-based devices. Fundamentally, the time required for electronic to optical signal conversion is given by the tunneling time, which is of the order of femtoseconds [96, 97]. Furthermore, owing to the small footprint of optical antennas, RC time constants governing the electronic response time can be very small. However, modulation experiments are currently hindered by low efficiencies/luminance of ACTJs. Nonetheless, we have carried out proof-of-principle experiments employing time-correlated single photon counting (TCSPC) techniques to demonstrate high modulation bandwidths. Setup and experimental results are summarized in Fig. 10.11. Modulation up to a frequency of 1 GHz has been demonstrated. Modulation frequencies are currently not limited by the device response time but by the accuracy of the measurement technique. As we show in Fig. 10.11c, the limit of this measurement technique can be greatly improved by reducing the timing jitter of the detection scheme.

10.4.2 LDOS and Impedance Matching Optimization

The efficiency of LEIT-devices can be improved by favorable antenna designs (c.f. (10.24)). Employing the model calculations shown in Fig. 10.10 we estimate the internal electron to MIM-SPP mode conversion efficiency to be approximately 1 %.

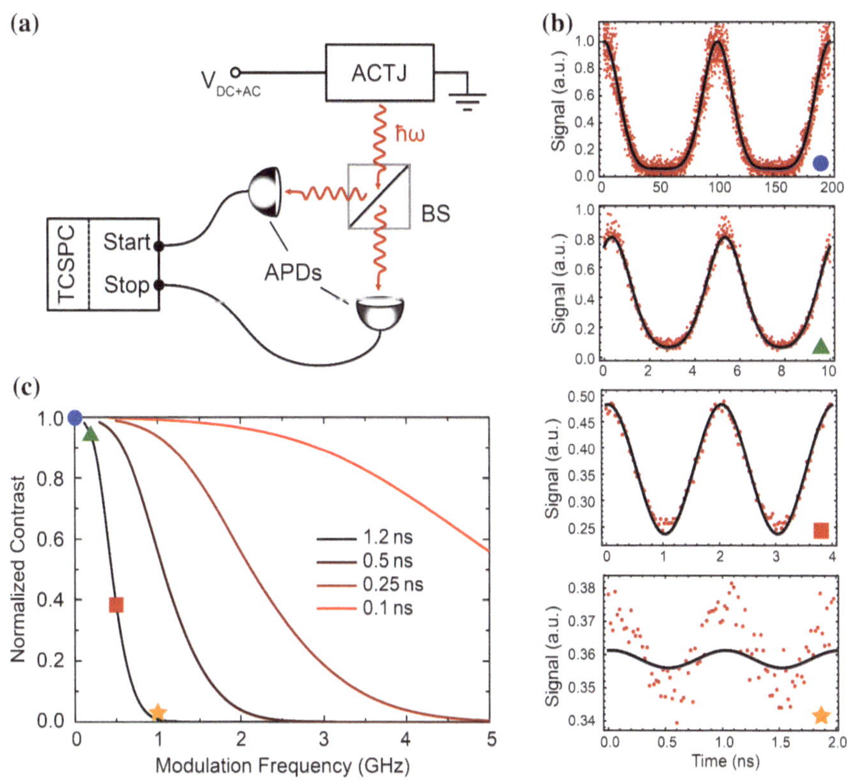

Fig. 10.11 **a** Experimental configuration for electrically modulating the light emission. The ACTJ is driven by a voltage of the functional form $V(t) = V_{DC} + V_{AC}\sin(2\pi ft)$. Emitted photons are directed to two avalanche photodiodes (APDs) via a beamsplitter (BS). The electrical photon-pulses are used to generate histograms of interphoton arrival times in a TCSPC module. **b** Averaged histograms $S_f(t)$ for several modulation frequencies from $f = 10\,\text{MHz}$ (*top*) to $f = 1\,\text{GHz}$ (*bottom*) for $V_{DC} = 1.50\,\text{V}$ and $V_{AC} = 0.50\,\text{V}$. The decrease in modulation contrast is caused by the timing jitter τ, i.e. the accuracy of the time delay measurement. $S_{10\text{MHz}}(t)$ is fitted by the series $S_{10\text{MHz}}(t) = \sum_{n=1}^{\infty} a_n + b_n \sin(2\pi nft + \phi_n)$ which determines the unaffected signal form $S_f^0(t)$ since τ is negligibly small. The signal evolution with increasing modulation frequency is well reproduced by folding $S_f^0(t)$ with a normal distribution with FWHM of $\tau = 1.2\,\text{ns}$. **c** Calculated contrast $C = (S_{max} - S_{min})/(S_{max} + S_{min})$ as a function of modulation frequency for several values of τ. The colored symbols indicate the experimentally determined contrast of the corresponding averaged histograms in panel (**b**)

The main difficulty in order to achieve higher external device efficiencies is the large impedance mismatch between MIM-SPPs and free space. Adiabatic mode conversion can be pursued to optimize MIM-SPP to free-space coupling [98].

10.4.3 Beyond MIM Devices

To increase the external device efficiency, one has to either suppress elastic tunneling or enhance inelastic tunneling, for example through molecular engineering [99]. The influence of the density of initial and final electronic densities on inelastic electron tunneling is well known within the STM community [63, 64, 100, 101]. Molecular junctions have already been studied for the purpose of light emission (see references in [73]). First solid-state light emitting devices based on molecular tunnel junctions have been recently demonstrated [102]. Furthermore, Van der Waals heterostructures hold promise to control the electronic properties of ACTJ-devices down to the atomic level [103].

10.4.4 Resonant Tunneling

Ten years after the initial discovery of LEIT, E.M. Belenov et al. theoretically predicted greatly improved quantum efficiencies for metal-barrier-metal structures containing quantum wells [104]. The resulting matrix elements for inelastic tunneling are resonantly enhanced if the energy levels of the quantum wells are suitably aligned. In a follow-up work, this phenomenon has been studied more elaborately in the context of plasmonic nanoantennas [27]. Although the efficiency of antenna excitation greatly depends on the exact properties of the barrier, efficiencies of the order of 10 % seem reachable. Efficiencies approaching unity can be achieved for the case of two quantum wells integrated into the barrier. If the energy levels of the two quantum wells differ by an energy $\hbar\omega$ in a geometry with a reasonably high optical mode density $\rho_p(\hbar\omega)$, inelastic electron tunneling becomes the dominant transfer channel [104].

10.4.5 Stimulated Emission

Our theoretical treatment in Sect. 10.2 considers inelastic electron tunneling as a spontaneous emission process. To the best of our knowledge this is well suited for the description of all experimental results reported so far. It would be highly interesting to devise experiments or devices which demonstrate the possibility to enter a regime where tunneling electrons and SPPs interact dynamically. Three publications in the late 70s have predicted that

1. Inelastic electron tunneling can be stimulated by SPPs; [105]
2. SPP propagation loss can be compensated by stimulated emission of SPPs; [106]
3. A MIM cavity of sufficiently high Q-factor can sustain stimulated SPP emission [107].

Recent experiments performed in the group of A. Meixner may provide evidence for such stimulated light-emission mechanisms in molecular junctions [108].

10.4.6 Beyond Visible Light Emission

LEIT experiments have been largely focused on light emission in the visible domain with the exception of one study on infrared STM light emission [109]. It might be interesting to explore device geometries featuring optical modes in the infrared and terahertz spectral domain. A promising material in the infrared to terahertz frequency range is graphene [110]. Several theoretical papers have already discussed inelastic electron tunneling in graphene-based tunnel devices [111–113].

10.5 Summary

We have introduced a theoretical description of LEIT which links the rate of electron-to-plasmon conversion to the optical density of states. We have discussed the chronological development of LEIT starting from the initial experiments on MIM tunnel devices, followed by light emission experiments using STM. Coupling tunnel junctions to optical antennas allows for the realization of electrically driven, nanoscale light emitters. The toolbox of antennas enables control over radiation patterns and polarization characteristics of such emitters.

Acknowledgements We thank J.B. Khurgin for valuable discussions. This work was supported by the NCCR-QSIT programme (grant no. 51NF40-160591) and the Swiss National Science Foundation (grant no. 200020_165841).

References

1. L. Novotny, B. Hecht, *Principles of Nano-Optics*, 2nd edn. (Cambridge University Press, Cambridge, 2012)
2. P. Bharadwaj, B. Deutsch, L. Novotny, Adv. Opt. Photonics **1**, 438 (2009)
3. P. Biagioni, J.-S. Huang, B. Hecht, Rep. Prog. Phys. **75**, 024402 (2012)
4. L. Novotny, N. van Hulst, Nat. Photonics **5**, 83 (2011)
5. J. Lambe, S.L. McCarthy, Phys. Rev. Lett. **37**, 923 (1976)
6. J. Kirtley, T. Theis, J. Tsang, D. DiMaria, Phys. Rev. B **27**, 4601 (1983)
7. L.C. Davis, Phys. Rev. B **16**, 2482 (1977)
8. D. Hone, B. Mühlschlegel, D.J. Scalapino, Appl. Phys. Lett. **33**, 203 (1978)
9. R.W. Rendell, D.J. Scalapino, Phys. Rev. B **24**, 3276 (1981)
10. R.W. Rendell, D.J. Scalapino, B. Mühlschlegel, Phys. Rev. Lett. **41**, 1746 (1978)
11. B. Laks, D.L. Mills, Phys. Rev. B **20**, 4962 (1979)
12. B. Laks, D.L. Mills, Phys. Rev. B **22**, 5723 (1980)
13. K. Arya, R. Zeyher, Phys. Rev. B **28**, 4080 (1983)

14. S. Ushioda, J.E. Rutledge, R.M. Pierce, Phys. Rev. B **34**, 6804 (1986)
15. A. Takeuchi, J. Watanabe, Y. Uehara, S. Ushioda, Phys. Rev. B **38**, 12948 (1988)
16. P. Johansson, R. Monreal, P. Apell, Phys. Rev. B **42**, 9210 (1990)
17. P. Johansson, R. Monreal, Z. Phys. B **84**, 269 (1991)
18. P. Johansson, Phys. Rev. B **58**, 10823 (1998)
19. J. Aizpurua, S.P. Apell, R. Berndt, Phys. Rev. B **62**, 2065 (2000)
20. Y. Uehara, Y. Kimura, S. Ushioda, K. Takeuchi, Jpn. J. Appl. Phys. **31**, 2465 (1992)
21. J. Bardeen, Phys. Rev. Lett. **6**, 57 (1961)
22. A.J. Bennett, C.B. Duke, S.D. Silverstein, Phys. Rev. **176**, 969 (1968)
23. C.B. Duke, *Tunneling in Solids* (Academic Press, New York, 1969)
24. B.N.J. Persson, A. Baratoff, Phys. Rev. Lett. **68**, 3224 (1992)
25. T. Schimizu, K. Kobayashi, M. Tsukada, Appl. Surf. Sci. **60–61**, 454 (1992)
26. A. Downes, M.E. Taylor, M.E. Welland, Phys. Rev. B **57**, 6706 (1998)
27. A.V. Uskov, J.B. Khurgin, I.E. Protsenko, I.V. Smetanin, A. Bouhelier. arXiv:1603.06721 [physics.optics] (2016)
28. H.J. Reittu, Am. J. Phys. **63**, 940 (1995)
29. C.J. Chen, *Introduction to Scanning Tunneling Microscopy*, 2nd edn. (Oxford University Press, New York, 1993)
30. D.A. Ryndyk, *Theory of Quantum Transport at Nanoscale* (Springer, New York, 2016)
31. E. Snoeks, A. Lagendijk, A. Polman, Phys. Rev. Lett. **74**, 2459 (1995)
32. D. Mills, M. Weber, B. Laks, in *Tunneling Spectroscopy: Capabilities, Applications, and New Techniques*, ed. by P.K. Hansma (Plenum Press, New York, 1982) Chapter 5
33. R. Berndt, in *Scanning Probe Microscopy: Analytical Methods*, ed. by. R. Wiesendanger (Springer, Berlin Heidelberg, 1998) Chapter 5
34. S.L. McCarthy, J. Lambe, Appl. Phys. Lett. **30**, 427 (1977)
35. N. Kroo, Z. Szentirmay, J. Felszerfalvi, Phys. Lett. A **81**, 399 (1981)
36. K. Parvin, W. Parker, Solid State Commun. **37**, 629 (1981)
37. D.G. Walmsley, H.F. Quinn, P. Dawson, Phys. Rev. Lett. **49**, 892 (1982)
38. J.R. Kirtley, T.N. Theis, J.C. Tsang, Appl. Phys. Lett. **37**, 435 (1980)
39. J.R. Kirtley, T.N. Theis, J.C. Tsang, Phys. Rev. B **24**, 5650 (1981)
40. S. Ushioda, J.E. Rutledge, R.M. Pierce, Phys. Rev. Lett. **54**, 224 (1985)
41. R.M. Pierce, J.E. Rutledge, S. Ushioda, Phys. Rev. B **36**, 1803 (1987)
42. J.B.D. Soole, H.P. Hughes, Surf. Sci. **197**, 250 (1988)
43. J. Watanabe, A. Takeuchi, Y. Uehara, S. Ushioda, Phys. Rev. B **38**, 12959 (1988)
44. P.D. Sparks, T. Sjodin, B.W. Reed, J. Stege, Phys. Rev. Lett. **68**, 2668 (1992)
45. P.K. Hansma, H.P. Broida, Appl. Phys. Lett. **32**, 545 (1978)
46. A. Adams, J.C. Wyss, P.K. Hansma, Phys. Rev. Lett. **42**, 912 (1979)
47. A. Adams, P.K. Hansma, Phys. Rev. B **23**, 3597 (1981)
48. R.K. Jain, S. Wagner, D.H. Olson, Appl. Phys. Lett. **32**, 62 (1978)
49. E.N. Economou, Phys. Rev. **182**, 539 (1969)
50. P. Dawson, D.G. Walmsley, H.A. Quinn, A.J.L. Ferguson, Phys. Rev. B **30**, 3164 (1984)
51. M. Hänisch, A. Otto, J. Phys.: Condens. Matter **6**, 9659 (1994)
52. M.J. Bloemer, J.G. Mantovani, J.P. Goudonnet, D.R. James, R.J. Warmack, T.L. Ferrell, Phys. Rev. B **35**, 5947 (1987)
53. S. Ushioda, Y. Uehara, M. Takada, K. Otsubo, J. Murota, Jpn. J. Appl. Phys. **31**, L870 (1992)
54. B.N. Kurdi, D.G. Hall, Phys. Rev. B **34**, 3980 (1986)
55. A.J.L. Ferguson, P. Dawson, D.G. Walmsley, J. Phys.: Condens. Matter **1**, 9021 (1989)
56. J.H. Coombs, J.K. Gimzewski, B. Reihl, J.K. Sass, R.R. Schlittler, J. Microsc. **152**, 325 (1988)
57. J.K. Gimzewski, B. Reihl, J.H. Coombs, R.R. Schlittler, Z. Phys. B **72**, 497 (1988)
58. J.K. Gimzewski, J.K. Sass, R.R. Schlittler, J. Schott, Europhys. Lett. **8**, 435 (1989)
59. R. Berndt, R. Gaisch, J.K. Gimzewski, B. Reihl, R.R. Schlittler, W.D. Schneider, M. Tschudy, Science **262**, 1425 (1993)
60. E. Ćavar, M.-C. Blüm, M. Pivetta, F. Patthey, M. Chergui, W.-D. Schneider, Phys. Rev. Lett. **95**, 196102 (2005)

61. R. Berndt, R. Gaisch, W.D. Schneider, J.K. Gimzewski, B. Reihl, R.R. Schlittler, M. Tschudy, Phys. Rev. Lett. **74**, 102 (1995)
62. C. Thirstrup, M. Sakurai, K. Stokbro, M. Aono, Phys. Rev. Lett. **82**, 1241 (1999)
63. G.V. Nazin, X.H. Qiu, W. Ho, Phys. Rev. Lett. **90**, 216110 (2003)
64. G. Hoffmann, T. Maroutian, R. Berndt, Phys. Rev. Lett. **93**, 076102 (2004)
65. X. Qiu, G. Nazin, W. Ho, Science **299**, 542 (2003)
66. Z.C. Dong, X.L. Guo, A.S. Trifonov, P.S. Dorozhkin, K. Miki, K. Kimura, S. Yokoyama, S. Mashiko, Phys. Rev. Lett. **92**, 086801 (2004)
67. Z.C. Dong, X.L. Zhang, H.Y. Gao, Y. Luo, C. Zhang, L.G. Chen, R. Zhang, X. Tao, Y. Zhang, J.L. Yang, J.G. Hou, Nat. Photonics **4**, 50 (2010)
68. C. Chen, P. Chu, C.A. Bobisch, D.L. Mills, W. Ho, Phys. Rev. Lett. **105**, 217402 (2010)
69. G. Schull, M. Becker, R. Berndt, Phys. Rev. Lett. **101**, 136801 (2008)
70. C. Chen, C.A. Bobisch, W. Ho, Science **325**, 981 (2009)
71. Y. Zhang, Y. Luo, Y. Zhang, Y.-J. Yu, Y.-M. Kuang, L. Zhang, Q.-S. Meng, Y. Luo, J.-L. Yang, Z.-C. Dong, J.G. Hou, Nature **531**, 623 (2016)
72. M. Sakurai, C. Thirstrup, M. Aono, Appl. Phys. A **80**, 1153 (2005)
73. F. Rossel, M. Pivetta, W.-D. Schneider, Surf. Sci. Rep. **65**, 129 (2010)
74. R. Berndt, J.K. Gimzewski, P. Johansson, Phys. Rev. Lett. **67**, 3796 (1991)
75. K. Takeuchi, Y. Uehara, S. Ushioda, S. Morita, J. Vac. Sci. Technol. B **9**, 557 (1991)
76. S. Ushioda, Y. Uehara, M. Kuwahara, Appl. Surf. Sci. **60**, 448 (1992)
77. T. Wang, E. Boer-Duchemin, Y. Zhang, G. Comtet, G. Dujardin, Nanotechnology **22**, 175201 (2011)
78. S. Divitt, P. Bharadwaj, L. Novotny, Opt. Express **21**, 27452 (2013)
79. P. Bharadwaj, A. Bouhelier, L. Novotny, Phys. Rev. Lett. **106**, 226802 (2011)
80. Y. Zhang, E. Boer-Duchemin, T. Wang, B. Rogez, G. Comtet, E. Le Moal, G. Dujardin, A. Hohenau, C. Gruber, J.R. Krenn, Opt. Express **21**, 13938 (2013)
81. F. Silly, A.O. Gusev, A. Taleb, F. Charra, M.-P. Pileni, Phys. Rev. Lett. **84**, 5840 (2000)
82. P. Myrach, N. Nilius, H.-J. Freund, Phys. Rev. B **83**, 035416 (2011)
83. E. Le Moal, S. Marguet, B. Rogez, S. Mukherjee, P. Dos Santos, E. Boer-Duchemin, G. Comtet, G. Dujardin, Nano Lett. **13**, 4198 (2013)
84. E. Le Moal, S. Marguet, D. Canneson, B. Rogez, E. Boer-Duchemin, G. Dujardin, T.V. Teperik, D.-C. Marinica, A.G. Borisov, Phys. Rev. B **93**, 035418 (2016)
85. L. Novotny, Phys. Rev. Lett. **98**, 266802 (2007)
86. A. Alù, N. Engheta, Phys. Rev. Lett. **101**, 043901 (2008)
87. J.C. Prangsma, J. Kern, A.G. Knapp, S. Grossmann, M. Emmerling, M. Kamp, B. Hecht, Nano Lett. **12**, 3915 (2012)
88. J. Kern, R. Kullock, J. Prangsma, M. Emmerling, M. Kamp, B. Hecht, Nat. Photonics **9**, 582 (2015)
89. M. Parzefall, P. Bharadwaj, A. Jain, T. Taniguchi, K. Watanabe, L. Novotny, Nat. Nanotechnol. **10**, 1058 (2015)
90. M. Buret, A.V. Uskov, J. Dellinger, N. Cazier, M.-M. Mennemanteuil, J. Berthelot, I.V. Smetanin, I.E. Protsenko, G. Colas-des-Francs, A. Bouhelier, Nano Lett. **15**, 5811 (2015)
91. N. Cazier, M. Buret, A. Uskov, L. Markey, J. Arocas, G. Colas-des-Francs, A. Bouhelier, Opt. Express **24**, 3873 (2016)
92. T. Taniguchi, K. Watanabe, J. Cryst. Growth **303**, 525 (2007)
93. G.-H. Lee, Y.-J. Yu, C. Lee, C. Dean, K.L. Shepard, P. Kim, J. Hone, Appl. Phys. Lett. **99**, 243114 (2011)
94. L. Britnell, R.V. Gorbachev, R. Jalil, B.D. Belle, F. Schedin, M.I. Katsnelson, L. Eaves, S.V. Morozov, A.S. Mayorov, N.M. Peres, A.H. Castro Neto, J. Leist, A.K. Geim, L.A. Ponomarenko, K.S. Novoselov, Nano Lett. **12**, 1707 (2012)
95. K. Watanabe, T. Taniguchi, H. Kanda, Nat. Mater. **3**, 404 (2004)
96. R. Landauer, T. Martin, Rev. Mod. Phys. **66**, 217 (1994)
97. D. Shafir, H. Soifer, B.D. Bruner, M. Dagan, Y. Mairesse, S. Patchkovskii, M.Y. Ivanov, O. Smirnova, N. Dudovich, Nature **485**, 343 (2012)

98. M.I. Stockman, Phys. Rev. Lett. **93**, 137404 (2004)
99. N.J. Tao, Nat. Nanotechnol. **1**, 173 (2006)
100. R. Berndt, J.K. Gimzewski, Phys. Rev. B **48**, 4746 (1993)
101. P. Chen, W. Wang, N. Lin, S. Du, Opt. Express **22**, 8234 (2014)
102. W. Du, T. Wang, H.-S. Chu, L. Wu, R. Liu, S. Sun, W.K. Phua, L. Wang, N. Tomczak, C.A. Nijhuis, Nat. Photonics **10**, 274 (2016)
103. A. Geim, I. Grigorieva, Nature **499**, 419 (2013)
104. É.M. Belenov, P. Luskinovich, V. Romanenko, A. Sobolev, A.V. Uskov, Sov. J. Quantum Electron. **17**, 1348 (1987)
105. D.P. Siu, R.K. Jain, T.K. Gustafson, Appl. Phys. Lett. **28**, 407 (1976)
106. D.P. Siu, T.K. Gustafson, Appl. Phys. Lett. **32**, 500 (1978)
107. D.M. Drury, T. Ishii, IEEE J. Quantum Electron. **16**, 58 (1980)
108. K. Braun, X. Wang, A.M. Kern, H. Adler, H. Peisert, T. Chassé, D. Zhang, A.J. Meixner, Beilstein J. Nanotechnol. **6**, 1100 (2015)
109. M.G. Boyle, J. Mitra, P. Dawson, Appl. Phys. Lett. **94**, 233118 (2009)
110. A. Grigorenko, M. Polini, K. Novoselov, Nat. Photonics **6**, 749 (2012)
111. S.A. Khorasani, IEEE J. Quantum Electron. **50**, 307 (2014)
112. D. Svintsov, Z. Devizorova, T. Otsuji, V. Ryzhii. arXiv:1509.03776 [cond-mat.mes-hall] (2015)
113. K.J.A. Ooi, H.S. Chu, C.Y. Hsieh, D.T.H. Tan, L.K. Ang, Phys. Rev. Appl. **3**, 054001 (2015)

Chapter 11
Spontaneous Emission in Nonlocal Metamaterials with Spatial Dispersion

Brian Wells, Pavel Ginzburg, Viktor A. Podolskiy and Anatoly V. Zayats

Abstract Recent successes in fabrication, characterization, numerical computations, and theory have brought to life a new class of composite materials with engineered optical properties, metamaterials. Uniaxial anisotropic artificially created structures based on plasmonic nanowire arrays have emerged as a versatile platform for negative refraction, subwavelength optics, biosensing, acoustic sensing, and nonlinearity engineering. It has been demonstrated, both experimentally and theoretically, that the optical response of plasmonic nanowire arrays is strongly affected by nonlocal electromagnetism, a phenomenon where permittivity of metamaterial strongly depends not only on the frequency, but also on wavevector of the plane wave interacting with this structure. Nonlocal dielectric response leads to excitation of additional electromagnetic wave that does not exist in conventional, local, metamaterials. The dispersion of this wave can be engineered by adjusting composition and geometry of metamaterial. In this chapter we present comprehensive review of nonlocal electromagnetic properties in plasmonic nanowire metamaterials. We begin by introducing the material platform, explain the theoretical approach for nonlocal homogenization, and finally discuss the implication of material nonlocality for emission of light in nonlocal environment.

B. Wells (✉)
Department of Physics, University of Hartford, West Hartford, CT 06117, USA
e-mail: brwells@hartford.edu

P. Ginzburg
Department of Electrical Engineering Physical Electronics, Tel Aviv University, Ramat Aviv, 69978 Tel Aviv, Israel
e-mail: pginzburg@post.tau.ac.il

V.A. Podolskiy
Department of Physics and Applied Physics, University of Massachusetts Lowell, One University Ave, Lowell, MA 01854, USA
e-mail: viktor_podolskiy@uml.edu

A.V. Zayats
Department of Physics, King's College London, Strand, London WC2R 2LS, UK
e-mail: a.zayats@kcl.ac.uk

S.I. Bozhevolnyi et al. (eds.), *Quantum Plasmonics*,
Springer Series in Solid-State Sciences 185,
DOI 10.1007/978-3-319-45820-5_11

11.1 Introduction

Metamaterials are a class of composite materials which are designed to have optical properties that are not readily found in nature. There are numerous applications in modern optics which can be realized only through photonic metamaterials. These range from imaging, sensing, and security, to solar power, optical information processing, and photonic circuits [1–12].

One of metamaterial platforms that has attracted significant attention is anisotropic uniaxial metamaterials which behave as a dielectric for one polarization of electric field and a metal for the other [13–20]. One of the most important features of those artificially created structures is their ability to support extremely high density of photonic states, if effective material parameters along orthogonal crystallographic directions have different signs. This regime is called hyperbolic, as the iso-frequency surfaces of bulk modes have hyperbolic shapes. Having high density of states promises substantial efficiency improvements of various light-matter interaction processes. The most basic and straightforward process is spontaneous emission, that arises from modified vacuum fluctuations. This effect is commonly referred by the name of Purcell enhancement [21]. It is worth noting, that higher order effects, such as scattering, nonlinear harmonic generation, two-photon emission [22], Compton scattering [23] and many others are also tailored by modifying the density of states. While various types of homogenization procedures are applied in order to retrieve effective metamaterial properties, local types of interactions could prevail quantum dynamics. In this case, the local density of states could be substantially different from the quantities, obtained with global modes in homogenized crystal. This effect implies the need of heavy routine numerical calculations for mapping the local density of states inside large scale nanostructure. In order to surpass this problem and gain physical understanding of properties, different type of homogenization, capable of addressing quantum processes, should be developed. The major goal of the subsequent analysis is to develop relevant tools for analysis of emission processes in hyperbolic metamaterials.

Two topologically different realizations of such metamaterials are possible with metal-dielectric multilayers and aligned nanowire arrays [7, 13, 24–26]. There are also naturally occurring materials where similar properties are observed near phonon resonances [27, 28]. In optical frequency range, the nanowire-based composites (Fig. 11.1) have been proposed for applications in negative refraction, nonlinear optics, cloaking, and sensing [29–34]. Such metamaterials can be successfully fabricated using electrochemical techniques over macroscopic sizes with the radius r for the individual nanowires between 10 and 100 nm, a separation distance a between 50 and 300 nm, and broad range of lengths [7, 35]. Various metals can be used such as Au, Al, Ni, Co and many others [36, 37].

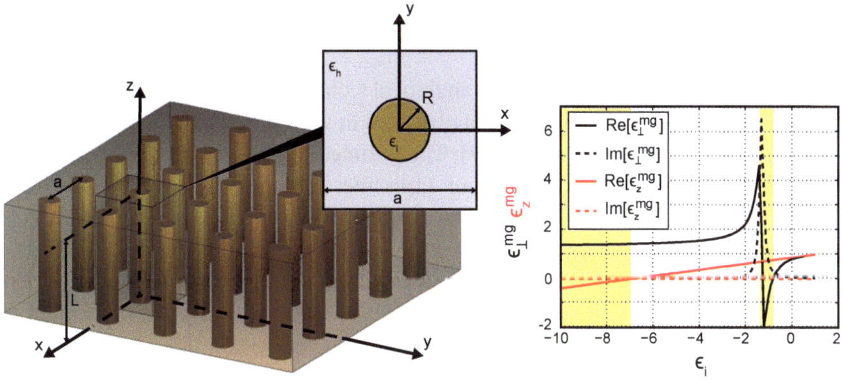

Fig. 11.1 *Left* schematic geometry and a unit cell of a nanowire composite. *Right* effective permittivity of the composite, $Im(\epsilon_i) = -0.1$; shaded areas represent the spectral range where the metamaterial operates in the hyperbolic regime. When $\epsilon_i \geq -1$, the metamaterial is local. An additional wave exists for $\epsilon_i \leq -1$. This wave propagates inside the metamaterial for $-7 \leq \epsilon_i \leq -1$ (the main wave has elliptical dispersion). When $\epsilon_i \leq -7$, the longitudinal wave exponentially decays along the wires, while the main wave has hyperbolic dispersion

With deep-subwavelength sizes involved, the effective medium approximation may be considered describing these metamaterials using anisotropic dielectric permittivity tensor, with the optical axis aligned in the direction of the nanowires. As such, their optical properties can be related to the dispersion of two waves with different polarizations, "ordinary" TE waves (electric field in the xy-plane) and "extraordinary" TM waves (magnetic field in the xy-plane). However, as shown in [38–40], in addition to these *transverse* electromagnetic waves, plasmonic nanowire metamaterials may also support *longitudinal* electromagnetic waves. In order to describe them, a nonlocal effective dielectric permittivity can be invoked incorporating spatial dispersion effects, while from a microscopic point of view they represent coupled cylindrical plasmon polariton (CPP) waves that propagate along the nanowires. The nonlocality significantly affects the "optical topology" of these metamaterials, drastically influencing the photonic density of states [41] in contrast with the predictions of local effective medium theory [22, 42–45], where Purcell enhancement is most pronounced in the hyperbolic regime.

In Sect. 11.2 we will introduce an analytical technique that provides an adequate description of electromagnetism in wire-based metamaterials that takes into account the nonlocal optical response originating from the homogenization procedure and develop an appropriate technique that significantly simplifies this calculation. In Sect. 11.3 we will overview the effects of optical nonlocality on the spontaneous radiation of emitters embedded inside the nanowire media. Experimental results on the Purcell effect with hyperbolic metamaterials will be revised in Sect. 11.4.

11.2 Nonlocal Effective Medium Theory

In this section, we will introduce the analytical technique that is needed to effectively describe electromagnetic phenomena observed in wire-based metamaterials that has been previously presented in [38, 39]. The outlined approach takes into account the nonlocal optical response originating from the homogenization procedure. This formalism combines the local and nonlocal effective-medium theories often used to describe the optics of nanowire composites in different limits and relates the origin of optical nonlocality to collective plasmonic excitation of wire composites. It will later be shown exactly how this formalism provides the required recipe to implement the needed additional boundary conditions for these composite structures.

We will develop this technique for the case of plasmonic nanowire metamaterials. These materials are formed by an array of aligned plasmonic nanowires embedded in a dielectric host. For simplicity, we fix the frequency of the electromagnetic excitations and the unit cell parameters of the system, and vary only the permittivity of the wire inclusions. It will be shown that the developed formalism can be readily applied for systems where both permittivity and frequency are changed at the same time. We assume that the system operates in the effective-medium regime (its unit cell $a \ll \lambda_0$ with λ_0 being the free-space wavelength) and that the surface concentration of plasmonic wires is small, $p \ll 1$ where $p = \pi R^2/a^2$. The parameters used in this chapter are $R = 20\,\text{nm}$, $a = 100\,\text{nm}$, $\epsilon_h = 1$, and $L = 1\,\mu\text{m}$ (see Fig. 11.1), which are typical for composites fabricated with anodized alumina templates [39].

The optical response of nanowire materials resembles that of uniaxial media with the optical axis parallel to the direction of the nanowires (z). Therefore, the dielectric permittivity tensor describing properties of the waves propagating in the wire media is diagonal with components $\epsilon_{xx} = \epsilon_{yy} = \epsilon_\perp$ and ϵ_z . It has been shown that at optical and near-IR frequencies, the transmission and reflection of these components is largely described by the Maxwell-Garnett type effective medium theory (EMT) [13, 46, 47]. In this approach, the microscopic distribution of the field is given by solutions of the Maxwell equations in the quasistatic limit

$$E_z^{mg} = e_z^{mg} \tag{11.1}$$

$$E_x^{mg} = e_x^{mg} \times \begin{cases} \frac{2\epsilon_h}{\epsilon_i + \epsilon_h}, r \le R \\ 1 + R^2 \frac{\epsilon_h - \epsilon_i}{\epsilon_i + \epsilon_h} \frac{y^2 - x^2}{(x^2 + y^2)^2}, r \ge R \end{cases} \tag{11.2}$$

with ϵ_i and ϵ_h being the permittivities of wire inclusions and of host material, respectively, and parameters e_x^{mg} and e_z^{mg} are the field amplitudes. Straightforward averaging of the jth component ($j = x, y, z$) of the fields over the unit cell yields the effective permittivity (Fig. 11.1)

$$\epsilon_z^{mg} = p\epsilon_i + (1-p)\epsilon_h \tag{11.3}$$

$$\epsilon_\perp^{mg} = \frac{2p\epsilon_i\epsilon_h + (1-p)\epsilon_h(\epsilon_i + \epsilon_h)}{2p\epsilon_h + (1-p)(\epsilon_i + \epsilon_h)} \tag{11.4}$$

When adjusting the composition of the metamaterial and operating wavelength, the optical response of the composite can be controlled between different optical regimes; elliptic $\left(\epsilon_\perp^{mg} > 0, \epsilon_z^{mg} > 0\right)$, epsilon-near-zero (ENZ, $\epsilon_z^{mg} \approx 0$) and hyperbolic $\left(\epsilon_\perp^{mg} > 0, \epsilon_z^{mg} < 0\right)$, as seen in Fig. 11.1 (right). For the ENZ and hyperbolic regimes, nanowire metamaterials support optical waves that have either small or very large effective modal index. This has been a motivating factor for a number of potential applications in light shaping [1, 9], cloaking [31], and subwavelength light manipulation [15, 18–20, 30].

At the same time, it has been shown that at lower frequencies where $-\epsilon_i \gg 1$, ϵ_z of wire composites becomes strongly nonlocal exhibiting a strong dependence on k_z [48–50]. Similar dependence has been recently shown to take place at visible frequencies in the ENZ regime [39]. Nonlocality, especially in the ENZ regime, has been shown to fundamentally alter the optical response of the wire composite, leading to the excitation of new types of optical waves, and requiring the use of additional boundary conditions for the analytical description of their excitation [51, 52].

Despite extensive previous research, until recently first-principal theoretical models describing the optics of wire composites [48, 49] could not be used at visible and (near)- IR frequencies, with remaining models requiring fitting [39, 50] or numerical solutions of Maxwell equations. The solution was first developed in [38]. Next we outline that formalism to ensure completeness throughout this work.

11.2.1 Calculation of E_z and H_z

The nanowire metamaterial can be separated into two regions, (i) inside and (ii) outside the wire material (Fig. 11.1). An expression for the fields will be derived for both by applying initial boundary conditions to the generalized field expression

$$\begin{aligned} F_z(r,\phi,z) = \sum_m & (c_{1m}J_m(\kappa r) + c_{2m}Y_m(\kappa r))\sin(m\phi)e^{ik_z z} \\ & + (c_{3m}J_m(\kappa r) + c_{4m}Y_m(\kappa r))\cos(m\phi)e^{ik_z z} \end{aligned} \tag{11.5}$$

where $F_z = \{E_z, H_z\}$.

First we will look at the solution inside the wire, where $\kappa_i = \sqrt{\epsilon_i\omega^2/c^2 - k_z^2}$ and as $r \to 0$, $Y_m(\kappa_i r) \to \infty$ thus c_{2m} and $c_{4m} = 0$, with k_z and ω being the wave vector in the z direction and the angular frequency of the plane wave, respectively, c being the speed of light in vacuum. Since the material is homogeneous along the z axis it is satisfactory to examine the fields just at $z=0$,

$$F_z^i = \sum_m J_m(\kappa_i r)\left(A_m^i \sin(m\phi) + B_m^i \cos(m\phi)\right) \tag{11.6}$$

with $E_z^l \propto \cos(m\phi)$ and $H_z^l \propto \sin(m\phi)$. Using these results, the inner wire fields can be written as

$$E_z^l(r \leq R) = \sum_m a_m J_m(\kappa_i r)\cos(m\phi) \tag{11.7}$$

$$H_z^l(r \leq R) = \sum_m b_m J_m(\kappa_i r)\sin(m\phi) \tag{11.8}$$

Outside the wire, the Bessel function of the second kind does not vanish as in the previous case. We will express this result in terms of Hankel functions, where $\kappa_h = \sqrt{\epsilon_h \omega^2/c^2 - k_z^2}$

$$F_z^h = \sum_m \left(c_{1m}^h H_m^+(\kappa_h r) + c_{2m}^h H_m^-(\kappa_h r)\right)\left(A_m^h \sin(m\phi) + B_m^h \cos(m\phi)\right) \tag{11.9}$$

and $H_m^+ = J_m + iY_m$ and $H_m^- = J_m - iY_m$. The outer wire fields can be written as

$$E_z^l(r > R) = \sum_m \left(\alpha_m^+ H_m^+(\kappa_h r) + \alpha_m^- H_m^-(\kappa_h r)\right)\cos(m\phi) \tag{11.10}$$

$$H_z^l(r > R) = \sum_m \left(\beta_m^+ H_m^+(\kappa_h r) + \beta_m^- H_m^-(\kappa_h r)\right)\sin(m\phi) \tag{11.11}$$

11.2.2 Calculation of E_r, H_r, E_ϕ, and H_ϕ

To obtain the other two components of the electromagnetic fields we must begin with Maxwell's equations $\nabla \times E = -\frac{1}{c}\frac{\partial H}{\partial t}$ and $\nabla \times H = \frac{\epsilon}{c}\frac{\partial E}{\partial t}$. Using the identity $\nabla \times A = \hat{r}\left(\frac{1}{r}\frac{\partial A_z}{\partial \phi} - \frac{\partial A_\phi}{\partial z}\right) + \phi\left(\frac{\partial A_r}{\partial z} - \frac{\partial A_z}{\partial r}\right) + \hat{z}\left(\frac{1}{r}\frac{\partial}{\partial r}(rA_\phi) - \frac{1}{r}\frac{\partial A_r}{\partial \phi}\right)$ we can write down the following relationships

$$\frac{\partial E_r}{\partial t} = \frac{c}{\epsilon}\left(\frac{1}{r}\frac{\partial H_z}{\partial \phi} - \frac{\partial H_\phi}{\partial z}\right) \tag{11.12}$$

$$\frac{\partial H_r}{\partial t} = c\left(\frac{\partial E_\phi}{\partial z} - \frac{1}{r}\frac{\partial E_z}{\partial \phi}\right) \tag{11.13}$$

$$\frac{\partial E_\phi}{\partial t} = \frac{c}{\epsilon}\left(\frac{\partial H_r}{\partial z} - \frac{\partial H_z}{\partial r}\right) \tag{11.14}$$

$$\frac{\partial H_\phi}{\partial t} = c\left(\frac{\partial E_z}{\partial r} - \frac{\partial E_r}{\partial z}\right) \tag{11.15}$$

$$\frac{\partial E_z}{\partial t} = \frac{c}{\epsilon}\left(\frac{1}{r}\frac{\partial}{\partial r}(rH_\phi) - \frac{1}{r}\frac{\partial H_r}{\partial \phi}\right) \tag{11.16}$$

$$\frac{\partial H_z}{\partial t} = c\left(\frac{1}{r}\frac{\partial E_r}{\partial \phi} - \frac{1}{r}\frac{\partial}{\partial r}(rE_\phi)\right) \tag{11.17}$$

The fields can generally be written as

$$E = E_o e^{ik_z z - i\omega t} \to \begin{cases} E_r e^{ik_z z - i\omega t} \\ E_\phi e^{ik_z z - i\omega t} \\ E_z e^{ik_z z - i\omega t} \end{cases} \tag{11.18}$$

$$H = H_o e^{ik_z z - i\omega t} \to \begin{cases} H_r e^{ik_z z - i\omega t} \\ H_\phi e^{ik_z z - i\omega t} \\ H_z e^{ik_z z - i\omega t} \end{cases} \tag{11.19}$$

After making the appropriate substitutions and some algebra we arrive at the following results

$$E_r = \frac{i}{\kappa^2}\left(k_z\frac{\partial E_z}{\partial r} + \frac{\omega}{rc}\frac{\partial H_z}{\partial \phi}\right) \to \begin{cases} E_r^{TE} = \frac{i\omega}{r\kappa^2 c}\frac{\partial H_z}{\partial \phi} \\ E_r^{TM} = \frac{ik_z}{\kappa^2}\frac{\partial E_z}{\partial r} \end{cases} \tag{11.20}$$

$$H_r = \frac{i}{\kappa^2}\left(-\frac{\epsilon\omega}{rc}\frac{\partial E_z}{\partial \phi} + k_z\frac{\partial H_z}{\partial r}\right) \to \begin{cases} H_r^{TE} = \frac{ik_z}{\kappa^2}\frac{\partial H_z}{\partial r} \\ H_r^{TM} = -\frac{i\epsilon\omega}{r\kappa^2 c}\frac{\partial E_z}{\partial \phi} \end{cases} \tag{11.21}$$

$$E_\phi = \frac{i}{\kappa^2}\left(\frac{k_z}{r}\frac{\partial E_z}{\partial \phi} - \frac{\omega}{c}\frac{\partial H_z}{\partial r}\right) \to \begin{cases} E_\phi^{TE} = -\frac{i\omega}{\kappa^2 c}\frac{\partial H_z}{\partial r} \\ E_\phi^{TM} = \frac{ik_z}{r\kappa^2}\frac{\partial E_z}{\partial \phi} \end{cases} \tag{11.22}$$

$$H_\phi = \frac{i}{\kappa^2}\left(\frac{\epsilon\omega}{c}\frac{\partial E_z}{\partial r} + \frac{k_z}{r}\frac{\partial H_z}{\partial \phi}\right) \to \begin{cases} H_\phi^{TE} = \frac{ik_z}{r\kappa^2}\frac{\partial H_z}{\partial \phi} \\ H_\phi^{TM} = \frac{i\epsilon\omega}{\kappa^2 c}\frac{\partial E_z}{\partial r} \end{cases} \tag{11.23}$$

Using (11.20)–(11.22) and our expression for $E_z \to E_z^l$ and $H_z \to H_z^l$ we can now write the fields in the r direction and ϕ direction. We denote that all $J_m(\kappa_i r) \to J_m$ and $H_m^{+/-}(\kappa_h r) \to H_m^{+/-}$ for simplicity. Then

$$E_r^l(r \le R) = \frac{ik_z^l}{\kappa_i} \sum_m a_m J_m^{'} \cos(m\phi) + \frac{i\omega}{r\kappa_i^2 c} \sum_m m b_m J_m \sin(m\phi) \tag{11.24}$$

$$E_r^l(r > R) = \frac{ik_z^l}{\kappa_h} \sum_m \left(\alpha_m^+ H_m^{'+} + \alpha_m^- H_m^{'-}\right) \cos(m\phi) + \frac{i\omega}{r\kappa_h^2 c} \sum_m \left(\beta_m^+ H_m^+ + \beta_m^- H_m^-\right) m \cos(m\phi) \tag{11.25}$$

where $H^{'\pm} = dH^{\pm}/d(\kappa_h r)$. For the radial magnetic fields

$$H_r^l(r \le R) = -\frac{i\epsilon_i \omega}{r\kappa_i^2 c} \sum_m a_m J_m m \sin(m\phi) + \frac{ik_z^l}{\kappa_{in}} \sum_m b_m J_m^{'} \sin(m\phi) \tag{11.26}$$

$$H_r^l(r > R) = -\frac{i\epsilon_h \omega}{r\kappa_h^2 c} \sum_m \left(\alpha_m^+ H_m^+ + \alpha_m^- H_m^-\right) m \sin(m\phi) + \frac{ik_z^l}{\kappa_h} \sum_m \left(\beta_m^+ H_m^{'+} + \beta_m^- H_m^{'-}\right) \sin(m\phi) \tag{11.27}$$

For the electric fields in the ϕ direction,

$$E_\phi^l(r \le R) = -\frac{ik_z^l}{\kappa_i^2 r} \sum_m a_m J_m m \sin(m\phi) - \frac{i\omega}{\kappa_i c} \sum_m b_m J_m^{'} \sin(m\phi) \tag{11.28}$$

$$E_\phi^l(r > R) = -\frac{ik_z^l}{\kappa_h^2 r} \sum_m \left(\alpha_m^+ H_m^+ + \alpha_m^- H_m^-\right) m \sin(m\phi) - \frac{i\omega}{\kappa_h c} \sum_m \left(\beta_m^+ H_m^{'+} + \beta_m^- H_m^{'-}\right) \sin(m\phi) \tag{11.29}$$

For the magnetic fields in the ϕ direction

$$H_\phi^l(r \le R) = \frac{i\epsilon_i \omega}{\kappa_i c} \sum_m a_m J_m^{'} \cos(m\phi) + \frac{ik_z^l}{r\kappa_i^2} \sum_m b_m J_m m \cos(m\phi) \tag{11.30}$$

$$H_\phi^l(r > R) = \frac{i\epsilon_h \omega}{\kappa_h c} \sum_m \left(\alpha_m^+ H_m^{'+} + \alpha_m^- H_m^{'-}\right) \cos(m\phi) + \frac{ik_z^l}{r\kappa_h^2} \sum_m \left(\beta_m^+ H_m^+ + \beta_m^- H_m^-\right) m \cos(m\phi) \tag{11.31}$$

11.2.3 Applying the Boundary Conditions at $r = R$

At the wire interface me must use the continuity of the tangential components of the electric and magnetic fields at $r = R$ to obtain an expression for $\{a_m, b_m\}$, $\{\alpha_m^+, \beta_m^+\}$ and $\{\alpha_m^-, \beta_m^-\}$. In the z direction

$$a_m J_m = \alpha_m^+ H_m^+ + \alpha_m^- H_m^- \tag{11.32}$$

$$b_m J_m = \beta_m^+ H_m^+ + \beta_m^- H_m^- \tag{11.33}$$

and in the ϕ direction

$$\frac{k_z^l}{\kappa_i^2 R} a_m J_m m + \frac{\omega}{\kappa_i c} b_m J_m^{'} = \frac{k_z^l}{\kappa_h^2 R} (\alpha_m^+ H_m^+ + \alpha_m^- H_m^-) m + \frac{\omega}{\kappa_h c} (\beta_m^+ H_m^{'+} + \beta_m^- H_m^{'-}) \tag{11.34}$$

$$\frac{\epsilon_i \omega}{\kappa_i c} a_m J_m^{'} + \frac{k_z^l}{R\kappa_i^2} J_m m = \frac{\epsilon_h \omega}{\kappa_h c} (\alpha_m^+ H_m^{'+} + \alpha_m^- H_m^{'-}) + \frac{k_z^l}{R\kappa_h^2} (\beta_m^+ H_m^+ + \beta_m^- H_m^-) \tag{11.35}$$

After solving (11.31) and (11.32) for $\{a_m, b_m\}$ and substituting these into (11.33) and (11.34), we see that only one of the three sets of coefficients $\{a_m, b_m\}$,$\{\alpha_m^+, \beta_m^+\}$, and $\{\alpha_m^-, \beta_m^-\}$ is independent. Explicitly, the linear relationship

$$\begin{pmatrix} \alpha_m^+ \\ \beta_m^+ \end{pmatrix} = \hat{S} \begin{pmatrix} \alpha_m^- \\ \beta_m^- \end{pmatrix} \tag{11.36}$$

can be derived from

$$\begin{bmatrix} \frac{k_z^l m}{R}\left(\frac{1}{\kappa_i^2} - \frac{1}{\kappa_h^2}\right) H_m^+ & \frac{\omega}{c}\left(\frac{J_m^{'}}{J_m}\frac{1}{\kappa_i} H_m^+ - \frac{1}{\kappa_h} H_m^{'+}\right) \\ \frac{\omega}{c}\left(\frac{\epsilon_i}{\kappa_i}\frac{J_m^{'}}{J_m} H_m^+ - \frac{\epsilon_h}{\kappa_h} H_m^{'+}\right) & \frac{k_z^l m}{R}\left(\frac{1}{\kappa_i^2} - \frac{1}{\kappa_h^2}\right) H_m^+ \end{bmatrix} \begin{pmatrix} \alpha_m^+ \\ \beta_m^+ \end{pmatrix} = \begin{bmatrix} \frac{k_z^l m}{R}\left(\frac{1}{\kappa_h^2} - \frac{1}{\kappa_i^2}\right) H_m^- & \frac{\omega}{c}\left(\frac{1}{\kappa_h} H_m^{'-} - \frac{J_m^{'}}{J_m}\frac{1}{\kappa_i} H_m^-\right) \\ \frac{\omega}{c}\left(\frac{\epsilon_h}{\kappa_h} H_m^{'-} - \frac{\epsilon_i}{\kappa_i}\frac{J_m^{'}}{J_m} H_m^-\right) & \frac{k_z^l m}{R}\left(\frac{1}{\kappa_h^2} - \frac{1}{\kappa_i^2}\right) H_m^- \end{bmatrix} \begin{pmatrix} \alpha_m^- \\ \beta_m^- \end{pmatrix} \tag{11.37}$$

where each of the four sub-matrices is a diagonal matrix with its elements corresponding to the Bessel function combinations evaluated at $r = R$. In this cylindrically-symmetric case, the S-matrix can be formally divided into four (diagonal) sub-matrices

$$\hat{S} = \begin{bmatrix} S_{11} & S_{12} \\ S_{21} & S_{22} \end{bmatrix} \tag{11.38}$$

The components S_{11} and S_{22} represent polarization-preserving TE-, TM-reflection, while the components S_{12}, S_{21} represent polarization-mixing coupling of TM to TE waves. Note that in the cylindrical geometry, polarization-preserving reflection is only possible when either $m = 0$ or $k_z = 0$, which yields $\det S_{12} = \det S_{21} = 0$.

Using (11.31) and (11.32) allows us to calculate the amplitudes $\{a, b\}$ based on the amplitudes of $\{\alpha^-, \beta^-\}$. The field equations will provide complete information about the field distribution inside the unit cell once the parameters k_z^l, and $\{\alpha^-, \beta^-\}$ are known. In order to obtain the additional expressions needed to solve for these unknowns the periodicity of the unit cell must now be investigated.

11.2.4 Dispersion of the Longitudinal Mode

We now focus on the problem of calculating the dispersion of the mode. This reduces to the problem of calculating a relationship between the internal structure of the unit cell and the set of parameters k_z^l, while $\{\alpha^-, \beta^-\}$ are known. For the square unit cell geometry, considered in this work, the latter combination will only contain cylindrical modes with $m = 0,4,8,\ldots$ The field of the eigenmode propagating in the periodic array of wires should satisfy the Bloch-periodicity condition

$$\left.\begin{matrix} E \\ H \end{matrix}\right|_{x=-\frac{a}{2},y} = \left.\begin{matrix} E \\ H \end{matrix}\right|_{x=+\frac{a}{2},y} \tag{11.39}$$

Using this requirement we can write down the field equations using the projection $\{E_y, H_y\} = \{E_r, H_r\}\cos\phi + \{E_\phi, H_\phi\}\sin\phi$

$$\begin{aligned} E_y = & \left[\frac{ik_z^l}{\kappa_h}\sum_m \left(\alpha_m^+ H_m^{'+} + \alpha_m^- H_m^{'-}\right)\cos(m\phi) + \frac{i\omega}{r\kappa_h^2 c}\sum_m \left(\beta_m^+ H_m^+ + \beta_m^- H_m^-\right) m\cos(m\phi)\right]\sin\phi \\ & - \left[\frac{ik_z}{\kappa_{out}^2 r}\sum_m \left(\alpha_m^+ H_m^+ + \alpha_m^- H_m^-\right) m\sin(m\phi) + \frac{i\omega}{\kappa_{out} c}\sum_m \left(\beta_m^+ H_m^{'+} + \beta_m^- H_m^{'-}\right)\sin(m\phi)\right]\cos\phi \end{aligned} \tag{11.40}$$

$$\begin{aligned} H_y = & \left[-\frac{i\epsilon_h\omega}{r\kappa_h^2 c}\sum_m \left(\alpha_m^+ H_m^+ + \alpha_m^- H_m^-\right) m\sin(m\phi) + \frac{ik_z^l}{\kappa_{out}}\sum_m \left(\beta_m^+ H_m^{'+} + \beta_m^- H_m^{'-}\right)\sin(m\phi)\right]\sin\phi \\ & + \left[\frac{i\epsilon_h\omega}{\kappa_h c}\sum_m \left(\alpha_m^+ H_m^{'+} + \alpha_m^- H_m^{'-}\right)\cos(m\phi) + \frac{ik_z^l}{r\kappa_h^2}\sum_m \left(\beta_m^+ H_m^+ + \beta_m^- H_m^-\right) m\cos(m\phi)\right]\cos\phi \end{aligned} \tag{11.41}$$

Here we enforce the periodicity of the y components of the electric and magnetic fields. Although this condition should ideally be satisfied for all values of the y coordinate within the interval $y \in \left[-\frac{a}{2}, \frac{a}{2}\right]$, in practice it suffices to enforce the Bloch-periodicity condition for a number of fixed points $\{x_j, y_j\}$ equal to the number of m values in (11.40) and (11.41). In the calculations we assume $y_j = \frac{a}{2N_m} j$, with N_m being the number of m terms. The analysis suggests that the choice of the exact location of the points does not significantly alter the dispersion of the mode, derived with the technique described below (see Fig. 11.2).

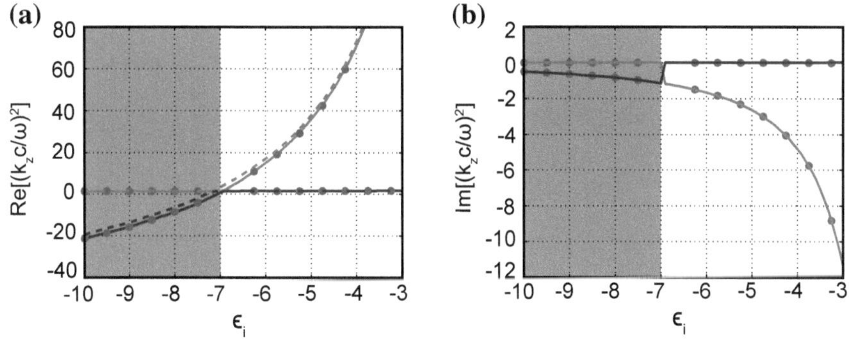

Fig. 11.2 Propagation constant of the TM-polarized waves in a nanowire composite as a function of wire permittivity. Lines and dashes represent (2.4.11) for $m = 0,4,8$ and $m = 0,4$ respectively; symbols represent the numerical solutions to Maxwell's equations. The *shaded regions* represents the hyperbolic regime

Noting that $\sin(\phi) = \sin(\pi - \phi)$, $\cos(\phi) = -\cos(\pi - \phi)$, $\sin(m\phi) = -\sin(m(\pi - \phi))$, and $\cos(m\phi) = \cos(m(\pi - \phi))$, it can be shown that the components of the electric and magnetic field possess the following symmetries:

$$E_y(x, y) = E_y(-x, y) \tag{11.42}$$

$$H_y(x, y) = -H_y(-x, y) \tag{11.43}$$

Therefore, (11.39) becomes equivalent to

$$\begin{pmatrix} \hat{0} & \hat{0} \\ \widetilde{H^+} & \widehat{H^+} \end{pmatrix} \begin{pmatrix} \alpha_m^+ \\ \beta_m^+ \end{pmatrix} + \begin{pmatrix} \hat{0} & \hat{0} \\ \widetilde{H^-} & \widehat{H^-} \end{pmatrix} \begin{pmatrix} \alpha_m^- \\ \beta_m^- \end{pmatrix} = \begin{pmatrix} 0 \\ 0 \end{pmatrix} \tag{11.44}$$

where the elements of the sub-matrices $H^{\pm}$ are evaluated based on the longitudinal field equations according to the following rules: the sub-matrices $\widetilde{H}$ and H represent the TM- and TE-polarized magnetic field respectively; the superscript of the expression $[\pm]$ corresponds to the superscript of the Hankel function; and the jm^{th} element of the sub-matrix represents the y component of the magnetic field due to the mth Hankel function, evaluated at the point $\{x_j, y_j\} = \{\frac{a}{2}, y_j\}$. With the help of the S matrix, (11.44) can be further simplified as

$$\begin{pmatrix} \hat{0} & \hat{0} \\ \widetilde{H^+}\widehat{S_{11}} + \widehat{H^+}\widehat{S_{21}} + \widetilde{H^-} & \widetilde{H^+}\widehat{S_{12}} + \widehat{H^+}\widehat{S_{22}} + \widehat{H^-} \end{pmatrix} \begin{pmatrix} \alpha_m^- \\ \beta_m^- \end{pmatrix} = \begin{pmatrix} 0 \\ 0 \end{pmatrix} \tag{11.45}$$

Finally, the amplitudes of the field of the longitudinal TM-polarized wave are represented as $\begin{pmatrix} \alpha_m^- \\ 0 \end{pmatrix}$ with values of the coefficients α given by non-trivial solutions of the linear relationship

$$(\widehat{\mathcal{H}_y})\alpha^- = \left(\widehat{\widehat{H^+}\widehat{S_{11}}} + \widehat{H^+}\widehat{S_{21}} + \widehat{H^-}\right)\alpha^- = 0 \tag{11.46}$$

with

$$\widehat{\widehat{H^+}} = \frac{-i\epsilon_h\omega}{r\kappa_h^2 c} H_m^+ m \sin(m\phi)\sin(\phi) + \frac{i\epsilon_h\omega}{\kappa_h^2 c} H_m^{'+} \cos(m\phi)\cos(\phi) \tag{11.47}$$

$$\widehat{H^\pm} = \frac{ik_z}{\kappa_h^2} H_m^{'\pm} \sin(m\phi)\sin(\phi) - \frac{ik_z}{r\kappa_h^2} H_m^\pm m\cos(m\phi)\cos(\phi) \tag{11.48}$$

representing the TM-polarized and TE-polarized waves respectively. It is now easy to see that the dispersion of this wave is given by

$$\det\left|\widehat{\mathcal{H}_y}\right| = 0 \tag{11.49}$$

To verify the validity of (11.49), the dispersion of the longitudinal wave corresponding to (i) $m = 0, 4$ and (ii) $m = 0, 4, 8$ is calculated and compared with the derived dispersions from numerical solutions of Maxwell's equations. From the results of these calculations, shown in Fig. 11.2, it is clearly seen that even the rough approximation with $m = 0, 4$ yields relatively good agreement with numerical solutions of Maxwell's equations. Including one extra harmonic makes agreement almost perfect.

Now that the longitudinal dispersion is realized the remaining coefficients can be solved and an analytical result of the wave profiles can be produced (Figs. 11.3 and 11.4).

Notice that since $E_z \neq 0$, this verifies that the solution represents the longitudinal wave with dispersion $\epsilon_z(k_z) = 0$.

Figure 11.4 demonstrates the excellent agreement between the numerical and analytical solutions corresponding to a three-term series $m = 0, 4, 8$, and clearly demonstrates the longitudinal character of this mode. This wave is strongly dispersive in the regime $\epsilon_i \to \epsilon_h$, corresponding to the surface plasmon oscillations on the metal-dielectric interface. On the other hand, when $-\epsilon_i \gg \epsilon_h$ (realized at mid-IR and lower frequencies for noble metals), the wavevector of the longitudinal mode approaches $n_l^\infty \omega/c$, and its transverse counterpart approaches the light line [38, 48, 50].

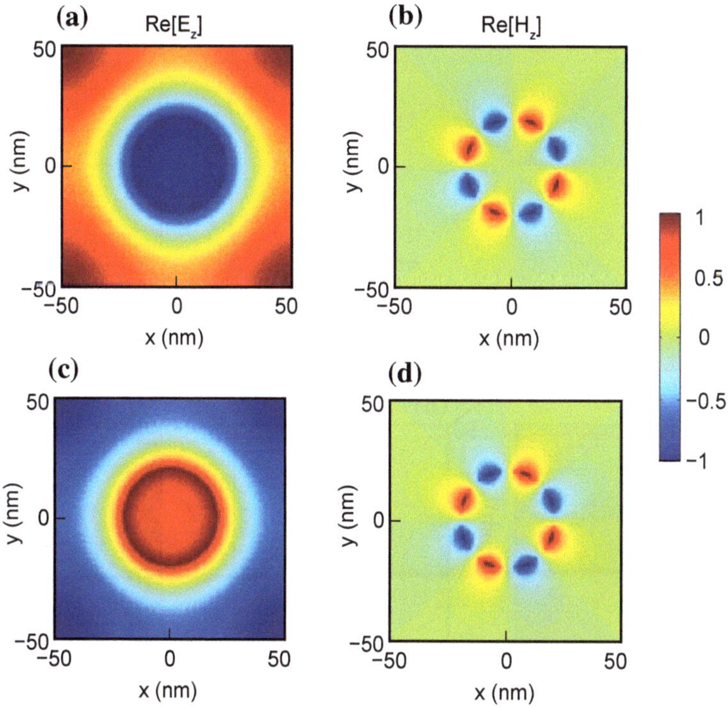

Fig. 11.3 Analytical electric and magnetic fields within the unit cell produced using the previously derived field expressions, (**a**, **b**) correspond to $\epsilon_i = -4 + 0.1i$ and (**c**, **d**) to $\epsilon_i = -9 + 0.1i$

11.2.5 Solutions at Oblique Angles

The next step is to obtain the dispersion and field profiles for angles other than normal. To achieve this we use the previously derived results for the longitudinal wave, which describes the microscopic field, and those obtained from effective medium theory (EMT), the macroscopic field. Comparing the dispersion relation corresponding to microscopic (11.49) and effective medium approximation $\epsilon_z(k_z) = 0$, a complete description of the nonlocal effective permittivity can be obtained. The functional dependence of nonlocal permittivity can be approximated as

$$\epsilon_z(k_z) = \xi\left(k_z^2 - k_z^{l\,2}\right)\frac{c^2}{\omega^2} \tag{11.50}$$

where k_z is the wavevector of the mode in the nonlocal effective medium approximation, k_z^l is the wavevector of the mode composite in the microscopic theory, and ξ is the factor which will be determined below.

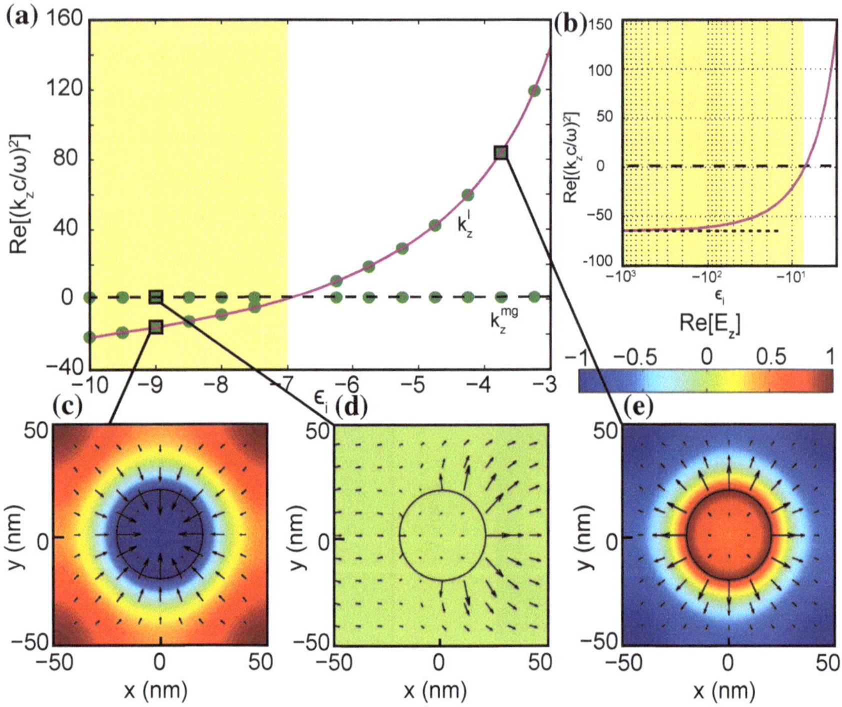

Fig. 11.4 **a**, **b** Dispersion in the nanowire composite as a function of wire permittivity. *Dashed* and *solid lines* represent propagation constants of transverse and longitudinal waves k_z^{mg} and k_z^l, (2.4.11) respectively, *symbols* represent numerical solutions to Maxwell's equations; for $-\epsilon_i \gg 1, k_z^l \to n_l^\infty \omega/c$ (*dotted line* in (**b**)). **c**, **d**, **e** Electric field in the unit cell; *surface plots* and *arrows* represent E_z and $\overrightarrow{E}_{x,y}$ components, respectively

Equation (11.50) can be used along with the above considerations to determine a complete nonlocal dispersion for propagation at any angle to the optical axis. For simplicity, we consider the case $k_y = 0$, $k_x \neq 0$. Using local effective medium theory where the propagation of the TM-polarized wave can be described by $\epsilon_z(k_z)\left(k_z^2 - \epsilon_\perp^{mg}\frac{\omega^2}{c^2}\right) = -\epsilon_\perp^{mg} k_x^2$ and substituting (11.50) into this equation, we obtain the following relation

$$\left(k_z^2 - k_z^{l2}\right)\left(k_z^2 - \epsilon_\perp^{mg}\frac{\omega^2}{c^2}\right) = -\frac{\epsilon_\perp^{mg}}{\xi}\frac{\omega^2}{c^2}k_x^2, \quad (11.51)$$

an exact solution to Maxwell's equations. Similar to other nonlocal materials, nanowire materials support two TM-polarized waves propagating with different indices

$$k_z = \sqrt{\frac{\left(k_z^{l2} + \epsilon_\perp^{mg}\frac{\omega^2}{c^2}\right) \pm \sqrt{\left(k_z^{l2} - \epsilon_\perp^{mg}\frac{\omega^2}{c^2}\right)^2 - 4\frac{\epsilon_\perp^{mg}}{\xi}\frac{\omega^2}{c^2}k_x^2}}{2}} \tag{11.52}$$

The final free parameter of the model, the multiplicative factor ξ, can be determined by requiring that in the limit of small k_x the properties of one of the two TM-modes follow elliptical or hyperbolic dispersion and has $k_z(k_x) = const$ dependence that is observed in the wire media when $\epsilon_z^{mg} > 0$, $\epsilon_z^{mg} \lesssim 0$, and $\epsilon_z^{mg} \ll -1$ respectively. The relationship

$$\xi = p\frac{\epsilon_i + \epsilon_h}{\epsilon_h - n_l^{\infty\,2}} \tag{11.53}$$

adequately describes the optics of wire media in all of these three limits. The excellent agreement between the predictions of (11.52) and the full-wave numerical solutions of Maxwell's equations are shown in Fig. 11.5.

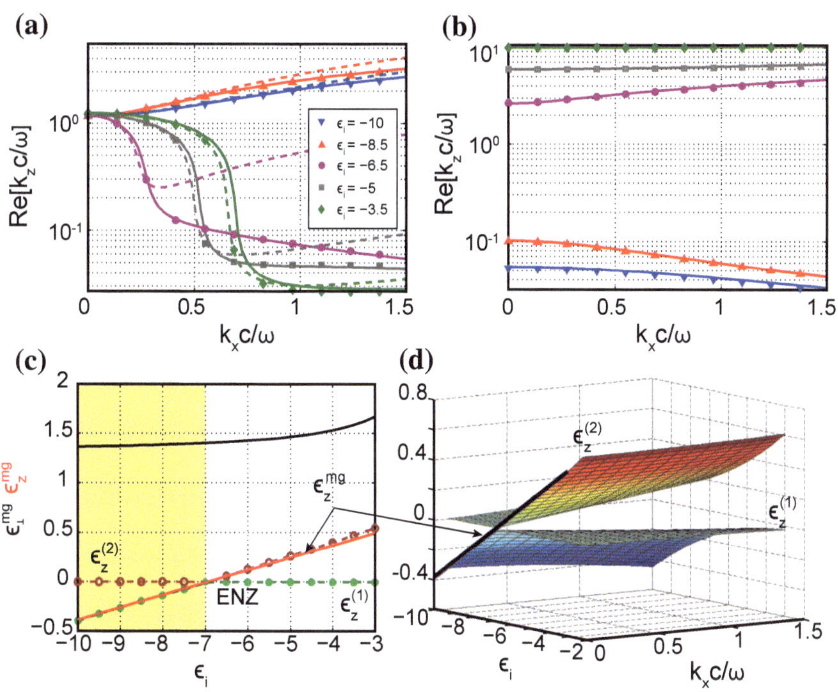

Fig. 11.5 **a**, **b** Isofrequency contours of TM-polarized modes in nanowire composites. *Solid lines* and *symbols* correspond to (11.52) and numerical solutions of Maxwell equations respectively; *dashed lines* represent local EMT. **c**, **d** The effective medium permittivity of the nonlocal nanowire composite for $k_x = 0$ (**c**) and for $k_x \neq 0$ (**d**); $\epsilon_z^{(1)}$, $\epsilon_z^{(2)}$ represent two solutions of (11.50)

As expected the isofrequency of the "main" TM-polarized wave resembles an ellipse or hyperbola which for small values of k_x and is well described by $\hat{\epsilon}^{mg}$. At the same time, the dependence $k_z(k_x)$ for the "additional" TM-wave is opposite to that of its "main" counterpart. It should be noted that when the main TM-wave is hyperbolic, the additional TM-wave is technically not propagating at all, even if it appears elliptical. The longitudinal wave decays along the wires, while the main wave has hyperbolic dispersion. The z-component of the permittivity can be described by (11.50) when $k_x = 0$, but for oblique angles exhibits strong spatial dispersion.

11.2.6 Wave Profiles at Oblique Angles

Now that the origin and dispersion of the modes propagating in nanowire systems is understood, we focus on the analysis of the optical properties of finite-size wire arrays. Since in the EMT approximation the fields of TE and TM-polarized modes are orthogonal to each other, and since propagation of TE polarized light through the wire-based system is only affected by x, y-components of the permittivity, this propagation can be successfully described by (11.3) and (11.4).

Here, we focus on the analysis of propagation of TM-polarized light. This analysis must describe the structure of electromagnetic waves propagating in the system, and determine what additional boundary conditions are needed to calculate the amplitudes of the two TM-polarized modes inside the wire system.

Consistency with the effective medium description requires that the unit-cell averaged fields satisfy both constituent relations $\epsilon_j = \langle \epsilon E_j / E_j \rangle$ and relations between the field components of the plane wave $\langle \epsilon E_z \rangle = -k_x/k_z \langle \epsilon E_x \rangle$. With these constraints, we start with (11.3) and (11.4) and the longitudinal field solutions recently derived. One can determine the parameters α_0^-, e_z^{mg}, and e_x^{mg} by normalizing $\langle E_z^l \rangle = \langle E_x^{mg} \rangle = \langle E_z^{mg} \rangle = 1$, and constructing the fields of the two waves propagating in the wire media as $\vec{E}(x, y) e^{i\omega t - ik_x x - ik_z z}$ with

$$E_x(x, y) = E_x^{mg} + (\gamma^{mg} + \gamma^l) E_x^l \big|_{z=0} \tag{11.54}$$

$$E_z(x, y) = \gamma^{mg} E_z^{mg} + \gamma^l E_z^l \big|_{z=0} \tag{11.55}$$

Using the above expressions we arrive at

$$\gamma^{mg} = -\frac{\epsilon_\perp^{mg} k_x}{\epsilon_z k_z} \frac{\epsilon_z - \epsilon^l}{\epsilon_z^{mg} - \epsilon^l} \tag{11.56}$$

$$\gamma^l = -\frac{\epsilon_\perp^{mg} k_x}{\epsilon_z k_z} \frac{\epsilon_z - \epsilon_z^e}{\epsilon^l - \epsilon_z^e} \tag{11.57}$$

In (11.56) and (11.57), $\epsilon_z \equiv \epsilon(k_z)$ which is given by (11.50), $k_z(k_x)$ by (11.52), and $\epsilon^l = \langle \epsilon(x,y) E_z^l(x,y) \rangle / \langle E_z^l(x,y) \rangle$.

Equation (11.55) represents a transition between full-wave solutions of Maxwell's equations in the nanowire array where the fields oscillate on the scale of the individual wires, and effective-medium solutions where plane waves propagate in the homogenized material. Since our model for $\vec{E}^{mg}$ assumes the quasistatic limit, (11.55) is technically valid in the limit $a \ll \lambda_0$ where $\lambda_0 = 2\pi/k_x$. However, Fig. 11.5 indicates that the developed formalism also provides an adequate approximation for the optics of wire systems for higher values of k_x.

11.2.7 Simplified Approach to Nonlocal Effective Medium Theory

We previously developed an analytical technique that provides an adequate description for the optical response of wire based metamaterials, once the dispersion for the longitudinal wave is known. This existing formalism maps the longitudinal dispersion to the solution of an eigenvalue problem (11.49) that is computationally intensive and, depending on implementation, may become numerically unstable.

Here we will derive a simplified analytical approach that can be used to approximate the dispersion of this longitudinal wave in the wire-based metamaterials. This approximation avoids using intensive numerical solutions to an eigenvalue problem. The results of nonlocal effective medium theory based on such approximate solutions will be shown to agree with the numerical solutions of Maxwell's equations and with predictions of the earlier developed formalism. This simplified approach will be used going forward and will prove to provide greater flexibility in a wide variety of applications.

It was previously stated that in the limit of small k_x the dispersion behaves in accordance with the Maxwell-Garnett approximation

$$k_z^{mg\,2} = \epsilon_\perp^{mg}\left(\frac{\omega^2}{c^2} - \frac{k_x^2}{\epsilon_z^{mg}}\right) \tag{11.58}$$

Using the nonlocal dispersion relation (11.51) and replacing $k_z^2 \rightarrow k_z^{mg\,2}$, we obtain

$$\left(\epsilon_\perp^{mg}\left(\frac{\omega^2}{c^2} - \frac{k_x^2}{\epsilon_z^{mg}}\right) - k_z^{l2}\right)\left(\epsilon_\perp^{mg}\left(\frac{\omega^2}{c^2} - \frac{k_x^2}{\epsilon_z^{mg}}\right) - \epsilon_\perp^{mg}\frac{\omega^2}{c^2}\right) = -\frac{\epsilon_\perp^{mg}}{\xi}\frac{\omega^2}{c^2}k_x^2 \tag{11.59}$$

Solving for k_z^{l2} and taking the limit as $k_x \rightarrow 0$ yields

$$k_z^{l2} = \left(\epsilon_\perp^{mg} - \frac{\epsilon_z^{mg}}{\xi} \right) \frac{\omega^2}{c^2} \tag{11.60}$$

This solution can also be validated doing a similar derivation but starting with the complete description of the nonlocal effective permittivity (11.50). By comparing this approximation with the exact solution (11.49) we see excellent agreement (Fig. 11.7).

Equation 11.60 is dependent on ξ, so we still need to solve an eigenvalue problem (11.49) to obtain n_l^∞ but only in the limit where the wires become perfect electric conductors (PEC's). Unfortunately there is no way to obtain a closed form solution as was done for k_z^l, but the scattering matrix and periodic boundary matrices involved can be greatly simplified, as will be shown below.

When the wire becomes PEC, $\epsilon_i \rightarrow -\infty$, and thus $\kappa_i^2 \rightarrow -\infty$. After applying this condition to (11.36) it can be shown that the S-matrix reduces to

$$\hat{S} = \begin{bmatrix} S_{11} & 0 \\ 0 & S_{22} \end{bmatrix} \tag{11.61}$$

where $S_{11} = S_{22}$ and for the $m = 0, 4, 8$ case becomes

$$S_{11} = \begin{bmatrix} H_0^- / H_0^+ & 0 & 0 \\ 0 & -1 & 0 \\ 0 & 0 & -1 \end{bmatrix} \tag{11.62}$$

As a consequence of the above simplification, the non-trivial solutions of the linear relationship from (11.46) reduces to

$$\widehat{\mathcal{H}_y} = \left(\widehat{H^+ \widehat{S_{11}}} + \widehat{H^-} \right) \tag{11.63}$$

To solve for n_l^∞, $\det|\mathcal{H}_y| = 0$ must be calculated for the above situation. This can only be calculated numerically and must be done for each dependent case where $n_l^\infty = n_l^\infty(p, \omega, \epsilon_h)$ (Fig. 11.6). It is worth noting that the solution can be substantially simplified when $m = [0, 4]$ case is considered.

11.2.8 Nonlocal Transfer Matrix Method

Now it is necessary to consider the problem of reflection/refraction of light at the interface of two (nonlocal) (meta-) materials, extending the well-developed transfer-matrix formalism (TMM) [53] to nonlocal composites. A complete study

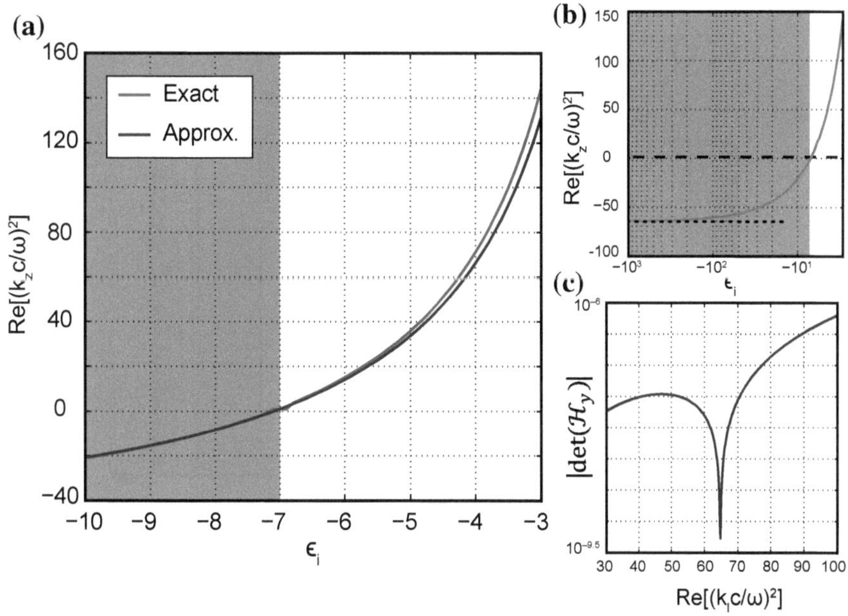

Fig. 11.6 **a** Comparison between exact (11.49) and approximate (11.60) solutions for longitudinal dispersion. **b** In the limit as the wire becomes PEC the longitudinal dispersion converges to n_l^{∞} which is calculated through an eigenvalue solution of (11.63) illustrated in (**c**)

of the transmission and reflection through a nonlocal metamaterial would not be possible using conventional TMM because now there is an additional TM mode propagating within the media. As a result of this additional mode, we must consider additional boundary conditions (ABC's) to solve for all the required amplitudes.

The typical geometry of light propagation through a finite-thickness slab of nanowire material is shown in Fig. 11.7. For this geometry, Maxwell's equations require continuity of (microscopic) E_x and D_z fields. The conventional effective-medium boundary conditions are obtained by averaging these relationships across the unit cell [14]. As previously shown, the nanowire media support two TM-polarized modes resulting in four TM-polarized waves propagating in the finite-thickness slab (see Fig. 11.7). Therefore, calculation of the amplitudes of these waves requires additional boundary conditions (ABC's). Different forms of ABC's, often based on heuristic arguments have been suggested in previous works [49–52]. As was first developed in [38], here we will present a first-principles approach to solve this long-standing problem in all possible configurations.

Maxwell's equations require continuity of (microscopic) E_x and D_z and the effective-medium boundary conditions must be obtained by averaging these relationships across the unit cell. Multiple linearly independent boundary conditions can be obtained by requiring the continuity of $E^n = \langle e^{2\pi i n \frac{x}{a}} E_x \rangle$ and $D^n = \langle e^{2\pi i n \frac{x}{a}} D_z \rangle$ with different integer n.

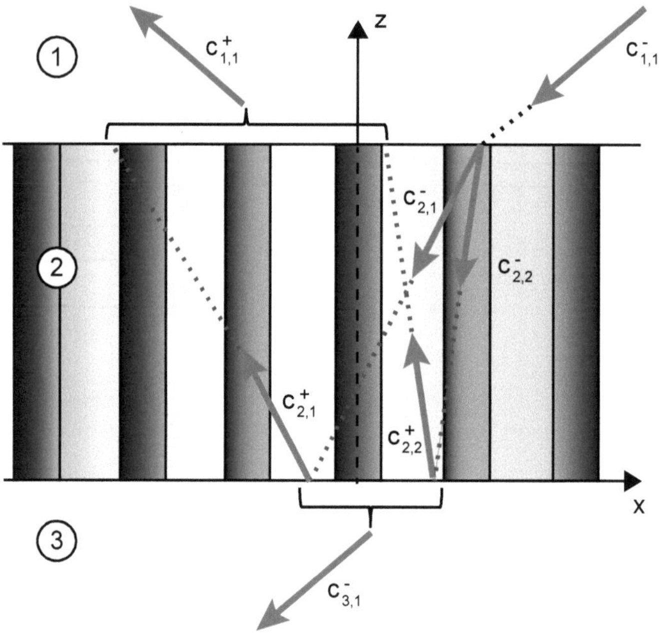

Fig. 11.7 Schematic of TMM for a nanowire composite. Amplitude of each mode is labeled to clarify notation as well as layer numbers

Continuity of E^0 and D^0 yields conventional EMT; continuity of D^1 (and, in the case of contact of two nonlocal media, of E^1) represents the set of ABC's. In general, the boundary conditions for any interface are implemented as:

$$\begin{cases} \sum_l c^+_{i-1,l}E^n_{i-1,l}e^{ik_{z_{i-1,l}}z_i} + c^-_{i-1,l}E^n_{i-1,l}e^{-ik_{z_{i-1,l}}z_i} = \sum_l c^+_{i,l}E^n_{i,l}e^{ik_{z_{i,l}}z_i} + c^-_{i,l}E^n_{i,l}e^{-ik_{z_{i,l}}z_i} \\ \sum_l c^+_{i-1,l}D^n_{i-1,l}e^{ik_{z_{i-1,l}}z_i} - c^-_{i-1,l}D^n_{i-1,l}e^{-ik_{z_{i-1,l}}z_i} = \sum_l c^+_{i,l}D^n_{i,l}e^{ik_{z_{i,l}}z_i} - c^-_{i,l}D^n_{i,l}e^{-ik_{z_{i,l}}z_i} \end{cases} \tag{11.64}$$

In the expressions above, the double-subscript represents the layer number in the system and the mode number within the layer, while the "±" superscript represent the direction of the wave propagation (see Fig. 11.7). The amplitudes of the wave propagating in the system represent the amplitude of E^0; therefore, the amplitudes of E_x are symmetric with respect to the change of propagation direction $k_z \to -k_z$, while the amplitudes of E_z, D_z are anti-symmetric.

Starting from the last layer and working forward, as conventionally done with TMM, there are a total of four possible last interface contacts; local-local, nonlocal-local, local-nonlocal, and nonlocal-nonlocal. To complete all possible scenarios the second to last interface cases must also be calculated for the above last contact interfaces. Once these cases are worked out and knowing the amplitude of the initial incoming wave, assuming it is entering from a local medium, the TMM can be extended for any n-layered system.

Using the above procedure we have developed a model for a full optical description of a nonlocal nanowire metamaterial slab. The transmission and reflection of the nanowire metamaterial, predicted by both local and nonlocal EMTs, are now compared (Figs. 11.8 and 11.9) with full-vectorial numerical solutions of Maxwell's equations. It is seen that the smaller the loss and the larger the angle of incidence the more important it becomes to take into account the nonlocal optics of nanowire composites. Interestingly, the nonlocal response strongly affects the optical response of the wire metamaterials across the broad range of the effective permittivities. This effect is most clearly seen in reflection, but is also visible in transmission, especially in the ENZ and elliptical regime.

We also calculate the transmission and reflection for a planar slab of nanowire material with the permittivity of the host $\epsilon_h = 2.25$. A comparison between the numerical solutions of Maxwell's equations, predictions of local EMT, and

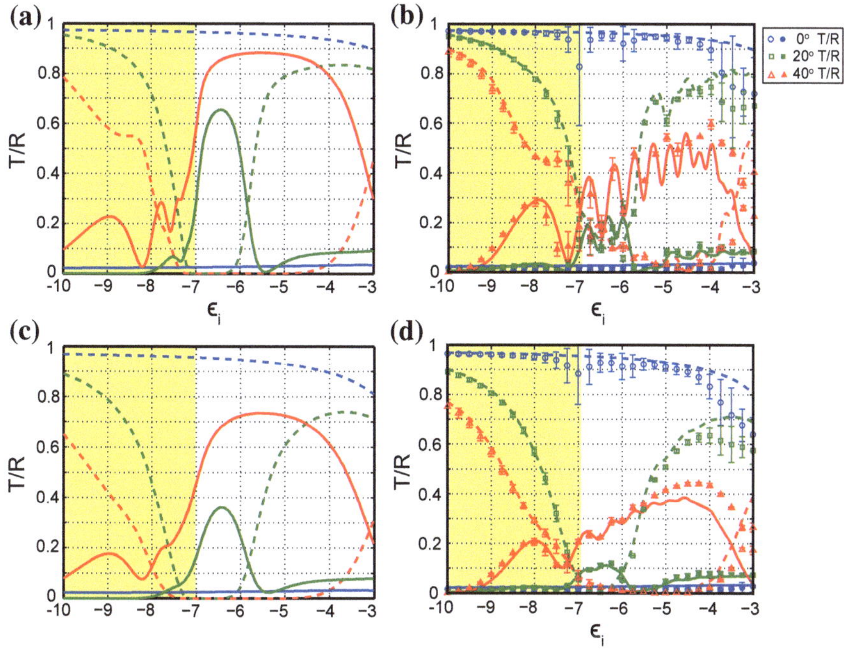

Fig. 11.8 Transmission and reflection of light through a parallel slab of nanowire media, suspended in air with $Im(\epsilon_i) = -0.1$ (**a**, **b**) and $Im(\epsilon_i) = -0.25$ (**c**, **d**). (**a**, **c**): local TMM calculations, (**b**, **d**): nonlocal EMT developed here (*lines*) and numerical solutions of Maxwell equations (*symbols*); *Solid lines* and *filled symbols* represent reflection, *dashed lines* and *empty symbols*—transmission

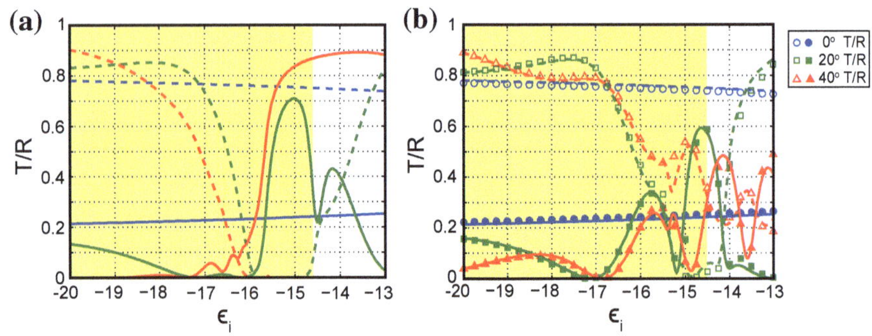

Fig. 11.9 Transmission and reflection of light through a parallel slab of nanowire material suspended in a substrate with $\epsilon_h = 2.25$ and $Im(\epsilon_i) = -0.1$. (**a**) Local calculations and (**b**) nonlocal EMT (*lines*) and numerical solutions to Maxwell's equations (*symbols*)

nonlocal EMT is shown in Fig. 11.9a, b. It can be seen that the nonlocal EMT provides a substantially more accurate description of the optics of nanowire systems than its local counterpart for this configuration as well.

11.3 Dipole Emission in Nonlocal Metamaterials

Now we will present an analytical and computational study of the light emission in nonlocal nanowire metamaterials. As it was previously shown, due to the nolocality of these materials there exists an additional TM-polarized optical mode that has hyperbolic-like properties at frequency ranges where the main TM wave has elliptic dispersion. Here we will analyze the effects of optical nonlocality on the emission of point dipoles embedded inside the nanowire media. It will be shown that the nonlocality significantly affects the "optical topology" of these metamaterials, enhancing the photonic density of states in the elliptic regime [41]. This is in contrast with the predictions of local effective medium theory [42, 44], where Purcell enhancement is limited to the hyperbolic regime.

In order to relate classical radiation to quantum effect of spontaneous emission, few important remarks should be made. The common approach for deriving rates of spontaneous emission in the weak coupling regime relies on the Fermi Golden Rule, suggesting that the rate is directly proportional to available density of electromagnetic (photonic) states. Those states are counted with solving a certain eigenvalue problem and deriving appropriate dispersion relations [54, 55]. However, lossy and dispersive nanostructures and, in particular, metamaterials are made of lossy dispersive components, prevent formulating pure eigenvalue problem [56]. There are other approaches, enabling to perform canonical quantization of modes in an arbitrary environment. Local Langevin operators could replace standard mode decomposition methods and lead to a local form of electromagnetic quantization [57]. In this case, the local density of states concept is introduced and the later quantity is directly proportional to the imaginary part of electromagnetic Green's function. The appearance of classical quantities, i.e. Green's functions, used for describing quantum effects, is rather expected. In fact, classical-quantum correspondence principle ensures relations between theories and, in particular, rates of spontaneous emission could be calculated by employing purely classical approaches, such as radiation reaction forces [62]. Specifically, quantum emitters could be represented by classical point dipoles. In this description, the strength of light-matter interaction depends on two key parameters—amplitude of electromagnetic field at a point of an emitter and the dipole moment of the emitter itself. In the frame of Antennae Theory [57] this procedure describes an impedance matching of a localized source to a far-field. The semi-classical approach, bridging the gap between classical (antenna) and quantum (vacuum fluctuations) theories, introduces additional damping force on a dipole in order to balance the outflow of radiated energy to the far-field with the one, stored in oscillations. This simple energy conservation principle enables deriving exactly the same decay rates, as could be calculated by employing second quantization of electromagnetic field and solving for quantum equations of motion. A complementary classical alternative is the calculation of so-called S-parameters of a small antenna, e.g. [58]. The full quantum-mechanical formula for the radiation reaction force in arbitrary electromagnetic environment brings this phenomenological description to the solid quantum-mechanical ground [59]. It is worth noting, that the above discussion is related to the weak coupling regime of light-matter interaction, while the strong coupling analysis requires more sophisticated approaches and solutions of time-dependent equations of motion. In all of the subsequent derivations, the beforehand mentioned approach on local density of stares will be used.

In order to calculate the local density of states and derive the resulting emitter's lifetime, we will represent the Green's function of an anisotropic metamaterial as a linear combination of plane waves. For loss-less homogeneous materials, the radiation rate enhancement due to the local density of optical states can be related to the imaginary part of Green's function representing the field E of the point dipole P inside the media [60]

$$\frac{\Gamma}{\Gamma_0} \simeq \frac{3}{2} \frac{Im(E \cdot P)}{\omega^3 |P|^2} \tag{11.65}$$

Since we will consider emission inside an anisotropic material, the field generated by the dipole needs to be separated into TE and TM-polarized components.

We will start with the derivation of the plane wave expansion and validation of a dipole emitting in a lossless homogeneous material. We can then expand this approach to analyze the dipole emission through the nanowire metamaterial. However, we must keep in mind that for realistic plasmonic media there is loss involved. It is therefore important to understand the effect of material absorption on light emission in the system. The quantization of the electromagnetic field in the presence of material bodies could be performed using a local Langevin approach, which however does not relate the eigenmodes of the structure [54]. Moreover, it is worth noting that the direct contact of a radiating dipole with lossy media causes an unphysical divergence that could be understood from the infinitely large energy, induced by the singular dipole [54, 61]. Various real and virtual cavity approaches could be employed in order to address the impact of surrounding media on a hosted emitter [62–64] but are still under debate. In terms of the Green's function description of the source region, the singularity could be removed by the introduction of a depolarization dyad, which however also depends on the shape of the extracted principle volume. To address these key issues, we will work through a four step strategy.

First, we will calculate the emission of the point dipole inside a hypothetical, lossless, metamaterial (Fig. 11.10a). Predictions of both local and nonlocal effective medium theories will be illustrated, with the two TM-polarized modes in nonlocal metamaterials taken as two competing decay channels.

Second, the emission of the point 3D dipole positioned in the small (vacuum-filled) slit surrounded by two infinite slabs of hypothetical lossless metamaterial will be analyzed (Fig. 11.10b). Combining this with the first step, these calculations will allow us to estimate the geometry-dependent local-field-correction effects. Applying a local field correction is a frequently employed technique to account for the arrangement of the emitter and the host matrix [54]. It is worth noting that the geometry of the artificially created cavity could affect the spontaneous emission rate; here we will use a slit geometry, reducing its physical dimensions to zero thickness to counter this possibility.

Third, we will calculate the emission of the point dipole positioned in the small vacuum slit surrounded by lossy metamaterial and use the local field correction, estimated above, to calculate the modulation of the dipole lifetime due to the metamaterial.

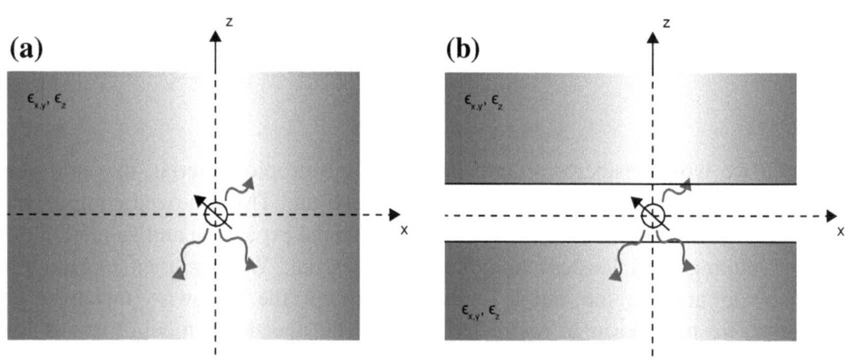

Fig. 11.10 Schematic of a point dipole inside a hypothetical, lossless, metamaterial (**a**) and the point 3D dipole positioned in a small (vacuum-filled) slit surrounded by two infinite slabs of hypothetical lossless metamaterial (**b**)

Finally, to generate a more realistic calculation, the small vacuum slit will be replaced with the lossless nanowire metamaterial, where the local field correction no longer needs to be considered. This will be done for both local and nonlocal approaches.

11.3.1 Plane Wave Expansion of Green's Function in Homogeneous Material

To calculate the field generated by the dipole separated into TE and TM-polarized components, we will start with Maxwell's equations in the frequency domain, and follow the recipes outlined in. [14, 65].

$$\nabla \cdot D(r,\omega) = 4\pi\rho(r,\omega) \tag{11.66}$$

$$\nabla \cdot B(r,\omega) = 0 \tag{11.67}$$

$$\nabla \times E(r,\omega) = \frac{i\omega}{c} H(r,\omega) \tag{11.68}$$

$$\nabla \times H(r,\omega) = -\frac{i\omega}{c}\hat{\epsilon}E(r,\omega) + \frac{4\pi}{c} j(r,\omega) \tag{11.69}$$

and the anisotropic permittivity tensor

$$\hat{\epsilon} = \begin{bmatrix} \epsilon_\perp & 0 & 0 \\ 0 & \epsilon_\perp & 0 \\ 0 & 0 & \epsilon_z \end{bmatrix} \tag{11.70}$$

We assume harmonic time dependence throughout, thus the current density can be written as

$$j(r) = -i\omega P\delta[r - r_0] \tag{11.71}$$

with the dipole moment P. We can rewrite Maxwell's equations as

$$\nabla \times E(r) = \frac{i\omega}{c} H(r) \tag{11.72}$$

$$\nabla \times H(r) = -\frac{i\omega}{c}\left[\hat{\epsilon}E(r) + 4\pi P\delta[r - r_0]\right] \tag{11.73}$$

By taking the curl of Faraday's equation we have

$$\nabla \times \nabla \times E(r) = \frac{\omega^2}{c^2}\left[\hat{\epsilon}E(r) + 4\pi P\delta[r - r_0]\right] \tag{11.74}$$

The Fourier transform of the electric field and delta function can be written as

$$E(r) = \int E_0(k)e^{ik\cdot r}d^3k \tag{11.75}$$

$$\delta(r - r_0) = \frac{1}{(2\pi)^3}\int e^{ik\cdot r}d^3k \tag{11.76}$$

Making these substitutions into (11.74)

$$\int\left[-k\times k\times E_0(k) - \frac{\omega^2}{c^2}\hat{\epsilon}E(k) - \frac{\omega^2}{c^2}\frac{P}{2\pi^2}\right]e^{ik\cdot r}d^3k = 0 \tag{11.77}$$

The integrand of (11.77) must go to zero, thus we just need to consider

$$-k\times k\times E_0(k) - \frac{\omega^2}{c^2}\hat{\epsilon}E(k) = \frac{\omega^2}{c^2}\frac{P}{2\pi^2} \tag{11.78}$$

After applying the identity $a\times b\times c = (a\cdot c)b - (a\cdot b)c$ we now have

$$-k(k\cdot E_0(k)) + k^2E_0(k) - \frac{\omega^2}{c^2}\hat{\epsilon}E(k) = \frac{\omega^2}{c^2}\frac{P}{2\pi^2} \tag{11.79}$$

This can be written as a matrix equation

$$M_{\beta\alpha}E_{0\alpha} = \frac{\omega^2}{c^2}\frac{P_\alpha}{2\pi^2} \tag{11.80}$$

where

$$M_{\beta\alpha} = \left(\delta_{\alpha\beta}k^2 - k_\alpha k_\beta - \frac{\omega^2}{c^2}\epsilon_{\alpha\beta}\right) \tag{11.81}$$

The Green's function $\overline{G}$ is essentially defined by the electric field E at the field point r generated by a radiating electric dipole P at the point source r'.

$$E = \frac{\omega^2}{2\pi^2c^2}\overline{G}P \tag{11.82}$$

This can be written in element form as

$$E_{0\alpha} = \frac{\omega^2}{2\pi^2 c^2} G_{\alpha\beta} P_\beta \tag{11.83}$$

where $G_{\alpha\beta} = M^{-1}_{\beta\alpha}$ and

$$M_{\beta\alpha} = \begin{bmatrix} k_y^2 + k_z^2 - \frac{\omega^2}{c^2}\epsilon_\perp & -k_x k_y & -k_x k_z \\ -k_x k_y & k_x^2 + k_z^2 - \frac{\omega^2}{c^2}\epsilon_\perp & -k_y k_z \\ -k_x k_z & -k_y k_z & k_x^2 + k_y^2 - \frac{\omega^2}{c^2}\epsilon_z \end{bmatrix} \tag{11.84}$$

The Green's tensor is the inverse of M

$$G_{\alpha\beta} = \frac{1}{|M|}(-1)^{\alpha+\beta} m_{\alpha\beta} \tag{11.85}$$

and to calculate the inverse, the determinant of M is

$$|M| = \frac{\omega^2}{c^2}\epsilon_z\left(k_x^2 + k_y^2 + k_z^2 - \epsilon_\perp \frac{\omega^2}{c^2}\right)\left(k_z^2 + \epsilon_\perp\left(\frac{k_x^2 + k_y^2}{\epsilon_z} - \frac{\omega^2}{c^2}\right)\right) \tag{11.86}$$

The Green's tensor can then be written in its final form as follows

$$G = \frac{1}{|M|}\begin{bmatrix} -k_y^2 k_z^2 + dTEy \cdot dEZ & k_x k_y dEM & k_x k_z dTE \\ k_x k_y dEM & -k_x^2 k_z^2 + dTEx \cdot dEZ & k_y k_z dTE \\ k_x k_z dTE & k_y k_z dTE & \left(k_z^2 - \epsilon_\perp \frac{\omega^2}{c^2}\right) dTE \end{bmatrix} \tag{11.87}$$

where $dEZ = k_x^2 + k_y^2 - \epsilon_z \frac{\omega^2}{c^2}$, $dTE = k_x^2 + k_y^2 + k_z^2 - \epsilon_\perp \frac{\omega^2}{c^2}$, $dTEy = k_x^2 + k_z^2 - \epsilon_\perp \frac{\omega^2}{c^2}$, $dTEx = k_y^2 + k_z^2 - \epsilon_\perp \frac{\omega^2}{c^2}$ and $dEM = k_x^2 + k_y^2 + k_z^2 - \epsilon_z \frac{\omega^2}{c^2}$.

Now that we developed the Green's tensor, we first consider the $\hat{z}$-polarized dipole. From (11.83) and (11.87)

$$E_{0z} = \frac{1}{2\pi^2 \epsilon_z}\begin{bmatrix} k_x k_z \\ k_y k_z \\ k_z^2 - \epsilon_\perp \frac{\omega^2}{c^2} \end{bmatrix} \frac{P}{k_z^2 - \epsilon_\perp\left(\frac{\omega^2}{c^2} - \frac{k_x^2 + k_y^2}{\epsilon_z}\right)} \tag{11.88}$$

The electric field is written as,

$$E_z(r) = \int E_0 e^{ik \cdot r} d^3k = \frac{i}{2\pi\epsilon_z} \int \begin{bmatrix} k_x k_z \\ k_y k_z \\ k_z^2 - \epsilon_\perp \frac{\omega^2}{c^2} \end{bmatrix} \frac{Pe^{ik \cdot r}}{2k_z} dk_x dk_y \tag{11.89}$$

where $k_z = \sqrt{\epsilon_\perp \left(\frac{\omega^2}{c^2} - \frac{k_x^2 + k_y^2}{\epsilon_z} \right)}$. Performing a coordinate transformation for $E_x, E_y \rightarrow E_r, E_\phi$ and evaluating the integral at $x = y = z = 0$ the field now becomes

$$E_z(0) = \frac{iP}{2\pi\epsilon_z} \int \begin{bmatrix} k\cos\theta \\ k\sin\theta \\ -\frac{k^2}{k_z} \frac{\epsilon_\perp}{z} \end{bmatrix} kdkd\theta = iP\epsilon_\perp \int \begin{bmatrix} 0 \\ 0 \\ -\frac{k^2}{k_z} \frac{\epsilon_\perp}{\epsilon_z} \end{bmatrix} kdk \tag{11.90}$$

Thus the radiation of the $\hat{z}$-polarized dipole is TM-polarized given explicitly by

$$E_z^{TM} = iP\epsilon_\perp \int_0^{k_{max}} \frac{k^3 dk}{k_z \epsilon_z^2} \tag{11.91}$$

with the integration parameter k representing the in-plane component of the wavevector. To appropriately deal with the limits of integration we must take into account that any effective medium theory fails in the limit when $k \cong \pi/a$. Therefore the integral is constrained inside the range of the first Brillouin zone, $k < k_{max} = \pi/a$.

On the other hand, the $\hat{x}$-polarized dipole emits both TE and TM-polarized waves, with amplitudes yielding the plane-wave expansion of the field as follows. Starting as we did for the $\hat{z}$ polarized field we can write out the $\hat{x}$-polarized field as

$$E_{0x} = \frac{P}{2\pi^2 |M|} \frac{\omega^2}{c^2} \begin{bmatrix} -k_y^2 k_z^2 + \left(k_x^2 + k_z^2 - \epsilon_\perp \frac{\omega^2}{c^2}\right)\left(k_x^2 + k_y^2 - \epsilon_z \frac{\omega^2}{c^2}\right) \\ k_x k_y dEM \\ k_x k_z dTE \end{bmatrix} \tag{11.92}$$

After some algebra we can separate this into TM and TE-polarized fields. We first investigate the TM-mode.

$$E_x^{TM}(r) = \frac{iP}{2\pi} \int \begin{bmatrix} \frac{-k_x}{k_x^2 + k_y^2} \frac{k_z}{\epsilon_\perp} \\ \frac{-k_y}{k_x^2 + k_y^2} \frac{k_z}{\epsilon_\perp} \\ \frac{1}{\epsilon_z} \end{bmatrix} k_x e^{ik \cdot r} dk_x dk_y \tag{11.93}$$

Performing another coordinate transformation and evaluating the integral at the origin yields

$$E_x^{TM}(0) = iP \int \begin{bmatrix} \frac{k_z}{2k\epsilon_\perp} \\ 0 \\ 0 \end{bmatrix} k^2 dk \tag{11.94}$$

The TM, $\hat{x}$-polarized dipole field is finally

$$E_x^{TM} = \frac{iP}{2\epsilon_\perp} \int_0^{k_{max}} k_z k dk \tag{11.95}$$

The last case to consider is the emission of the $\hat{x}$-polarized dipole for the TE-mode

$$E_x^{TE}(r) = \frac{iP}{2\pi} \frac{\omega^2}{c^2} \int \begin{bmatrix} k_y \\ k_x \\ 0 \end{bmatrix} \frac{k_y}{k_x^2 + k_y^2} \frac{e^{ik \cdot r}}{k_z} dk_x dk_y \tag{11.96}$$

Again, using the same procedure as was done for the last two cases the final result for the $\hat{x}$- polarized TE-mode is

$$E_x^{TE} = iP \frac{\omega^2}{c^2} \int_0^{k_{max}} \frac{k}{2k_z} dk \tag{11.97}$$

Due to the uniaxial symmetry of the problem, the $\hat{y}$-component of the electric field generated by a $\hat{y}$-polarized dipole is identical to the E_x generated by a $\hat{x}$-polarized dipole.

11.3.2 Spontaneous Decay Rates Near Planar Interfaces

Now that we are able to calculate the fields emitted by a point dipole we can use transfer matrix method (TMM) to determine the enhancement of the decay rate for a variety of configurations. The emission of the point dipole is calculated according to the plane wave-expansion, described above, incorporated into TMM formalism, as suggested in [66]. Before looking at the emission from the nanowire metamaterial, we verify that the above setup works by testing it for a well-known configuration. In [59] exact calculations were performed for the configuration in Fig. 11.11.

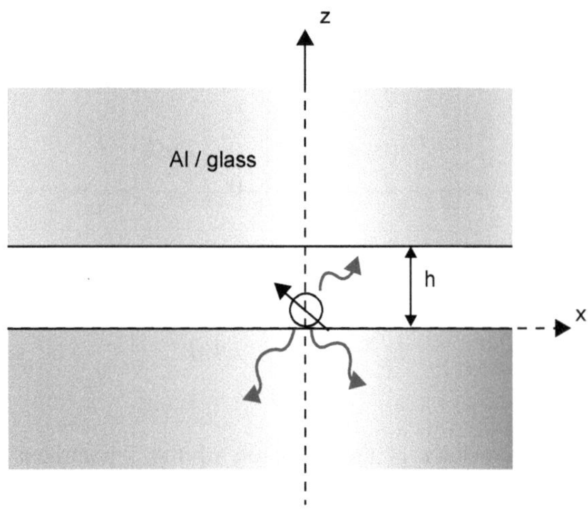

Fig. 11.11 Single point dipole near planar interfaces. The dipole is located on the surface of a dielectric substrate and a metal ($\epsilon = -34.5 + 8.5i$) or a glass ($\epsilon = 2.25$) interface is brought towards the dipole. The emission wavelength is $\lambda = 488$nm

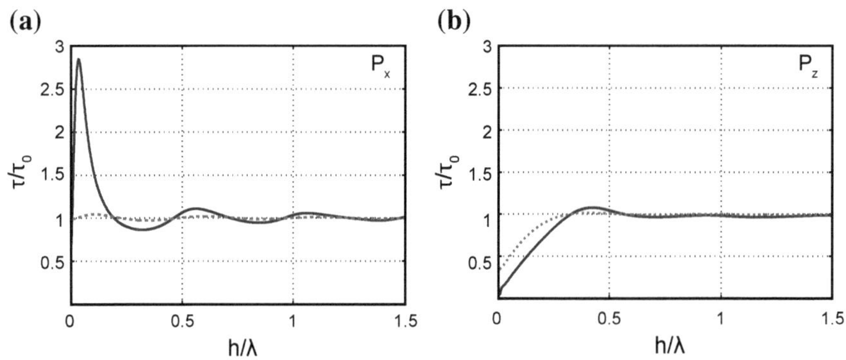

Fig. 11.12 Point dipole lifetime as a function of gap h. The *solid curves* are obtained for an approaching metal interface and the dashed curves represent the approaching dielectric interface

The normalized lifetime, $\tau/\tau_0 = (\Gamma/\Gamma_0)^{-1}$, of the dipole is a function of the separation h between a substrate and an approaching interface (Fig. 11.11). The normalization τ_0 refers to the situation where the dipole is located on the glass substrate and the second interface is at $h \to \infty$.

The modulations seen in Fig. 11.12 originate from the interference between the propagating plane waves from the point dipole and the reflected fields from the approaching interface. As expected, the modulations are more emphasized for the P_x orientation. The results that are obtained using the derived fields from the previous section and TMM are in exact agreement with [59]. This is a validation that the technique works to replicate exact-form calculations for a dipole near planar interfaces.

11.3.3 Emission in Lossless Metamaterials and Local Field Corrections

Now let's begin analyzing the emission of a point dipole positioned inside an infinite material (Fig. 11.10a) described by local effective medium theory using (11.3) and (11.4). The results are shown in Fig. 11.13a are in agreement with previous studies [42–45]. The local effective medium theory suggests that the enhancement of the photonic density of states is linked to the plasmon resonance in the wires $(\epsilon_i \cong 1)$, the broadband region corresponding to epsilon near zero (ENZ) $(\epsilon_i \cong -7)$, and the hyperbolic response of the metamaterial $(\epsilon < -7)$. The ENZ regime provides the maximum reduction in lifetime in the local response.

Calculations of the lifetime dynamics modulation is also repeated for a situation when a point dipole is positioned inside the homogeneous material described by the nonlocal EMT model [38]. In this case, we assume that the dipole can independently emit into two TM-polarized channels.

As seen in Fig. 11.13b, the predictions of nonlocal EMT differ drastically from local EMT. The main difference is the dramatic enhancement of decay rate that is observed in the elliptic regime. This can be accredited to the existence of the additional TM mode. It may be useful to note that the hyperbolic-like CPP-rich mode dominates the emission in across both the elliptic and hyperbolic regimes

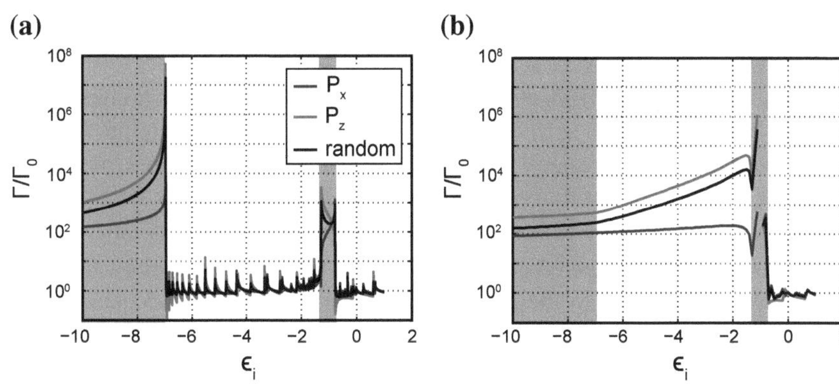

Fig. 11.13 Enhancement of the decay rate in bulk lossless metamaterials as the function of rods' permitivity, (**a**) local and (**b**) nonlocal response. (The different *colors* represent different orientation of the dipole and the *shaded areas* represent the spectral range where the metamaterial operates in the hyperbolic regime)

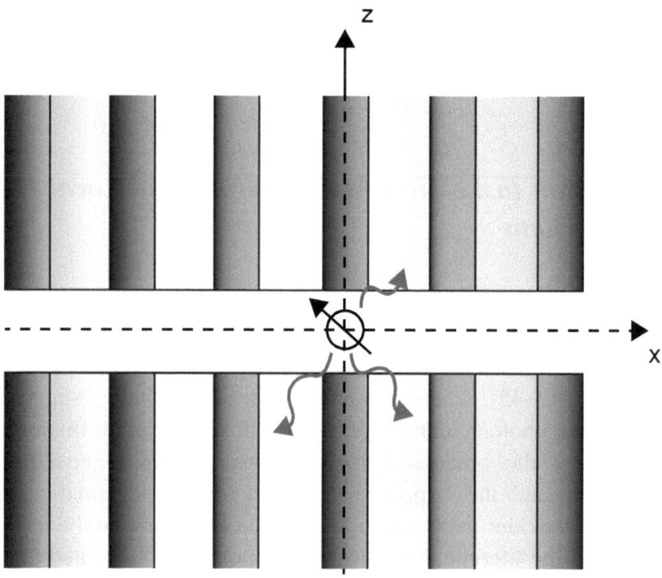

Fig. 11.14 Schematic of a point dipole positioned in a small slit vacuum surrounded by a nanowire metamaterial

[41]. The nonlocality also removes the strong enhancement that is observed at ENZ in the local calculations.

To analyze the emission from a dipole in realistic lossy metamaterials, it becomes necessary to place the dipole inside a substantially small vacuum cavity that is carved in the homogeneous material. As result of the methods that we are implementing, using TMM, the cavity takes the shape of a small planar slit that is oriented perpendicular to the wires (Fig. 11.14). The dipole emission is calculated according to the plane wave expansion described above, incorporated in the transfer-matrix formalism, as suggested in [38, 66].

For a slit below $\sim 10\,\mathrm{nm}$ the enhancement of the decay rate becomes independent of the slit size as a result of the high-wavenumber cut-off that was described earlier. Furthermore, for a local metamaterial it becomes possible to eliminate local field correction effects by renormalizing emission in the $\hat{z}$-polarized dipole by $1/\epsilon_z^{mg\,2}$ (Fig. 11.15). Similarly, renormalization of the $\hat{z}$-polarized dipole in the nonlocal EMT response can be obtained by $1/\epsilon_z(k)^2$. The results using this local correction applied to the point dipole positioned inside a vacuum slit surrounded by a nonlocal nanowire medium qualitatively agree with the total decay rate calculated for a dipole positioned inside the nonlocal medium (Fig. 11.13b). Therefore, we use the above normalization to take into account local field correction effects.

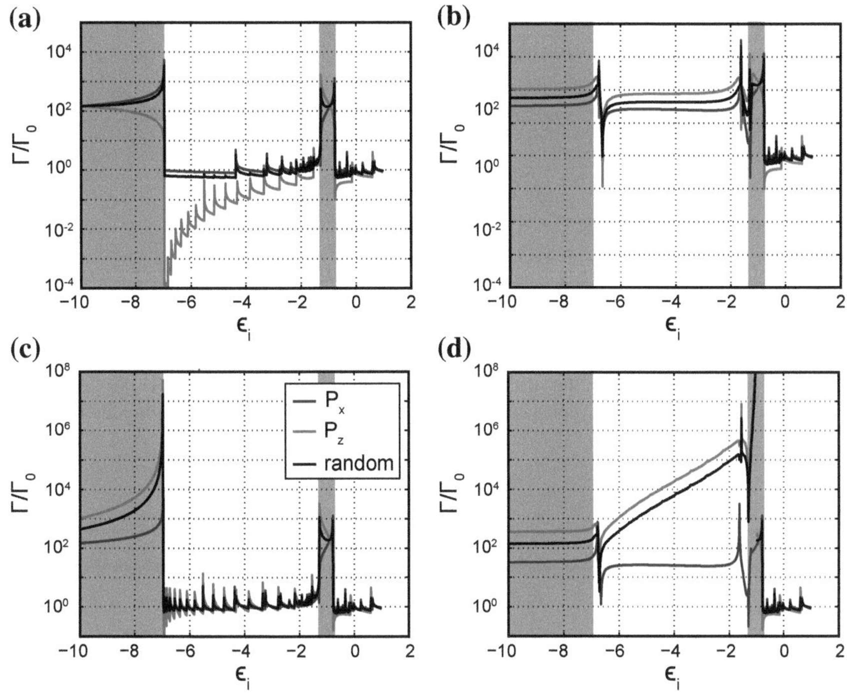

Fig. 11.15 Emission of the point dipole positioned in the vacuum slit inside the lossless nanowire metamaterial; (**a**, **b**) original data with no local field corrections, (**c**, **d**) data accounting for a local field correction

11.3.4 Effects of Finite Material Absorption

In the third phase, we calculate the decay rate of the dipole positioned in the slit between two slabs of lossy nanowire metamaterial with $Im(\epsilon_i) = 0.1$. The results for these calculations are shown in Fig. 11.16. As expected, the presence of material absorption does not fundamentally alter the dynamics of the emission. In general, increase in $Im(\epsilon_i)$ leads to further enhancement of the decay rate.

11.3.5 Non-Local Field Correction Approach

The final phase is to show that replacing the vacuum slit with the lossless nanowire metamaterial will improve the calculated accuracy and establish a technique that can potentially adequately deal with highly absorbing materials. Figure 11.17 illustrates these results for the configurations for the lossless and lossy nanowires

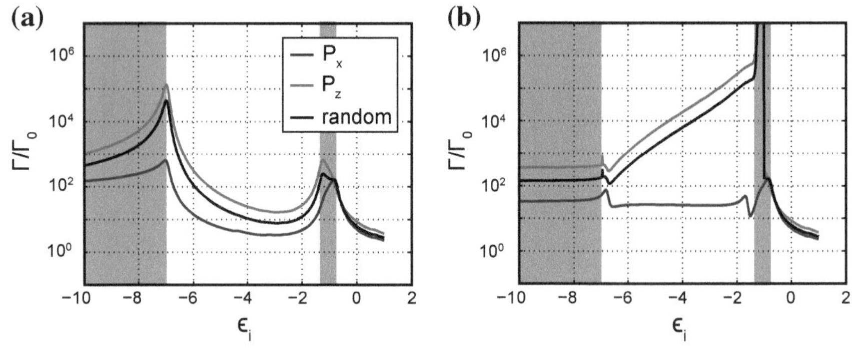

Fig. 11.16 Enhancement of the decay rate in nanowire metamaterials for (**a**) local and (**b**) nonlocal calculations with finite losses; the results account for local field corrections

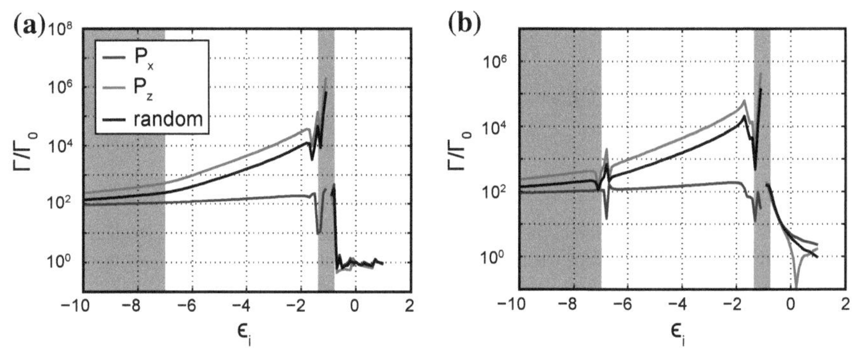

Fig. 11.17 Enhancement of decay rate in nanowire metamaterials for (**a**) lossless and (**b**) finite losses; the results use the lossless nanowire metamaterial in the slit, no local field corrections are used

that were considered in the previous sections. We see that the comparison is quantitatively similar.

By using this approach we can more accurately calculate systems that are realized in nature.

To summarize, here we have analyzed the enhancement of the decay rate for a dipole emitting inside the plasmonic nanowire composite. The predictions from nonlocal effective medium theory suggests that there is a significant enhancement of the decay rate across the spectral range where the nanowires are plasmonic, and in particular across the spectral range where the composite as a whole exhibits elliptic local response.

11.4 Experimental Results on Collective Purcell Enhancement

The distinctive advantage of hyperbolic metamaterials over plasmonic antennas and micro-cavities, is that the former could provide a collective enhancement for a macroscopic number of fluorophores. While the regular Purcell enhancement with nanostructures is highly sensitive to relative position and orientation of a fluorescent molecule in respect to a structure, metamaterials, in principle, do not have those limitations. It should be noted, that practical realizations of hyperbolic metamaterials frequently favor near-field interactions over macroscopic photonic effects, however, substantial Purcell enhancement could be obtained for any spatial orientation of a florescent molecule inside a unit cell of nanowires composites [67].

In order to obtain position dependent Purcell factor, numerical methods for calculating classical Green's functions could be employed. Those tools are required in order to account for near-field effects and perform a comparison with the beforehand developed theoretical approach. The reported results were obtained with the finite element method (FEM) implemented in Comsol Multiphysics software. The Purcell factor at various positions inside the nanowire metamaterial was calculated as a ratio of power flow from a point dipole placed there and the corresponding value for a dipole in the uniform dielectric (Purcell enhancement in homogeneous material). The arbitrary orientation of the dipole was taken into account by averaging the Purcell factor over all dipole orientations, which in this case presents a linear combination of the corresponding values for emitters with a dipole moment along the three coordinate axes. For the numerical simulations, in order to simulate infinite number of nanowires in the arrays, the number of rods was gradually enlarged in the finite-size arrays with periodic boundary conditions on their sides. The convergence of the Purcell factor with the number of the nanowires within the simulation domain confirmed that a 10×10 nanowire array with periodic boundary conditions can be used to simulate the behaviour of an infinite metamamterial. The averaging over the dipole position within the primitive cell of the array of the metamaterial was performed, assuming a uniform distribution of the emitters with a position dependent decay rate and local excitation efficiency, by a pump light illuminating the metamaterial from the substrate side.

The results of numerical analysis appear on Fig. 11.18. Both convergence issues and position-depended Purcell enhancement are discussed in the figure caption. Since the numerical procedure could be performed for the finite set of dipole's positions, the Purcell enhancement in the entire domain was obtained by performing a linear interpolation.

Time-resolved photoluminescence analysis is usually performed using time-correlated single photon counting (TCSPC) [68]. Experimental results, obtained with this kind of apparatus, provide time-dependent fluorescent signal. In an ideal case, this signal could be fitted with a small number of exponential decays, with each one attributed to a certain emission process. However, in a complex electromagnetic environment this process could fail owing to a complex nature of

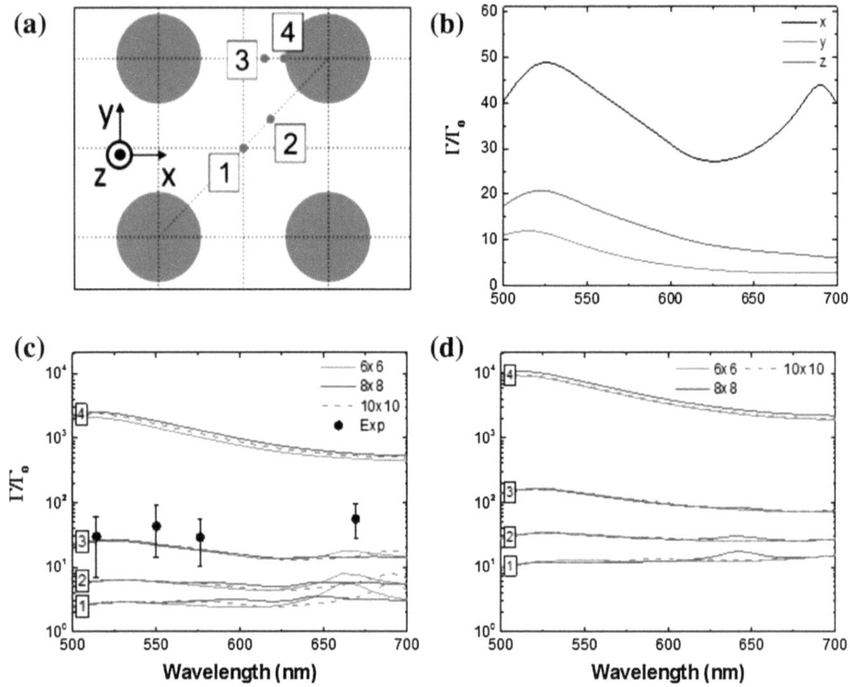

Fig. 11.18 Numerical simulations of the Purcell factor in metamaterials. (**A**) Schematic of the metamaterial with the position of the emitters used in the simulations. All the emitters are situated in the middle of the nanowire length. **B**, **C**, **D** Spectral dependence of the Purcell factor for (**B**) a dipole with different orientations at position 3 inside the metamaterial, (**C**, **D**) a randomly oriented dipole at different positions inside the nanowire metamaterial (as indicated in (**A**)) for the metamaterials with (B.C) period a = 100 nm and nanowire radius r = 25 nm (as in the experiment) and (**D**) a = 50 nm and r = 12.5 nm, corresponding to the same local effective medium parameters. The *colored lines* correspond to different sizes of the finite nanowire array used in the simulations (as indicated in the legends) showing the convergence to the behavior of the infinite metamaterial; *symbols* in (**c**) represent the experimental data with points indicating the maximum of the lifetime distributions (unpublished data) and the bar corresponding to the width of the distribution at 0.1 amplitude. In all simulations the internal quantum yield of the emitter was considered to be 1

the Purcell enhancement. In this case it is convenient to introduce the quantity, called lifetime distributions, indicating a continuous span of exponential times, forming the process. This distribution could be obtained by taking an inverse Laplace transform of the following expression:

$$I(t) = \int_0^\infty F(s)e^{-st}ds \tag{11.98}$$

Where I(t) is the time dependent intensity decay (measured quantity), deconvoluted from instrumental response function (also measured), and F(s) is the

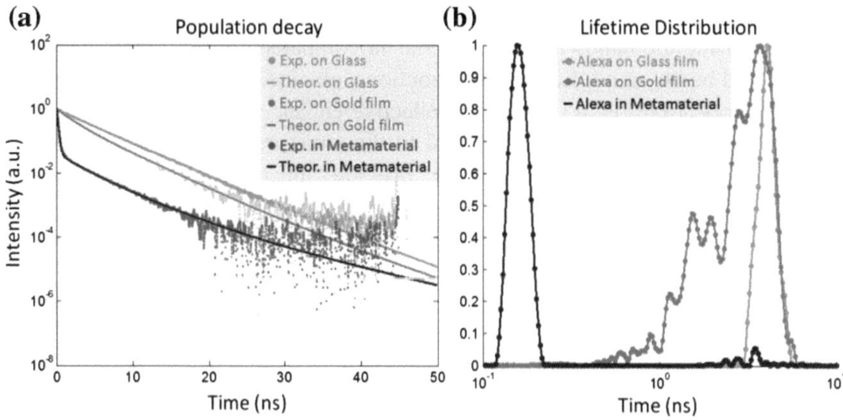

Fig. 11.19 Collective Purcell enhancement of Alexa dye, dissolved in water and situated inside nanowires metamaterial. (**a**) Decay of the fluorescent signal as the function of time. *Blue* data—the dye solution on top of a glass slide, *Red*—gold film, *Green*—nanowires metamaterial. (**b**) Lifetime distribution for 3 scenarios—glass, metal, and nanowires. The distributions were obtained from the time-dependent signal by applying an inverse Laplace transform. Scattered data—experiment, solid line—theoretical fit

relative weight of actual exponential decay components (lifetime distribution). Since the inverse Laplace transform is known to be an ill-defined problem (especially for analysis of noisy data), an iterative fitting procedure was applied in order to achieve stable results. Solution of emitters (Alexa dye) situated on the glass slide show smooth localized lifetime distributions, peaked in the nanosecond range (Fig. 11.19b, blue curve). It is worth noting that the glass-ethanol interface has only a small refractive index contrast, and hence could be neglected in the analysis. Next, a plasmonic (Au) interface was introduced for testing the concept of collective lifetime manipulation. The dynamics of fluorophores' lifetimes, modified by the presence of the nearby surfaces, in particular those made of noble metals, is well understood. The key contributors to lifetime modification are mirror-reflected waves, excitation of surface waves and quenching [69]. Using this theoretical formulation and the experimentally measured lifetime distribution of dyes on a glass slide, the theoretical prediction of the distribution on the gold film could be obtained (unpublished data). An almost perfect fit between the theoretical prediction taking into account spatial averaging of the emitter position with respect to the metal film, and experimental data could be shown and it suggests the validity of the approach for studies of more complex nanostructures. The spatial average takes into account a random distribution of positions and orientations of florescent molecules in the solution. Taking into account the obtained good correspondence between the experiment and the model, contributions of various possible collective fluorophore concentration-dependent effects, e.g. super-fluorescence [70, 71], can then be ruled out.

The results of time-dependent decays appear on Fig. 11.19a. Collective lifetime reduction inside the nanowire metamaterial in comparison to gold and glass substrates could be observed. Lifetime distributions, obtained with the inverse Laplace technique appear on Fig. 11.19b. The collective enhancement owing to nanowires reaches the factor of ~30 and is typical to the structure. It is worth noting, that layered realizations of metamaterials provide much smaller enhancements (e.g. [72]).

11.5 Conclusion

The goal of this chapter is to review the effect of spontaneous emission inside an artificially created highly nonlocal medium. In particular, theoretical models for nonlocal homogenization were revised in detail and related to recent numerical and experimental reports. The emphasis was put on the effect of structural nonlocality, that were shown to have a major impact on the quantum electrodynamics inside structured media. In particular, it dominates the spontaneous emission behavior, setting fundamental limits on the Purcell enhancement. It also enables the design of the local density of optical states via geometrical parameters of the metamaterial composite, opening a new road for engineering density of optical states with (meta) materials.

Acknowledgements This work has been funded in part by ESPRC (UK), the ERC iPLASMM project (321268), and the US Army Research Office (Grant No. W911NF-12-1-0533). A.Z. acknowledges support from the Royal Society and the Wolfson Foundation. P.G. acknowledges TAU Rector Grant and German-Israeli Foundation (GIF, grant number 2399).

References

1. Nader Engheta, Circuits with light at nanoscales: optical nanocircuits inspired by metamaterials. Science **317**(5845), 1698–1702 (2007)
2. John B. Pendry, David R. Smith, Reversing light with negative refraction. Phys. Today **57**, 37–43 (2004)
3. V.M. Shalaev, Optical negative-index metamaterials. Nat. Photonics **1**(1), 41–48 (2007)
4. B. John Pendry, D. Schurig, R. David Smith, Controlling electromagnetic fields. Science **312** (5781), 1780–1782 (2006)
5. W. Cai, et al. Optical cloaking with metamaterials. *Nature photonics* 1.4, pp. 224–227 (2007)
6. R. Atkinson, et al. Anisotropic optical properties of arrays of gold nanorods embedded in alumina. Phys. Rev. B **73.23**, 235402 (2006)
7. J. Yao, et al. Optical negative refraction in bulk metamaterials of nanowires. Science **321.5891**, 930–930 (2008)
8. B.D.F. Casse et al., Super-resolution imaging using a three-dimensional metamaterials nanolens. Appl. Phys. Lett. **96**(2), 023114 (2010)

9. Mário Silveirinha, Nader Engheta, Tunneling of electromagnetic energy through subwavelength channels and bends using ε-near-zero materials. Phys. Rev. Lett. **97**(15), 157403 (2006)
10. C. Erik Garnett, L. Mark Brongersma, Yi Cui, D. Michael Mc Gehee, Nanowire solar cells. Annu. Rev. Mater. Res. **41**, 269–295 (2011)
11. A. Aubry, D. Lei, A. Fernandez-Dominguez, Y. Sonnefraud, S.A. Maier, J.B. Pendry, Plasmonic light harvesting devices over the whole visible spectrum. Nano Lett. **10**, 2574 (2010)
12. S.A. Maier, P. Kik, H.A. Atwater, S. Meltzer, E. Harel, B. Koel, A. Requicha, Local detection of electromagnetic energy transport below the diffraction limit in metal nanoparticle plasmon waveguides. Nat. Mater. **2**, 229 (2003)
13. V.A. Podolskiy, E.E. Narimanov, Strongly anisotropic waveguide as a nonmagnetic left-handed system. Phys. Rev. B **71**, 201101(R) (2005)
14. M.A. Noginov and V.A. Podolskiy (eds.), *Tutorials in Metamaterials* (CRC Press, Boca Raton, FL, 2012)
15. N. Fung, H. Lee, C. Sun, X. Zhang, Sub-diffraction-limited optical imaging with a silver superlens. Science **308**, 534 (2005)
16. L.V. Alexeyev, E.E. Narimanov, Slow light and 3D imaging with non-magnetic negative index systems. Opt. Exp. **14**, 11184 (2006)
17. Z. Jacob, L.V. Alekseyev, E. Narimanov, Optical hyperlens: far-field imaging beyond the diffraction limit. Opt. Exp. **14**, 8247–8256 (2006)
18. Z. Liu, H. Lee, Y. Xiong, C. Sun, X. Zhang, Far-field optical hyperlens magnifying sub-diffraction-limited objects. Science **315**, 1686 (2007)
19. A. Salandrino, N. Engheta, Far-field subdiffraction optical microscopy using metamaterial crystals: theory and simulations. Phys. Rev. B **74**, 075103 (2006)
20. I.I. Smolyaninov, Y.J. Hung, C.C. Davis, Science **315**, 1699 (2007)
21. E.M. Purcell, Phys. Rev. **69**, 681 (1946)
22. A.N. Poddubny, P.A. Belov, P. Ginzburg, A.V. Zayats, Y.S. Kivshar, Microscopic model of Purcell enhancement in hyperbolic metamaterials. Phys. Rev. B **86**, 035148 (2012)
23. I.V. Iorsh, A.N. Poddubny, P. Ginzburg, P.A. Belov, Y.S. Kivshar compton-like polariton scattering in hyperbolic metamaterials. Phys Rev. Lett. **114**, 185501 (2015)
24. A.J. Hoffman, L. Alekseyev, S.S. Howard, K.J. Franz, D. Wasserman, V.A. Podolskiy, E.E. Narimanov, D.L. Sivco, C. Gmachl, Negative refraction in semiconductor metamaterials. Nat. Mater. **6**, 946 (2007)
25. A. Kabashin, P. Evans, S. Pastkovsky, W. Hendren, G.A. Wurtz, R. Atkinson, R. Pollard, V. Podolskiy, A.V. Zayats, Plasmonic nanorod metamaterials for biosensing. Nat. Mater. **8**, 867 (2009)
26. J. Kim, V.P. Drachev, Z. Jacob, G.V. Naik, A. Boltasseva, E.E. Narimanov, V.M. Shalaev, Improving the radiative decay rate for dye molecules with hyperbolic metamaterials. Opt. Express **20**, 7 (2012)
27. L. Alekseyev, V.A. Podolskiy, E. Narimanov, Homogeneous hyperbolic systems for terahertz and far-infrared frequencies. Adv. Optoelectron. **2012**, 267564 (2012)
28. V.A. Podolskiy, L. Alekseyev, E.E. Narimanov, Strongly anisotropic media: the THz perspectives of left-handed materials. J. Mod. Opt. **52**(16), 2343 (2005)
29. N. Vasilantonakis, G.A. Wurtz, V.A. Podolskiy, A.V. Zayats, Refractive index sensing with hyperbolic metamaterials: strategies for biosensing and nonlinearity enhancement. Opt. Exp. **23**, 14329 (2015)
30. S. Ishii, A.V. Kildishev, E. Narimanov, V.M. Shalaev, V.P. Drachev, Sub-wavelength interference pattern from volume plasmon polaritons in a hyperbolic medium. Laser Photon. Rev. **7**, 2 (2013)
31. W. Cai, U.K. Chettiar, A.V. Kildishev, V.M. Shalaev, Optical cloaking with metamaterials. Nat. Phot. **1**, 224 (2007)
32. E.E. Narimanov, A.V. Kildishev, Optical black hole: Broadband omnidirectional light absorber. Appl. Phys. Lett. **95**, 041106 (2009)

33. D.A. Genov, S. Zhang, X. Zhang, Mimicking celestial mechanics in metamaterials. Nat. Phys. Adv. online pub. NPHYS1338 (2009)
34. G.A. Wurtz, R. Pollard, W. Hendren, G.P. Wiederrecht, D.J. Gosztola, V.A. Podolskiy, A.V. Zayats, Designed nonlocality-enhanced sub-picosecond nonlinearities in plasmonic nanorod metamaterial. Nat. Nanotechnol. **6**, 107 (2011)
35. G.A. Wurtz, W. Dickson, D. O'Connor, R. Atkinson, W. Hendren, P. Evans, R. Pollard, A.V. Zayats, Guided plasmonic modes in nanorod assemblies: strong electromagnetic coupling regime. Opt. Exp. **16**, 7460 (2008)
36. S. Melle, J.L. Menendez, G. Armelles, D. Navas, M. Vanzquez, K. Nielsch, R.B. Wehrspohn, U. Gozele, Magneto-optical properties of nickel nanowire arrays. Appl. Phys. Lett. **83**, 4547 (2003)
37. Y. Veniaminova, A.A. Stashkevich, Y. Roussigne, S.M. Cherif, T.V. Murzina, A.P. Murphy, R. Atkinson, R.J. Pollard, A.V. Zayats, Brillouin light scattering by spin waves in magnetic metamaterials based on Co nanordos. Opt. Mat. Exp **2**, 1260 (2012)
38. B. Wells, A.V. Zayats, V.A. Podolskiy, Nonlocal optics of plasmonic nanowire metamaterials. Phys. Rev. B **89**, 035111 (2014)
39. R.J. Pollard, A. Murphy, W.R. Hendren, P.R. Evans, R. Atkinson, G.A. Wurtz, A.V. Zayats, V.A. Podolskiy, Optical nonlocalities and additional waves in epsilon-near-zero metamaterials. Phys. Rev. Lett. **102**, 127405 (2009)
40. T. Geng, S. Zhuang, J. Gao, X. Yang, Nonlocal effective medium approximation for metallic nanorod metamaterials. Phys. Rev. B. **91**, 245128 (2015)
41. V.A. Podolskiy, P. Ginzburg, B. Wells, A.V. Zayats, Light emission in nonlocal plasmonic metamaterials. Faraday Discuss. **178**, 61 (2015)
42. E.E. Narimanov, Infinite at any frequency: the photonic density of states in (meta)materials with hyperbolic dispersion and related phenomena, in *Proceedings of SPIE Optics and Photonics*, pp. 7754–09, 54 (2010); full manuscript is in press
43. H.K. Krishnamoorthy, Z. Jacob, E. Narimanov, I. Kretzschmar, V.M. Menon, Topological transitions in metamaterials. Science **336**, 205 (2012)
44. O. Kidwai, S. Zhukovsky, J.E. Sipe, Opt. Lett. **36**, 2530 (2011)
45. C.I. Cortes, W. Newman, S. Molesky, Z. Jacob, J. Opt. **14**, 063001 (2012)
46. J.C.M. Garnett, Philos. Trans. R. Soc. London, Ser. B **203**, 385 (1904)
47. G.W. Milton, *The theory of composites* (Cambridge U. Press, Cambridge, UK, 2002)
48. A.L. Pokrovsky, A.L. Efros, Nonlocal electrodynamics of two-dimensional wire mesh photonic crystals. Phys. Rev. B **65**, 045110 (2002)
49. P.A. Belov, R. Marques, S.I. Maslovski, I.S. Nefedov, M. Silveirinha, C.R. Simovski, S.A. Tretyakov, Strong spatial dispersion in wire media in the very large wavelength limit. Phys. Rev. B **67**, 113103 (2003)
50. M. Silveirinha, Nonlocal homogenization model for a periodic array of epsilon-negative rods. Phys. Rev. E **73**, 046612 (2006)
51. V.L. Ginzburg, Electromagnetic waves in isotropic and crystalline media characterized by dielectric permittivity with spatial dispersion. JETP **34**, 1096 (1958)
52. V.M. Agranovich, V.L. Ginzburg, *Crystal Optics with Spatial Dispersion and Excitons, Springer Series in Solid-State Sciences*, vol. 42 (Springer, Berlin, 1984)
53. M. Born, E. Wolf, *Principles of Optics* (Press, Cambridge U, 1999)
54. M.O. Scully, M. S. Zubairy, *Quantum Optics* (Cambridge University Press, 1997)
55. P. Ginzburg, Accelerating spontaneous emission in open resonators. *Annalen der Physik* (2016)
56. J.E. Sipe, Photons in dispersive dielectrics. J. Opt. A: Pure Appl. Opt. **11**, 114006 (2009)
57. C.A. Balanis, *Antenna Theory: Analysis and Design* (Wiley, New York, 2005)
58. A.P. Slobozhanyuk, P. Ginzburg, D.A. Powell, I. Iorsh, A.S. Shalin, P. Segovia, A.V. Krasavin et al., Phys. Rev. B **92**, 195127 (2015)
59. P. Ginzburg, A.V. Krasavin, A.N. Poddubny, P.A. Belov, Y.S. Kivshar, A.V. Zayats, Phys. Rev. Lett. **111**, 036804 (2013)
60. Vogel, W. Welsch, D-G, *Quantum Optics, 3rd, Revised and Extended Edition* (Wiley 2006)

61. S.M. Barnett, B. Huttner, R. Loudon, Phys. Rev. Lett. **68**, 3698 (1992)
62. H.T. Dung, S.Y. Buhmann, D.G. Welsch, Phys. Rev. A **74**, 023803 (2006)
63. S. Scheel, L. Knoll, D.G. Welsch, Phys. Rev. A **60**, 4094 (1999)
64. S. Scheel, L. Knoll, D.G. Welsch, Phys. Rev. A **61**, 069901 (2000)
65. L. Novotny, B. Hecht, *Principles of Nano-Optics* (Cambridge U Press, Cambridge UK, 2006)
66. S.R.J. Brueck, IEEE J. Sel. Top. Quantum Elctron. **6**, 899 (2000)
67. A.P. Slobozhanyuk, P. Ginzburg, D.A. Powell, I. Iorsh, A.S. Shalin, P. Segovia, A.V. Krasavin, G.A. Wurtz, V.A. Podolskiy, P.A. Belov, A.V. Zayats, Purcell effect in hyperbolic metamaterial resonators. Phys. Rev. B **92**, 195127 (2015)
68. J.R. Lakowicz, *Principles of Fluorescence Spectroscopy,* 3rd edn. (Springer, New York, 2006)
69. G.W. Ford, W.H. Weber, Electromagnetic effects on a molecule at a metal surface: I. Effects of nonlocality and finite molecular size. Surf. Sci. **109**, 451–481 (1981)
70. R. Bonifacio, L.A. Lugiato, Cooperative radiation processes in two-level systems: superfluorescence. Phys. Rev. A **11**, 1507 (1975)
71. D. Martín-Cano, L. Martín-Moreno, F.J. García-Vidal, E. Moreno, Resonance energy transfer and superradiance mediated by plasmonic nanowaveguides. Nano Lett. **10**, 3129–3134 (2010)
72. T. Tumkur, G. Zhu, P. Black, Y.A. Barnakov, C.E. Bonner, M.A. Noginov, Control of spontaneous emission in a volume of functionalized hyperbolic metamaterial. Appl. Phys. Lett. **99**, 151115 (2011)

Chapter 12
Nonlocal Response in Plasmonic Nanostructures

Martijn Wubs and N. Asger Mortensen

Abstract After a brief overview of nanoplasmonics experiments that defy explanation with classical electrodynamics, we introduce nonlocal response as a main reason for non-classical effects. This concept is first introduced phenomenologically, and afterwards based on the semi-classical hydrodynamic Drude model (HDM) that indeed exhibits nonlocal response. In particular, we discuss recent generalizations and extensions of the HDM, to include both convection and diffusion dynamics of the induced charges. This generalized nonlocal optical response (GNOR) model allows for the first time unified semi-classical explanations of known experimental phenomena for both monomers and dimers that previously seemed to require microscopic theory. Finally, we turn to Landau damping and discuss the microscopic origin of the size-dependent damping captured by the classical diffusion mechanism in the GNOR model.

12.1 Introduction

In most chapters of this book, "*quantum plasmonics*" is to be understood as "*quantum optics with plasmonic structures*", where the quantumness is in the non-classical states of the electromagnetic field, or in the quantum properties of the emitters. The plasmonic environment on the other hand is typically described in the same way as in classical electrodynamics, with the same modes or the same classical Green function [1–7].

The approach of this chapter is different, in that we consider how plasmonic properties are influenced by quantum properties of the free-electron plasma, properties

M. Wubs (✉) · N.A. Mortensen
Department of Photonics Engineering and Center for Nanostructured Graphene, Technical University of Denmark, 2800 Kongens Lyngby, Denmark
e-mail: mwubs@fotonik.dtu.dk
URL: http://www.martijnwubs.nl

N.A. Mortensen
e-mail: asger@mailaps.org
URL: http://www.mortensen-lab.org

S.I. Bozhevolnyi et al. (eds.), *Quantum Plasmonics*,
Springer Series in Solid-State Sciences 185,
DOI 10.1007/978-3-319-45820-5_12

that are neglected in the usual treatments of plasmon dynamics. Yet these quantum properties of the plasma, in particular nonlocal response, have measurable effects. Prominent resonance frequency shifts and field enhancement reductions occur for example for few-nanometer sized plasmonic particles but also for larger particles separated by nanometer-sized gaps. This is quantum plasmonics in the sense of "*classical optics with quantum plasmonic structures*", and we will give ample examples below how this can be accounted in electrodynamics modeling.

Historically, the field of plasmonics has developed with a solid foundation in classical electrodynamics employing semi-classical descriptions of the interactions of light with matter [8–10]. In particular, the collective oscillation of conduction electrons subject to driving optical fields has been conceptually analyzed within Drude theory and the local-response approximation (LRA), where the material response occurs only in the point of space of the perturbation, while there is no response at even short distances. For dielectrics, this is of course a well-established and accurate approach, while it is traditionally being applied to metals too. Despite its simplifications, the LRA framework has fostered both predictions and experimental confirmations of novel plasmonic phenomena, such as the squeezing of light beyond the diffraction limit [11, 12], the tunability of the optical properties of metallic structures with size and shape [13], and large enhancement of the electric field in metal nanoparticles of close proximity [14] as well as in metal geometries with sharp surface corrugation [15].

In accounts of light-matter interactions we commonly rely on linear-response theory and for insulating materials the further simplification associated with the LRA is usually excellent all the way down to the atomic scale. The success of the LRA in nanoplasmonics is perhaps more intriguing. As an example, novel experimental explorations of gold bow-tie dimers with few-nanometer gaps were published by a Maier and Yang collaboration [16] under the headline "*Nanoplasmonics: Classical down to the Nanometer Scale*". On the other hand, when light interacts with the free conduction electrons in metals the optical response could conceptually be nonlocal with spatial dispersion in the response function [17]. In passing, we note that similar nonlocal effects are also anticipated in the context of both exciton dynamics and the free-carrier plasmonic response in doped two-dimensional materials and semiconductors (see recent review [18] and references therein). So, what makes the LRA so reasonable even for metals?

The underlying quantum wave dynamics of the electron gas manifests itself at a length scale being intrinsic to the metal: the Fermi-wavelength scale which is in the nanometer-to-Ångström scale regime for most metals [19]. This largely constitutes the success of the LRA and the application of Drude theory to plasmonics even in nanoscale metallic structures, while also hinting to the mesoscopic size regime where we can anticipate a departure from the predictions within the LRA of classical electrodynamics. In fact, the neglect of nonlocal effects is causing field-singularities in the LRA response, e.g. for arbitrarily sharp changes in the metal-surface topography or in dimers with vanishing gaps [20].

In the following, we discuss nonlocal response, which was recently revived [21–24] in relation to experimental developments where plasmonics is explored in

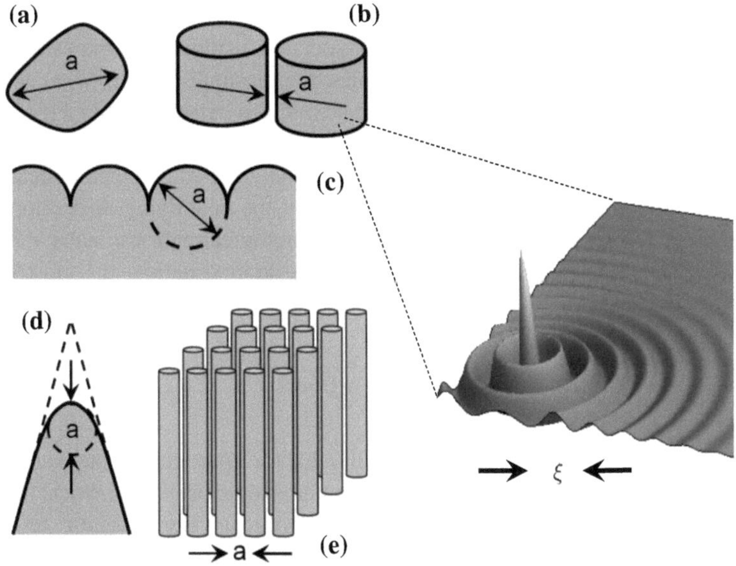

Fig. 12.1 Examples of metallic nanostructures, indicating characteristic length scales a for (a) isolated metal particles, (b) metal-particle dimers with gaps, (c) corrugated metal surfaces, (d) sharp metal tips, and (e) metal-nanowire metamaterials. While the local-response approximation is typically adequate for large structures, the nonlocal correction becomes important in the mesoscopic regime where $a \longrightarrow \xi$, with ξ being an intrinsic length scale associated with the nonlocal dynamics of the electron gas

structures of still smaller dimensions [25–33]. These experiments share at least one common feature: nanometer-to-Ångström scale characteristic dimensions appear to promote a plasmonic response that can only qualitatively be accounted for the LRA, while a more quantitative description seem to require developments beyond classical electrodynamics. Figure 12.1 shows schematic representations of a number of generic mesoscopic geometries with competing extrinsic and intrinsic length scales, such as characteristic geometrical dimensions a and the finite range ξ of nonlocal response. In order to address such problems, we will discuss developments of a real-space formulation [18, 24] and numerical implementations [34] of long-existing nonlocal hydrodynamic theory [35–37] as well as a more recent extension to a generalized account of drift-diffusion dynamics (the so-called GNOR theory) [38] along with discussions of Landau damping in connection to nonlocal response [39–41].

12.2 Linear-Response Theory

To simplify our discussion we will first consider the linear-response theory for scalar fields, before turning to a vectorial account. We consider a generic system's response

$\mathcal{Z}$ associated with a system input $\mathcal{F}$. Here, $\mathcal{F}$ could be a classical force field in a mechanical system, while $\mathcal{Z}$ would be the associated displacement in position of the system. In our later discussion of plasmonics, the plasma is driven by an electrical field $\mathbf{E}$ and the response is given in terms of an induced current density $\mathbf{J}$ (or alternatively it can be formulated as a displacement field $\mathbf{D}$). It is clear that $\mathcal{Z}$ is a functional of $\mathcal{F}$, written as $\mathcal{Z}[\mathcal{F}]$, but in general the exact functional could be complicated and nonlinear, depending on the detailed dynamics of the underlying microscopic system. In order to proceed along a more phenomenological path we make a Volterra series of $\mathcal{Z}[\mathcal{F}]$. Focusing on the linear-response approximation, the most general expression becomes

$$\mathcal{Z}(t,\mathbf{r}) = \int_{-\infty}^{t} dt' \int d\mathbf{r}' \, \chi(t-t',\mathbf{r},\mathbf{r}')\mathcal{F}(t',\mathbf{r}') \tag{12.1}$$

with χ being the linear-response susceptibility that for a particular system needs to be determined by some additional means, e.g. a microscopic or semi-classical theory or through carefully conducted linear-response experiments. Here, $\mathbf{r}$ and $\mathbf{r}'$ are spatial coordinates, while t and t' are different time instances. As an inherent assumption in physics, the principle of causality implies that there is only a response at times following the action, i.e. $t > t'$. The linear response function $\chi(t-t',\mathbf{r},\mathbf{r}')$ contains both temporal memory effects and spatial nonlocal effects. In passing, we note that in Fourier space the temporal response is associated with frequency dispersion (ω dependence) while nonlocal response gives rise to spatial dispersion (k dependence) in the response function. Assuming a *homogeneous* medium, the translational invariance implies that $\chi(t-t',\mathbf{r},\mathbf{r}') = \chi(t-t',\mathbf{r}-\mathbf{r}')$. This consequently turns both the temporal and the spatial integrals in (12.1) into convolutions, so that in ω and k-space the linear response becomes a simple product, i.e.

$$\mathcal{Z}(\omega,k) = \chi(\omega,k)\mathcal{F}(\omega,k). \tag{12.2}$$

This motivates that the phrases *nonlocal response* and *spatial dispersion* represent two sides of the same coin. In the following, we deliberately emphasize a real-space notation to facilitate our later application to finite-size structures (see Fig. 12.1), rather than to spatially extended systems. Our starting point is a phenomenological approach first outlined by Mortensen [42]. In particular, we will explore

$$\mathcal{Z}(\omega,\mathbf{r}) = \int d\mathbf{r}' \, \chi(\omega,\mathbf{r},\mathbf{r}')\mathcal{F}(\omega,\mathbf{r}') \tag{12.3}$$

in the situation where nonlocal response is a correction to the common LRA associated with a homogeneous system, i.e.

$$\chi(\omega,\mathbf{r},\mathbf{r}') = \chi_{\mathrm{LRA}}(\omega)\delta(\mathbf{r}-\mathbf{r}') + f(\omega,\mathbf{r}-\mathbf{r}'). \tag{12.4}$$

Here, we have introduced f as a nonlocal response function that we assume to be of short range. For the ease of presentation, we also assume a scalar response, leaving a vectorial response for Sect. 12.3. If the system is isotropic, as it is the case for the free-electron gas, then f is symmetric, i.e. $f(\omega, \mathbf{r} - \mathbf{r}') = f(\omega, |\mathbf{r} - \mathbf{r}'|)$. Thus, if we now turn to the moments of this function, then without further assumptions we in general have that

$$\int dr\, rf(\omega, r) = 0 \tag{12.5}$$

and

$$\int dr\, r^2 f(\omega, r) \equiv 2\xi^2. \tag{12.6}$$

Here, we have introduced the phenomenological parameter ξ as the characteristic range of the nonlocal response function. What is the physics associated with this length scale? As we shall see later, in a plasmonic context ξ is roughly the distance that an electron can travel during the time of an optical cycle.

Now, let us return to (12.3). If the perturbing field $\mathcal{F}$ varies slowly on the scale of ξ, we may proceed by Taylor expanding $\mathcal{F}$ around the point $\mathbf{r}$,

$$\mathcal{F}(\mathbf{r}') \simeq \mathcal{F}(\mathbf{r}) + [\nabla\mathcal{F}(\mathbf{r})] \cdot (\mathbf{r}' - \mathbf{r}) + \frac{1}{2}(\mathbf{r}' - \mathbf{r})^T \cdot [H\mathcal{F}(\mathbf{r})] \cdot (\mathbf{r}' - \mathbf{r}), \tag{12.7}$$

where the Hessian matrix H has the elements $H_{ij} = \partial^2/(\partial_i \partial_j)$ with $i,j = x, y, z$. Next, we substitute this expansion into (12.3) and perform the integrals with the help of (12.5) and (12.6). The resulting constitutive relation then becomes

$$\mathcal{Z}(\omega, \mathbf{r}) \simeq \left[\chi_{\text{LRA}}(\omega) + \xi^2\nabla^2\right] \mathcal{F}(\omega, \mathbf{r}) \tag{12.8}$$

where we have absorbed the zeroth-order moment integral of f into our definition of the local term $\chi_{\text{LRA}}(\omega)$. We note that the Laplacian is a consequence of the scalar model. In Sect. 12.3 we turn to a vectorial description and consequently we find contributions from other second-order spatial derivatives too.

What are the immediate implications of this nonlocal-corrected constitutive relation? Interestingly, (12.8) shows that scalar nonlocal response manifests itself through the Laplacian term, seemingly irrespective of the microscopic or semi-classical origin, and with a strength given by the scalar ξ, which relates to the width of the nonlocal response function through (12.6). This result also suggests the possibility of several nonlocal mechanisms playing in concert and adding up to an effective ξ^2. We shall return to this point below. We also note a quite striking property of (12.8) when compared to (12.3): With very few generic assumptions we have transformed the nonlocal two-point integral relation into a constitutive law that is mathematically of a local-response form, i.e. only involving a single spatial coordinate. Thus, if we can cope with an additional Laplacian term (note that such differential operators are already present in Maxwell's wave equation), then we have conceptually returned to a formulation that we already routinely employ in local-response computational photonics and plasmonics. This is no small feat as it will ease both our conceptual understanding and computational explorations of nonlocal effects in nanoplasmonic systems.

Which additional physics can we expect from the Laplacian term? As in a diffusion equation, the Laplacian term also here gives a spatial smearing of fields and thus we can already at this stage anticipate its importance for field-enhancement phenomena, by its regularization of field singularities associated with the usual local-response term in (12.8). This brings us to also note a more practical side to this Laplacian term. As it is already well known from efforts to numerically solve the Navier–Stokes equation for hydrodynamic problems, the pragmatic addition of a small artificial diffusion is often useful and even essential to stabilize numerical solutions [43]. Here, the Laplacian term actually benefits numerical solutions of plasmonic problems in a similar way, even though we emphasize the physical origin rather than being an artificially introduced term. With this smearing term in the equation, electrodynamic simulations of metal structures with even arbitrarily sharp features in the surface topography will not suffer the usual convergence issues due to the underlying singular response [34].

Finally, since we have already interchangeably used the terms *nonlocal response* and *spatial dispersion*, we now provide an explicit link. Fourier transforming (12.8) it is immediately clear that in the context of (12.2) we have $\chi(\omega, k) \simeq \chi_{\text{LRA}}(\omega) - (k\xi)^2$ with a leading quadratic correction for $k\xi \ll 1$.

12.3 Linear-Response Electrodynamics

The theoretical modeling of plasmonic phenomena is for the most part resting on the macroscopic Maxwell equations [8]. In particular, the optical response of metals is described through the constitutive relations, which relate the response of the material to the applied field. Following our above discussion, the displacement field D caused in response to a perturbing electric field E is in accordance with (12.3) given by

$$\mathbf{D}(\omega, \mathbf{r}) = \varepsilon_0 \int d\mathbf{r}' \varepsilon(\omega, \mathbf{r}, \mathbf{r}')\mathbf{E}(\omega, \mathbf{r}'), \tag{12.9}$$

and from Maxwell's equations we arrive at the following integro-differential wave equation

$$\nabla \times \nabla \times \mathbf{E}(\omega, \mathbf{r}) = \left(\frac{\omega}{c}\right)^2 \int d\mathbf{r}' \varepsilon(\omega, \mathbf{r}, \mathbf{r}')\mathbf{E}(\omega, \mathbf{r}'). \tag{12.10}$$

Obviously, this equation is not too appealing, neither for numerical implementations nor for further analytical exercises. In fact, in order to make any progress we would as a first prerequisite need some kind of microscopic or semi-classical account for the actual response function $\varepsilon(\omega, \mathbf{r}, \mathbf{r}')$. Here, Ginzburg and Zayats were among the first to suggest a more phenomenological avenue where one in the lack of more microscopic insight boldly assumes a simple short-range function (say, a Gaussian) as a modification to the usual Drude delta-function response [44]. However, adapting the approach that lead us to (12.8), we may transform the integro-differential equation into a more attractive regular partial-differential equation (PDE). With only few notational complications associated with vectorial fields the result is the same as for the scalar considerations above and in a hydrodynamic model one arrives at [18, 38, 45]

$$\nabla \times \nabla \times \mathbf{E}(\omega, \mathbf{r}) = \left(\frac{\omega}{c}\right)^2 \left[\varepsilon_{\text{LRA}}(\omega) + \xi^2 \nabla(\nabla \cdot)\right] \mathbf{E}(\omega, \mathbf{r}). \tag{12.11}$$

To better appreciate that the nonlocal correction term is qualitatively of a Laplacian form, as discussed above for scalar fields, we first note that for any vectorial field $\mathcal{F}$ we have that $\nabla \times \nabla \times \mathcal{F} = \nabla(\nabla \cdot)\mathcal{F} - \nabla^2 \mathcal{F}$. Thus, for our conceptual understanding we might turn the gradient-of-divergence term into a Laplacian term by absorbing the double-curl term into the already-existing double-curl term on the left-hand side of (12.11). Doing so would be even further motivated if $\xi^2\omega^2/c^2 \ll 1$, in which case the double-curl correction could be entirely neglected. Introducing the free-space wavelength $\lambda = 2\pi c/\omega$, we see that this is equivalent to the condition $\xi \ll \lambda$, which is easily met in noble-metal plasmonics. There are however good reasons to simply proceed with the explicit form in (12.11). Most importantly, the gradient-of-divergence form will respect the vectorial nature of the problem, by addressing transverse and longitudinal field components differently [18] which is important to accurately account for both transverse resonances below the plasma frequency and longitudinal resonances above the plasma frequency [24, 46].

12.4 Hydrodynamic Drift-Diffusion Theory

We have already anticipated above that the overall nonlocal response could have contributions from several underlying physical mechanisms. In this section we offer a semi-classical theory which gives two distinct contributions to ξ from celebrated classical transport mechanisms: convection and diffusion of charge carriers. The detailed derivation rests on a classical hydrodynamic-diffusion problem of an electron responding to an external electric field. We start from the linearized hydrodynamic equation-of-motion for an electron in an electric field (for simplicity, we suppress spatial and temporal variables) [24, 36]

$$\frac{\partial}{\partial t}\mathbf{v} = -\gamma \mathbf{v} + \frac{(-e)}{m}\mathbf{E} - \frac{\beta^2}{n_0}\nabla n_1. \tag{12.12}$$

Here, v is the nonequilibrium velocity correction to the static sea of electrons and γ is the Drude damping parameter also appearing within LRA. The right-hand side contains a semi-classical correction where the pressure term ∇n_1 is classical in spirit, while its strength originates from a quantum description of pressure effects in the electron gas. Thus, β is a characteristic velocity for pressure waves associated with the finite compressibility of the electron gas. In the high-frequency limit, $\omega \gg \gamma$, Thomas–Fermi theory gives $\beta^2 = 3/5v_F^2$, with v_F being the Fermi velocity [18]. For the electron density $n(\mathbf{r},t) = n_0 + n_1(\mathbf{r},t)$, the latter assumed small term ($n_1 \ll n_0$) is the induced density variation associated with the E field that drives the system away from the equilibrium density n_0. The commonly employed LRA is motivated by the fact that $\beta \ll c$ and leaving out the pressure term on the right-hand side we indeed recover the usual text-book equation-of-motion for the response of an electron to an electrical field [8, 19].

Of course, (12.12) cannot stand alone and it should be complemented by the principle of charge conservation. However, rather than turning to the usual simple form for the continuity equation we extend our considerations to include both convective and diffusive transport of charge. Thus, we consider the linearized convection-diffusion equation

$$\frac{\partial}{\partial t}\{(-e)n_1\} = D\nabla^2(-e)n_1 - \nabla\cdot\{(-e)n_0\mathbf{v}\} = -\nabla\cdot\mathbf{J} \tag{12.13}$$

where the current density is then given by Fick's law

$$\mathbf{J} = (-e)n_0\mathbf{v} - D\nabla(-e)n_1, \tag{12.14}$$

with D being the diffusion constant (of course not to be confused with the previously discussed displacement field).

We proceed by multiplying (12.12) by the equilibrium density n_0, which will eventually allow us to arrive at a governing equation for the current density $\mathbf{J}$ rather than for the velocity field $\mathbf{v}$. In addition, we take the time-derivative and after re-arranging the terms we get

$$\left(\frac{\partial}{\partial t}+\gamma\right)\frac{\partial}{\partial t}\{(-e)n_0\mathbf{v}\} = \frac{n_0e^2}{m}\frac{\partial}{\partial t}\mathbf{E} - \beta^2\nabla\left\{\frac{\partial}{\partial t}(-e)n_1\right\}. \tag{12.15}$$

Fick's law, (12.14), can now be applied to the left-hand side to eliminate the convective part of the current

$$\left(\frac{\partial}{\partial t}+\gamma\right)\left[\frac{\partial}{\partial t}\mathbf{J} + D\nabla\left\{\frac{\partial}{\partial t}(-e)n_1\right\}\right] = \frac{n_0e^2}{m}\frac{\partial}{\partial t}\mathbf{E} - \beta^2\nabla\left\{\frac{\partial}{\partial t}(-e)n_1\right\}. \tag{12.16}$$

Next, we use the diffusion-convection equation, (12.13), to eliminate the induced-density terms (the two terms in curly brackets), thereby arriving at a governing equation that explicitly connects the current density $\mathbf{J}$ with the driving electrical field $\mathbf{E}$,

$$\left(\frac{\partial}{\partial t}+\gamma\right)\left[\frac{\partial}{\partial t}\mathbf{J} - D\nabla(\nabla\cdot\mathbf{J})\right] = \frac{n_0e^2}{m}\frac{\partial}{\partial t}\mathbf{E} + \beta^2\nabla(\nabla\cdot\mathbf{J}). \tag{12.17}$$

Finally, Fourier transforming with respect to time and re-arranging the terms we arrive at a generalized constitutive equation, where the local-response Ohm's law ($\mathbf{J} = \sigma_D \mathbf{E}$) is corrected for nonlocal dynamics

$$\xi^2 \nabla(\nabla \cdot \mathbf{J}) + \mathbf{J} = \sigma_{\mathrm{D}} \mathbf{E}. \tag{12.18}$$

Here, we have introduced the nonlocal parameter

$$\xi^2 = \frac{\beta^2}{\omega(\omega + i\gamma)} + \frac{D}{i\omega} \tag{12.19}$$

while

$$\sigma_{\mathrm{D}} = \frac{\frac{e^2 n_0}{m}}{-i\omega + \gamma} \tag{12.20}$$

is the usual frequency-dependent Drude conductivity appearing also in the LRA which we immediately recover in the limit $\xi \longrightarrow 0$ where (12.18) reduces to Ohm's law.

Of course, (12.18) should be accompanied by the Maxwell equation (neglecting interband transitions for the ease of discussion)

$$\nabla \times \nabla \times \mathbf{E} = \left(\frac{\omega}{c}\right)^2 \mathbf{E} + i\omega\mu_0 \mathbf{J} \tag{12.21}$$

and it still remains to be shown that this will indeed lead to (12.11). Rather than following the intuitive step of substituting (12.18) into (12.21), we will do the opposite to eliminate the current density from the wave equation [45]. At first glance, we would have to pay the price by the appearance of fourth-order derivatives. However, since $\nabla \cdot (\nabla \times \mathcal{F}) = 0$ for any vector field $\mathcal{F}$, we eventually arrive at an equation with only second-order derivatives

$$\nabla \times \nabla \times \mathbf{E}(\omega, \mathbf{r}) = \left(\frac{\omega}{c}\right)^2 \left[\varepsilon_{\mathrm{D}} + \xi^2 \nabla(\nabla \cdot)\right] \mathbf{E}(\omega, \mathbf{r}) \tag{12.22}$$

where $\varepsilon_{\mathrm{D}} = \varepsilon_0 + i\sigma/(\varepsilon_0\omega)$ is the Drude dielectric function. As seen, this is indeed of the form in (12.11) and thus (12.19) does indeed give a semi-classical account for the length scale that was first introduced phenomenologically. Furthermore, it is from (12.19) also immediately clear how the two co-existing transport mechanisms of convection and diffusion are playing in concert and adding up to give an effective nonlocal length scale with contributions from both mechanisms. Mathematically, it is somewhat remarkable that they together form a complex-valued nonlocal parameter, with convection giving predominantly a real-valued contribution, while diffusion is contributing to the imaginary part. Physically, there is a simple explanation for the difference: The diffusion of induced charge over time will serve to degrade the plasmonic polarization and as such it represents a damping mechanism. In contrast, the convection is associated with propagation of non-dissipative pressure waves which to a first approximation do not cause additional broadening. Equation 12.22 is derived in the spirit of nonlocal correction to the LRA picture. Nevertheless, the correction term (proportional to ξ^2) can near surfaces (where field derivatives are significant) give rise to pronounced corrections. As such, the diffusive damping can locally exceed the standard Drude damping proportional to γ.

12.5 Boundary Conditions

In this section we address a seemingly trivial point, that has nevertheless been subject to different treatments in the literature. Perhaps no need to say, but solving any PDE in a finite geometry calls for boundary conditions. Within the LRA one solves the $\xi \longrightarrow 0$ version of (12.22) with the usual boundary conditions for $\mathbf{E}$ that follow as a consequence of Maxwell's equations themselves, in the sense that the derivation of the boundary conditions only involves Maxwell's equations plus mathematics (the Gauss and Stokes theorems). On the other hand, solving the coupled equations for the nonlocal problem, (12.17) and 12.21, would imply the need for an *additional boundary condition* (ABC) that explains how $\mathbf{J}$ behaves at the boundary. Another perspective on this is that the nonlocal problem involves both transverse and longitudinal field components, in contrast to the LRA where only transverse fields are excited. Intuitively, a problem involving more field components would call for additional boundary conditions. One could incorrectly be left with the impression that there is some ambiguity (or even freedom) in choosing this ABC. Here, we note that the ABC should reflect underlying physical assumptions, rather than being associated with convenient mathematical choices. Thus, once the physical assumptions have been stated, boundary conditions simply follow from our governing equations. A detailed discussion of these points has been given elsewhere both in the context of the coupled wave equations [24, 47] (12.17) and (12.21) and the generalized wave equation [38, 45] (12.22).

Our starting point, the linearized equation-of-motion in (12.12), is resting on an inherent assumption of a spatially homogeneous equilibrium electron density n_0 and it is only the induced charge n_1 that exhibits temporal and spatial dynamics, i.e. $n(\mathbf{r}, t) = n_0 + n_1(\mathbf{r}, t)$. Thus, n_0 is assumed constant throughout the metal while it drops abruptly to zero beyond the surface of the metal. When assuming such a simple step-like behavior of n_0 at the dielectric-metal interface, this unambiguously leads to one and only one required ABC, namely the continuity of the normal component of the free-electron current density $\mathbf{J}$ [24, 48]. To see this, we first note that the assumption of an infinite work function implies that there are no free electrons outside the metal and consequently $\mathbf{J} = 0$, while $\mathbf{J}$ can still be finite inside the metal. In other words, no electrons are moving across the metal surface, while they are of course free to flow parallel to the surface. Mathematically, the normal component of $\mathbf{J}$ is zero at the surface (i.e. $\mathbf{n} \cdot \mathbf{J} = 0$, with $\mathbf{n}$ being the normal vector) which one can also derive more stringently from (12.13) with the aid of Gauss' theorem.

In terms of (12.22), $\mathbf{n} \cdot \mathbf{J} = 0$ implies that $\mathbf{n} \cdot \mathbf{E}$ is only continuous across the boundary in the absence of interband transitions and for vacuum surroundings, while there is a normal-component electric-field discontinuity if the interband contribution on the metal side is not compensated by similar dielectric contributions on the dielectric side of the interface [47].

How appropriate are our physical assumptions? The assumed ground state of the electron gas corresponds to a situation with an infinite work function, thereby preventing quantum-spill out of electrons beyond the surface of the metal, while we at the same time also ignore density variations inside the metal, such as Friedel oscillations that occur on the Fermi wavelengths scale near metal surfaces. Is this a severe limitation? The pragmatic answer is that this picture is close in spirit to the LRA where the equilibrium electron density is assumed uniform too! More importantly, this turns out to be an appropriate description of noble metals commonly employed in plasmonics while spill-out effects are important in less common metals like sodium [41, 49–51]. In passing, we emphasize that there have been attempts of relaxing the assumption of a homogeneous equilibrium density [52–54] and it has recently demonstrated how to include density-gradient corrections to also account for quantum-spill out in a hydrodynamic model [55–57].

12.6 Numerical Implementations

Obviously, there is a need to be able to solve the nonlocal wave problem, either addressing its coupled-wave formulation [(12.17) and (12.21)] or by invoking the generalized wave equation [(12.22)]. Here, computational nanophotonics is rich on numerical methods [59] and in principle the nonlocal problem can be addressed by many approaches ranging from Mie-scattering descriptions applied to cylindrical and spherical geometries [24, 60, 61] (see [62] for a freely-available numerical implementation) and Fourier-modal methods to periodic systems [63] to finite-element implementations [34, 64] and boundary-element methods [65] applicable to arbitrarily shaped metallic geometries.

Here, we briefly discuss our finite-element implementation [34]. For our numerical solution of the system of equations [(12.17) and (12.21)] we rely on a commercially available code dedicated to solving partial differential equations based on the finite-element method (FEM). We rely on a weak-form implementation [66] that allows us to draw on built-in routines for electromagnetic scattering and the code is also offering built-in meshing and mesh-refinement routines for arbitrarily shaped geometries. Our code is made freely available [67] and works as an add-on to COMSOL Multiphysics 4.1. The performance of the implementation is being supported by examples and it has been documented in detail, including convergence tests and rigorous benchmarking for geometries where semi-analytical accounts allow for solutions with arbitrary numerical accuracy [58].

Figure 12.2 illustrates an example of the scattering of a plane wave from a metallic nanowire with a nontrivial cross section. Despite its three-fold cylindrical rota-

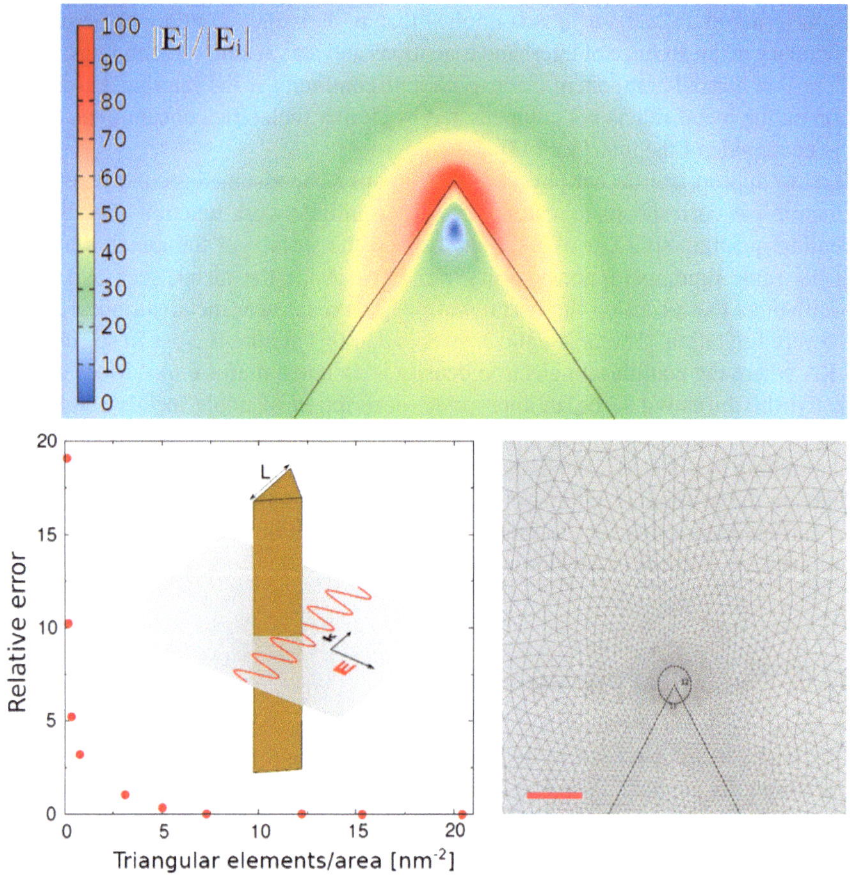

Fig. 12.2 Numerical finite-element evaluation of nonlocal response of a triangular structure with infinitely sharp corners. Despite the abrupt changes in surface topography, the nonlocal smearing facilitates numerical convergence upon appropriate mesh refinement near the corners. Courtesy of Giuseppe Toscano [58]

tional symmetry, the structure is still too complex to allow significant analytical progress and it thus serves the purpose of illustrating the capabilities of our numerical approach. More importantly, we consider a cross section with arbitrarily sharp corners which holds field singularities within the LRA. No matter the mesh refinement, there would be no convergence in the fields due to the singular nature of the problem. Turning to the nonlocal description, the appearance of the nonlocal length scale ξ changes this completely and the problem converges with a mesh refinement that allows spatially resolving variations on the scale of ξ. This is illustrated in the top panel where the electrical field varies smoothly in space while attaining large but finite values, despite the underlying arbitrarily sharp change in the surface topography of the metal.

Finally, we note that the implementation was initially targeting the common hydrodynamic model where ξ^2 reflects only convection dynamics. However, the code is sufficiently general to also include effects of diffusion, since this only renders the length scale in (12.19) complex valued, while leaving all other equations unchanged [38]. In fact, this conclusion also applies to any other method relying on the original hydrodynamic equations, including the other methods and approaches [24, 34, 60, 61, 64, 65] briefly mentioned above.

12.7 Characteristic Material Parameters

Table 12.1 lists characteristic length scales and parameters for the common plasmonic metals Au and Ag. We have also included data for Na which has recently received considerable attention in the ab initio quantum plasmonic community [51, 68, 69]. The entries for different metals are based on Fermi wavelengths λ_F, Fermi velocities v_F, and plasma frequencies ω_p taken from standard tables [19], while the values for τ originate from various references as indicated in the right-most column of the table. The mean-free path is obtained from the scattering time using the classical expression $\ell = v_F\tau$ and likewise, the diffusion constant is given by $D = \ell v_F = v_F^2\tau$.

A quick glance immediately reveals that all length scales are nanometric and in most cases they are even approaching atomic dimensions. This perfectly illustrates the success of the LRA when applied to larger plasmonic structures; the nonlocal correction to the delta-function response of the LRA is indeed negligible on longer lengths scales.

Landau and Lifshitz [17] already emphasized both convection and diffusion length scales in their discussion of spatial dispersion, pointing out that spatial dis-

Table 12.1 Characteristic length scales and parameters for Au, Ag, and Na

	Fermi wavelength λ_F (nm)	Mean-free path ℓ (nm)	Convec. length v_F/ω_p (nm)	Dif. length $\sqrt{D/\omega_p}$ (nm)	Scat. time $\omega_p\tau$
Au	0.52	103	0.10	1.9	1000[a]
	0.52	50	0.11	1.3	465[b]
Ag	0.52	103	0.10	1.9	1000[a]
	0.52	40	0.10	1.1	421[b]
Na	0.68	2.6	0.12	0.32	21[b]
	0.68	4.4	0.12	0.42	37[c]
	0.68	2.3	0.12	0.31	20[d]

[a]Data for crystalline materials taken from Ashcroft and Mermin [19]
[b]Data tabulated by Blaber et al. [89]
[c]Data used in simulations by Teperik et al. [51]
[d]Data used in simulations by Stella et al. [68]

persion would be dominated by one of these two transport mechanisms, namely the one that manifests itself on the longest length scale. Often, diffusion would be considered a slow process, suggesting that $D \ll v_F^2/\omega_p$. However, Table 12.1 illustrates that diffusion and convection are indeed playing in concert. The two length scales are comparable and thus the mechanisms should be treated on an equal footing as our (12.19) indeed does!

12.8 A Unifying Description of Monomers and Dimers

Monomers and their dimer counterparts are archetypal plasmonic structures and a versatile theory could rightfully be expected to offer new insights for both individual monomers as well as for assemblies of such building blocks.

At this point we mention that the phenomenological theory of Kreibig and coworkers captures the size-dependent spectral broadening observed for few-nanometer sized close-to-spherical monomers [70], while it does not immediately apply to non-spherical monomers or to small-gap dimers, e.g. being composed of monomers that are not necessarily small themselves. Likewise, the phenomenological quantum-corrected model [71] invokes quantum tunneling to explain the non-classical optical response of a small-gap dimer, while tunneling-based theories naturally have no explanations to offer for the non-classical optical response of its monomer building blocks. This is where the GNOR proves versatile and with predictive power reaching far beyond the case of spherical monomers; despite its simplicity and semi-classical nature, the generalized nonlocal response model actually unifies the phenomena observed for monomers and dimers and it underlines a close connection between size-dependent damping in monomers and gap-dependent broadening in dimers. Compared to previous hydrodynamic efforts, the key point is the additional effect of diffusion. In this way, convection-diffusion dynamics does the job for both monomers and dimers.

We start with a brief discussion of the monomer case. For simplicity, we consider the plasmonic response of a spherical particle (of radius R) where the complex-valued nonlocal length scale ξ in (12.19) leads to $1/R$ corrections in both the dipole resonance frequency and its linewidth [38]. This observation immediately links the diffusion constant D to the A-coefficient in Kreibig's phenomenological theory for size-dependent broadening where the damping in spherical nanoparticles is given by $\gamma + Av_F/R$ where A is experimentally found to be of the order unity for most considered metals. Theoretical studies of Landau damping also leads to $A \approx 1$ [72], as we will return to. Interestingly, the estimated D-values and the associated diffusion lengths are in qualitative good agreement with the independent estimates in Table 12.1. Figure 12.3 illustrates the differences between the GNOR and the LRA models for a $R = 1.5$ nm spherical metal particle, highlighting in particular the LDOS, EELS, and extinction spectra. As seen in the extinction spectrum, the dipole resonance within the GNOR model exhibits a blueshift and a further broadening as compared to the LRA.

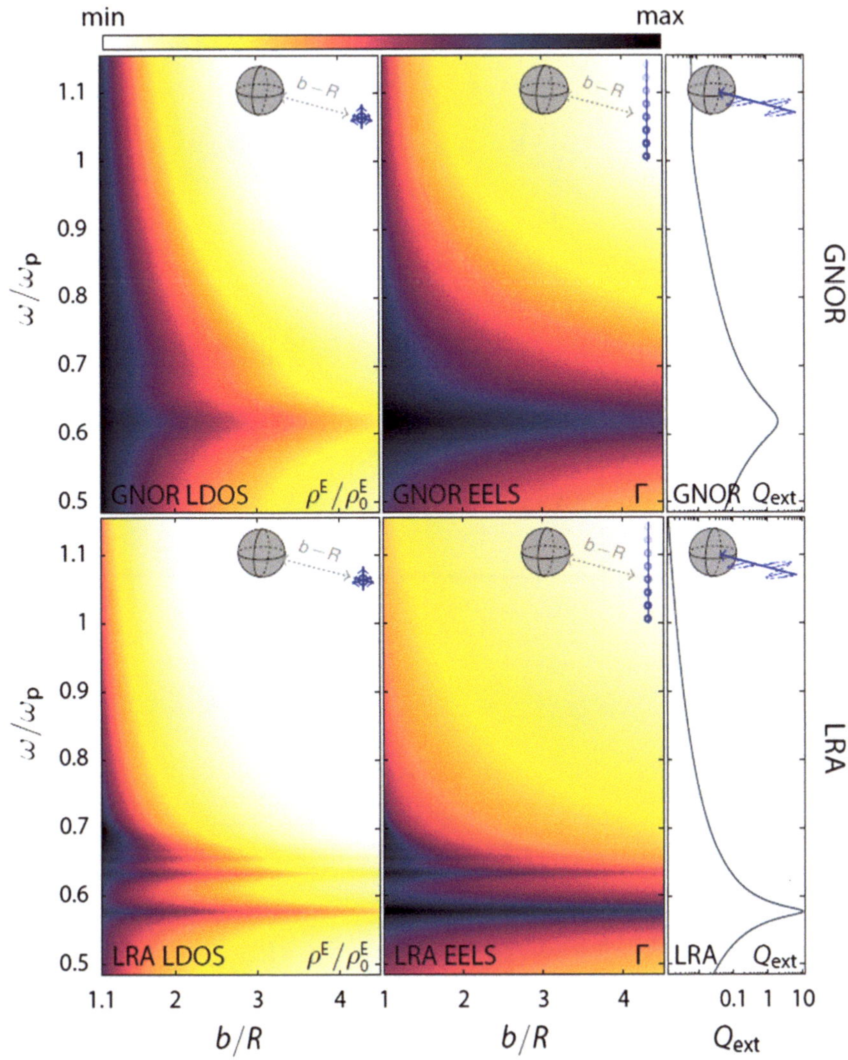

Fig. 12.3 GNOR and LRA calculations for the LDOS, EELS, and extinction spectra for a $R = 1.5$ nm metal particle. Courtesy of Thomas Christensen [73]

Next, the critical check is of course if the very same choice of D-value could explain spectra of the dimer too. While experimental results are available [27, 74] one may also compare the nonlocal results directly to spectra obtained with ab initio approaches [51]. Figure 12.4 illustrates our simulation of extinction spectra of a Na dimer with a radius of $R = 10$ nm and a gap size g varying from 50 Å to −50 Å in steps of 5 Å. Note that the negative gap values correspond to overlapping particles. We emphasize that our results are obtained with an entirely semi-classical model

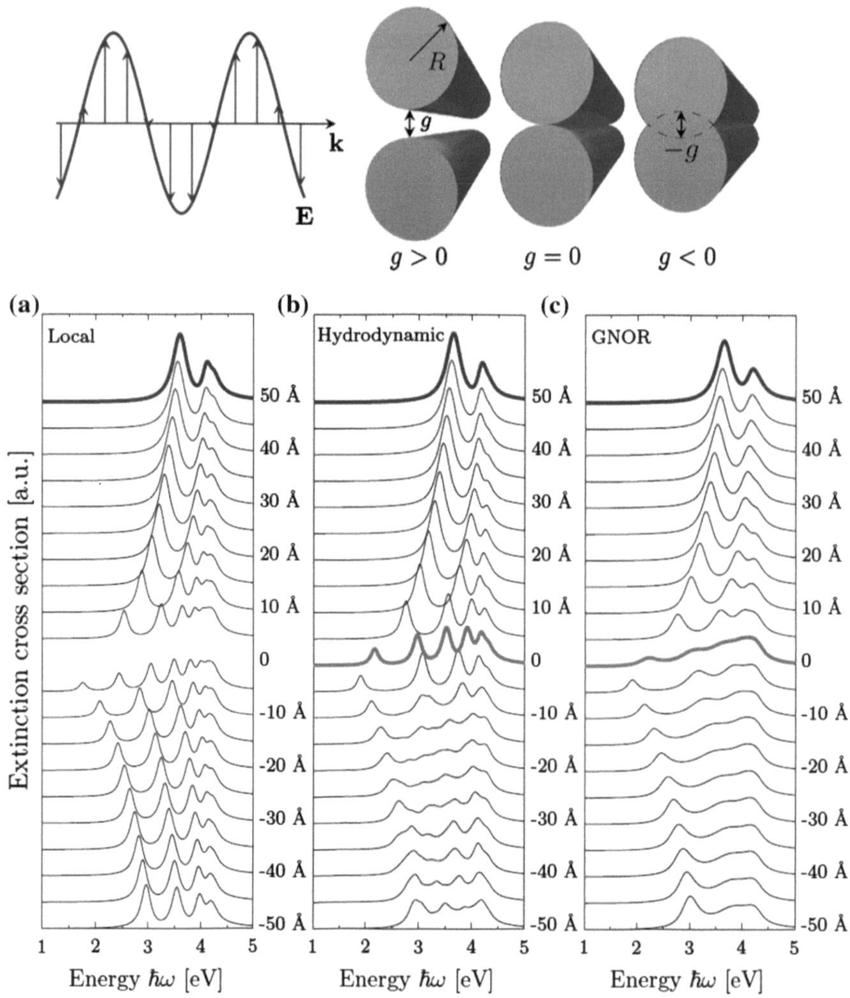

Fig. 12.4 Extinction cross section as a function of energy for a Na dimer with $R = 10\,\text{nm}$ and gap size g varying from 50 Å to -50 Å in steps of 5 Å. Simulations are done using (a) LRA, (b) hydrodynamic model, and (c) GNOR model. The values next to the spectra denote the corresponding values for the gap size g. Courtesy of Søren Raza [75]

where the simplifications and the resulting boundary conditions (which we have discussed in detail in Sect. 12.5) do not leave room for a quantum-tunneling interpretation; the work function is considered infinite and the metal surfaces act as hard walls for the electrons so that charge transport across the dimer gap is prohibited. Instead, the diffusion contribution to nonlocal response causes size-dependent broadening which becomes pronounced as the gap approaches the magnitude of the nonlocal length scale. In other words, gap-dependent broadening sets in for nanometric dimer gaps as observed by both experiments [27, 74] and in ab inito studies [51].

So, what is the important common feature of monomers and dimers? For sure, it cannot be tunneling (since it has no meaning for monomers)! In this context, the semi-classical prediction of gap-dependent broadening in dimers apparently leaves us with the following intriguing question: Should the experimentally observed broadening [27, 74] be attributed to quantum-tunneling mediated short-circuiting of the capacitive junction (as described in the phenomenological quantum-corrected model [71] and more recent extensions [76, 77]) or is it rather due to diffusive damping (we will return to a microscopic explanation of the diffusion below)? Of course, co-existing mechanisms could even play in concert!

To facilitate a qualitative discussion we turn to a pedestrian circuit model that addresses the relative importance of diffusive damping (characterized by a resistance R_{dif} in a circuit model) and the damping associated with the relaxation of a possible quantum tunneling current (characterized by R_{tun}) short-circuiting the classically impenetrable capacitive gap (characterized by a capacitance C). In a simple picture, C and R_{tun} constitute a parallel circuit [78] connected in series with R_{dif}, see Fig. 12.5. The circuit impedance is then given by

$$Z = R_{\text{dif}} + \frac{R_{\text{tun}}}{1 + i\omega\tau_{\text{tun}}} = R_{\text{dif}} - \frac{i}{\omega C} + \mathcal{O}[1/(\omega\tau_{\text{tun}})^2] \tag{12.23}$$

where $\tau_{\text{tun}} = R_{\text{tun}}C$ is interpreted as the tunneling RC time [78].

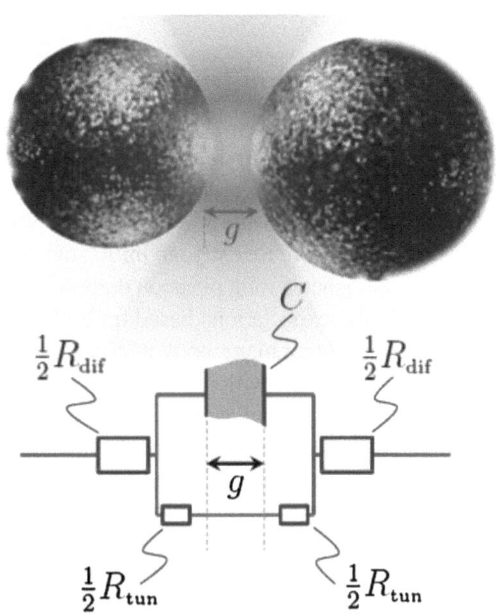

Fig. 12.5 Equivalent circuit model for the dissipation dynamics in plasmonic dimers. The gap is represented by a capacitance C in parallel with a resistor R_{tun} that mimics possible tunneling currents, while further relaxation by diffusion dynamics inside the metal particles is mimicked by in-series coupled resistors R_{dif}

The RC-model has previously been applied in analysis of the ultra-fast response of a scanning-tunneling microscope [78] and here it has been extended to include dissipation associated with the diffusive dynamics. The tunneling dynamics simplifies in the slow adiabatic-following regime and the limit of fast external driving [79]. This analysis suggests that the high-frequency dimer dynamics can become entirely dominated by the diffusive broadening and the junction capacitance, see the second equality in (12.23). Tunneling dynamics has been explored in the context of the mesoscopic capacitance [80] and ultra-fast tunneling experiments have reported tunneling RC times in the picosecond range [78, 81]. Thus, at optical frequencies the plasmon response may be too fast on the scale of the characteristic RC time. If so, then $\omega\tau_{\text{tun}} \gg 1$ and consequently, the relaxation would be dominated by diffusive broadening rather than the short-circuiting tunneling current.

We note that the above circuit analysis is completely independent of the question whether tunneling relaxation occurs within the gap, as it is the case in the quantum-corrected model [71] or takes place inside the metal surfaces as in a more recent model [77] (in agreement with the common understanding of relaxation processes in mesoscopic quantum electron transport). We emphasize that given the qualitative nature of the circuit model combined with the lack of a rigorous measure of the RC time in plasmonic dimers, the discussion is of course not conclusive. In fact, if one approaches the single-Ångström gaps, then one could imagine a different and perhaps short RC time, in which case tunneling would play a more significant role [41]. Below, we briefly turn to a microscopic study where more rigorous insight consolidates the importance of diffusive broadening as a unifying feature of both monomers and dimers with nanometric smallest dimensions.

12.9 The Origin of Diffusion: Insight from ab Initio studies

The phenomenological model for size-dependent broadening has already been linked to quantum mechanical calculations of Landau damping associated with electron-hole pair generation in the metal near the surface, see recent work [39] and references therein. More recently, ab initio studies [41] and electron spectroscopy [33, 82] have established strong plasmon damping at the very surface of metals; damping that exceeds the expectations based on bulk-damping parameters. Indeed, it is this very same underlying many-body interaction within the electron gas that the GNOR model seeks to capture with one single and entirely classical parameter: the diffusion constant D. Since the induced charge is always residing near the surface, the diffusion is effective near the surface too. In fact, the nonlocal correction term in (12.22) mainly contributes near the surface where the $\mathbf{E}$-field changes the most. This is why our drift-diffusion model mimics both longitudinal pressure waves and Landau damping so well. The enhanced damping near the surface is illustrated in Fig. 12.6 where we show the local effective permittivity $\varepsilon_{\text{eff}}(\mathbf{r}, \omega)$ extracted from GNOR simulations via $\mathbf{D}(\mathbf{r}, \omega) \equiv \varepsilon_0 \varepsilon_{\text{eff}}(\mathbf{r}, \omega)\mathbf{E}(\mathbf{r}, \omega)$. Note that due to the abrupt termination of the surface and the associated hard-wall boundary condition for the normal

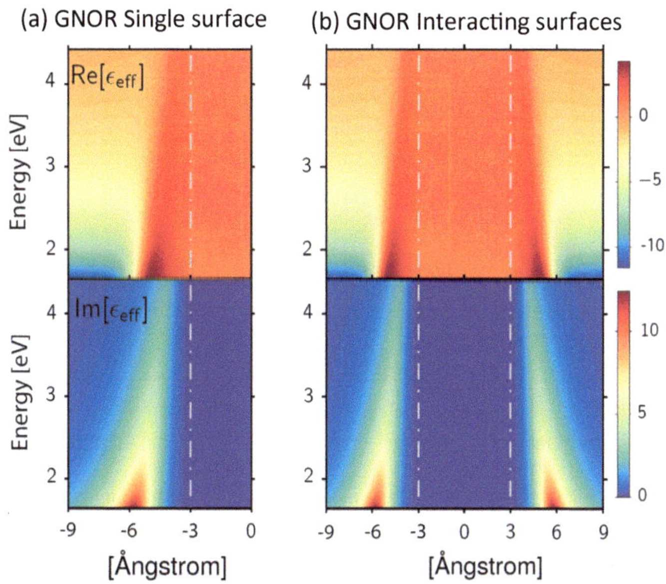

Fig. 12.6 GNOR results for the real part (top panels) and imaginary part (lower panels) of ε_{eff} for different frequencies throughout the optical range. Panel (a) is for a single interface with an infinite work function and a homogeneous equilibrium electron density that vanishes abruptly outside the surface. Panel (b) is for a corresponding dimer with a 0.5 nm vacuum gap separating the two surfaces. Courtesy of Wei Yan

component of the current, the additional damping is 'forced' to occur slightly inside the surface. This links up to the Feibelman parameter and the importance of the *actual position of the surface of the electron plasma* was recently excellently discussed by Teperik and co-workers [51, 83] in the context of plasmonic ruler effects of sub-nanometer gap dimers. Here, we note that the hard-wall boundary condition can be relaxed to also include density-gradient and spill-out effects in a hydrodynamic model [55].

A different perspective on enhanced Landau damping and its relation to the diffusion in GNOR theory has been given by Khurgin and Sun [72]. Essentially, Landau damping sets in when the plasmon wave vector q exceeds ω/v_F [47], i.e. $q \gtrsim \text{Re}\{\xi\}$, thus compensating for exceedingly high field gradients near the surface of metal nanostructures. The role of Landau damping for both monomers and dimers can actually be illustrated quite beautifully by turning to a time-dependent study of the optical response of the electron gas near the surface of the metal [41]. For simplicity we describe a simple metal, such as Na, within a jellium approximation using time-dependent density-functional theory to find the response to a time-dependent electrical field. In this way, the calculation provides us with both the equilibrium density $n_0(\mathbf{r})$ (now space dependent), the induced charge density $n_1(\mathbf{r})$, and the displacement field $\mathbf{D}$ that occurs in response to the perturbing $\mathbf{E}$ field. Once the $\mathbf{D}$ field

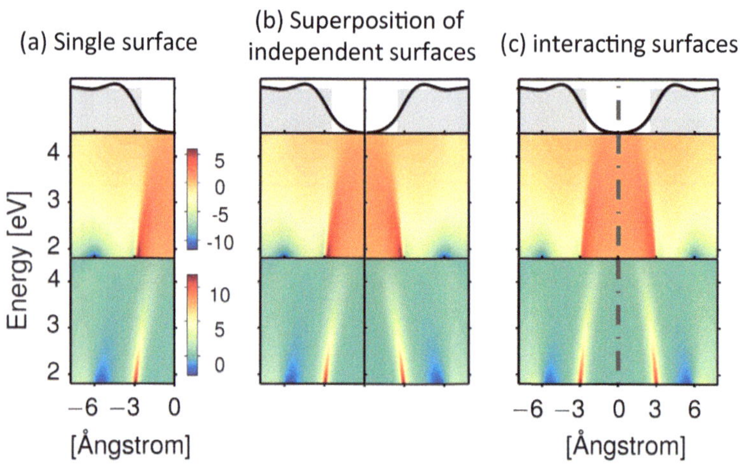

Fig. 12.7 Equilibrium density in the jellium model (top panels), exhibiting both Friedel oscillations and quantum spill-out, along with the TD-DFT results for the real part (middle panels) and imaginary part (lower panels) of $\varepsilon_{\rm eff}$ for different frequencies throughout the optical range. Panel (a) is for a single interface. Panel (b) is for a dimer with a 0.5 nm gap separating two independent interfaces based on the results in (a). Finally, panel (c) is a full calculation for interacting surfaces

is obtained one may similarly to the GNOR discussion above again infer an effective relative dielectric function $\varepsilon_{\rm eff}(\mathbf{r},\omega)$ where in particular the imaginary part holds key information about the damping and where it is spatially occurring.

The top panel of Fig. 12.7a illustrates the equilibrium density (exhibiting both Friedel oscillations and quantum spill-out) along with the real part (middle panel) and imaginary part (lower panel) of $\varepsilon_{\rm eff}$ for different frequencies throughout the optical range. In particular, inspecting the imaginary part of the response, we note how large Landau damping (exceeding the bulk damping) is clearly visible in the near vicinity of the surface. In the context of the work by Khurgin and Sun [72], the enhanced Landau damping in the vicinity of the surfaces (lower panel) penalizes the rapid field variation associated with the epsilon-near-zero behavior (middle panel), so that the material response remains finite rather than causing a field singularity as observed in a LRA study with an inhomogeneous equilibrium density [53].

Next, the very same calculation can be done for dimers too, which in our 1D consideration corresponds to two opposing jellium surfaces separated only by a sub-nanometric vacuum gap. Turning to a dimer with a 0.5 nm gap we arrive at the results in Fig. 12.7c where we recover the same physics found for a single interface, but now with significant Landau damping occurring at both interfaces while there is no appreciable damping occurring inside the gap. In fact, the response of the dimer is very well represented by a simple superposition of the response of two opposing independent surfaces, see Fig. 12.7b, thus driving home the point suggested by the circuit analysis; even for a 0.5 nm the dissipation is dominated by Landau damping (captured in our diffusion model), while the presence of tunneling currents does not effectively change the plasmon-energy dissipation. We that note that the typical ab initio simulations of the electron-gas dynamics include no energy-relaxation processes (e.g. coupling to a thermalizing phonon bath) and as such possible tunneling currents are necessarily relaxed only through the generation of electron-hole pairs. In closing this section, we emphasize that the observed importance of Landau damping does not rule out the existence of quantum tunneling currents at optical frequencies; in the present context of 0.5 nm gaps such currents start to appear, but they just do not seem to significantly influence the gap-dependent broadening of dimers before one enters true Ångström-scale problems where the individual atoms start to matter; an effect not captured by the presented continuum electrodynamic models!

12.10 Conclusions and Outlook

To summarize, a semi-classical description of quantum-related nonlocality has been developed in this chapter. According to the hydrodynamic description, as explained in earlier sections of this work, electrons are localized in real space and the dielectric function changes in the vicinity of the surface as a result of convection and diffusion. Not unlike the diffraction limit, this diffusion in the end sets an ultimate limit on the degree of electromagnetic-field confinement achievable in nanoplasmonic structures.

A main reason why this chapter appears in this book is that the quantum properties of the free-electron plasma in the metal will also affect the emission properties of any nearby quantum emitters. With quantumness in the emitters as well as their environment, this could be dubbed Doubly-Quantum Plasmonics (DQP). The emerging subfield DQP is driven by two types of researchers: those who are interested in interacting efficiently with single quantum emitters, and those who would like to probe efficiently the quantum properties of a free-electron plasma. While until now both aspects have been studied largely independently, a few studies have appeared that could be called quantum plasmonics in a double sense, where the quantum properties of the free-electron plasma affect the properties of nearby quantum emitters [41, 47, 84, 85]. We expect many more of such theory studies and experiments in the near future, not because they are fancy but because strong light-matter interaction required for quantum optics can be offered by plasmonics, but especially so in the near field of plasmonic nanostructures where light-matter interactions are strong but at the same time nonlocal-response effects could be large. Groove waveguides [86] constitute one such system that has been explored in the context of both nonlocal response [45] and quantum emitters [87].

Finally, while we have here limited ourselves to considerations inherently assuming a linear response, recent work suggest that nonlocal dynamics and nonlinear response could enter a fruitful marriage [88].

Acknowledgements We acknowledge funding from the Danish National Research Foundation (CoE Project DNRF103) and the Danish Council for Independent Research–Natural Sciences (Project 1323-00087). Over the course of years we have benefited tremendously from numerous interactions with now former PhD students and postdocs as well as with our peers in the plasmonic community. In particular, many of the figures are by courtesy of PhD theses by Giuseppe Toscano [58], Søren Raza [75], and Thomas Christensen [73]. We also thank Wei Yan, Christin David, Nicolas Stenger, Sanshui Xiao, Yu Luo, Alexander Uskov, Uriel Levy, Javier García de Abajo, Antti-Pekka Jauho, Jacob Khurgin and Sergey Bozhevolnyi for long-term interactions.

References

1. T. Gruner, D.-G. Welsch, Phys. Rev. A **53**, 1818 (1996)
2. L.G. Suttorp, M. Wubs, Phys. Rev. A **70**, 013816 (2004)
3. D.E. Chang, A.S. Sørensen, P.R. Hemmer, M.D. Lukin, Phys. Rev. Lett. **97**, 053002 (2006)
4. D.E. Chang, A.S. Sørensen, E.A. Demler, M.D. Lukin, Nat. Phys. **3**(11), 807 (2007)
5. M.S. Tame, K.R. McEnery, S.K. Özdemir, J. Lee, S.A. Maier, M.S. Kim, Nat. Phys. **9**(6), 329 (2013)
6. A. Delga, J. Feist, J. Bravo-Abad, F.J. García-Vidal, Phys. Rev. Lett. **112**, 253601 (2014)
7. P.T. Kristensen, S. Hughes, ACS Photonics **1**(1), 2 (2014)
8. S.A. Maier, *Plasmonics: Fundamentals and Applications* (Springer, New York, 2007)
9. M.L. Brongersma, Faraday Discuss. **178**, 9 (2015)
10. A. Baev, P.N. Prasad, H. Ågren, M. Samoć, M. Wegener, Phys. Rep. **594**, 1 (2015)
11. D.K. Gramotnev, S.I. Bozhevolnyi, Nat. Photonics **4**(2), 83 (2010)
12. D.K. Gramotnev, S.I. Bozhevolnyi, Nat. Photonics **8**(1), 14 (2014)
13. L.M. Liz-Marzán, Langmuir **22**(1), 32 (2006)
14. E. Hao, G.C. Schatz, J. Chem. Phys. **120**(1), 357 (2004)
15. M.I. Stockman, Phys. Rev. Lett. **93**(13), 137404 (2004)
16. H. Duan, A.I. Fernández-Domínguez, M. Bosman, S.A. Maier, J.K.W. Yang, Nano Lett. **12**(3), 1683 (2012)
17. L.D. Landau, E.M. Lifshitz, L.P. Pitaevskii, in *Electrodynamics of Continuous Media*, *Landau and Lifshitz Course on Theoretical Physics*, vol. 8, 2nd edn. (Butterworth Heinemann, Oxford, 1984)
18. S. Raza, S.I. Bozhevolnyi, M. Wubs, N.A. Mortensen, J. Phys. Condens. Matter **27**, 183204 (2015)
19. N.W. Ashcroft, N.D. Mermin, *Solid State Physics* (Saunders College Publishing, Fort Worth, 1976)
20. I. Romero, J. Aizpurua, G.W. Bryant, F.J. García de Abajo, Opt. Express **14**(21), 9988 (2006)
21. F.J. García de Abajo, J. Phys. Chem. C **112**(46), 17983 (2008)
22. J.M. McMahon, S.K. Gray, G.C. Schatz, Phys. Rev. Lett. **103**, 097403 (2009)
23. C. David, F.J. García de Abajo, J. Phys. Chem. C **115**(40), 19470 (2011)
24. S. Raza, G. Toscano, A.P. Jauho, M. Wubs, N.A. Mortensen, Phys. Rev. B **84**, 121412(R) (2011)
25. J. Kern, S. Grossmann, N.V. Tarakina, T. Häckel, M. Emmerling, M. Kamp, J.S. Huang, P. Biagioni, J. Prangsma, B. Hecht, Nano Lett. **12**(11), 5504 (2012)
26. C. Ciracì, R.T. Hill, J.J. Mock, Y. Urzhumov, A.I. Fernández-Domínguez, S.A. Maier, J.B. Pendry, A. Chilkoti, D.R. Smith, Science **337**(6098), 1072 (2012)
27. K.J. Savage, M.M. Hawkeye, R. Esteban, A.G. Borisov, J. Aizpurua, J.J. Baumberg, Nature **491**(7425), 574 (2012)
28. J.A. Scholl, A.L. Koh, J.A. Dionne, Nature **483**(7390), 421 (2012)
29. S. Raza, N. Stenger, S. Kadkhodazadeh, S.V. Fischer, N. Kostesha, A.P. Jauho, A. Burrows, M. Wubs, N.A. Mortensen, Nanophotonics **2**(2), 131 (2013)

30. S. Raza, N. Stenger, A. Pors, T. Holmgaard, S. Kadkhodazadeh, J.B. Wagner, K. Pedersen, M. Wubs, S.I. Bozhevolnyi, N.A. Mortensen, Nat. Commun. **5**, 4125 (2014)
31. S.F. Tan, L. Wu, J.K. Yang, P. Bai, M. Bosman, C.A. Nijhuis, Science **343**(6178), 1496 (2014)
32. D. Paria, K. Roy, H.J. Singh, S. Kumar, S. Raghavan, A. Ghosh, A. Ghosh, Adv. Mater. **27**(10), 1751 (2015)
33. S. Raza, S. Kadkhodazadeh, T. Christensen, M. Di Vece, M. Wubs, N.A. Mortensen, N. Stenger, Nat. Commun. **6**, 8788 (2015)
34. G. Toscano, S. Raza, A.P. Jauho, N.A. Mortensen, M. Wubs, Opt. Express **20**(4), 4176 (2012)
35. F. Bloch, Z. Phys, A **81**(5–6), 363 (1933)
36. A. Boardman, *Electromagnetic surface modes. Hydrodynamic theory of plasmon-polaritons on plane surfaces.* (John Wiley and Sons, Chichester, 1982)
37. J.M. Pitarke, V.M. Silkin, E.V. Chulkov, P.M. Echenique, Rep. Prog. Phys. **70**, 1 (2007)
38. N.A. Mortensen, S. Raza, M. Wubs, T. Søndergaard, S.I. Bozhevolnyi, Nat. Commun. **5**, 3809 (2014)
39. A.V. Uskov, I.E. Protsenko, N.A. Mortensen, E.P. O'Reilly, Plasmonics **9**, 185 (2014)
40. J.B. Khurgin, Faraday Discuss. **178**, 109 (2015)
41. W. Yan, M. Wubs, N.A. Mortensen, Phys. Rev. Lett. **115**, 137403 (2015)
42. N.A. Mortensen, Phot. Nanostr. **11**(4), 303 (2013)
43. J. Monaghan, Rep. Prog. Phys. **68**(8), 1703 (2005)
44. P. Ginzburg, A. Zayats, ACS Nano **7**(5), 4334 (2013)
45. G. Toscano, S. Raza, W. Yan, C. Jeppesen, S. Xiao, M. Wubs, A.P. Jauho, S.I. Bozhevolnyi, N.A. Mortensen, Nanophotonics **2**(3), 161 (2013)
46. M. Wubs, Opt. Express **23**(24), 31296 (2015)
47. W. Yan, M. Wubs, N.A. Mortensen, Phys. Rev. B **86**(20), 205429 (2012)
48. P. Jewsbury, J. Phys. F: Met. Phys. **11**, 195 (1981)
49. A. Liebsch, Phys. Rev. B **48**(15), 11317 (1993)
50. R.C. Monreal, T.J. Antosiewicz, S.P. Apell, New J. Phys. **15**(8), 083044 (2013)
51. T.V. Teperik, P. Nordlander, J. Aizpurua, A.G. Borisov, Phys. Rev. Lett. **110**, 263901 (2013)
52. M. Ichikawa, J. Phys. Soc. Jpn. **80**(4), 044606 (2011)
53. Z.F. Öztürk, S. Xiao, M. Yan, M. Wubs, A.P. Jauho, N.A. Mortensen, J. Nanophot. **5**, 051602 (2011)
54. C. David, F.J. García de Abajo, ACS Nano **8**(9), 9558 (2014)
55. G. Toscano, J. Straubel, A. Kwiatkowski, C. Rockstuhl, F. Evers, H. Xu, N.A. Mortensen, M. Wubs, Nat. Commun. **6**, 7132 (2015)
56. W. Yan, Phys. Rev. B **91**, 115416 (2015)
57. X. Li, H. Fang, X. Weng, L. Zhang, X. Dou, A. Yang, X. Yuan, Opt. Express **23**(23), 29738 (2015)
58. G. Toscano, Semiclassical theory of nonlocal plasmonic excitation in metallic nanostructures. Ph.D. thesis, Technical University of Denmark (2013)
59. B. Gallinet, J. Butet, O.J.F. Martin, Laser Phot. Rev. **9**(6), 577 (2015)
60. S. Raza, G. Toscano, A.P. Jauho, N.A. Mortensen, M. Wubs, Plasmonics **8**(2), 193 (2013)
61. T. Christensen, W. Yan, S. Raza, A.P. Jauho, N.A. Mortensen, M. Wubs, ACS Nano **8**(2), 1745 (2014)
62. D. Chulhai, L. Jensen, Generalized nonlocal optical response (2016). doi:10.4231/D3V97ZS7X
63. A. Yanai, N.A. Mortensen, U. Levy, Phys. Rev. B **88**, 205120 (2013)
64. K.R. Hiremath, L. Zschiedrich, F. Schmidt, J. Comp. Phys. **231**(17), 5890 (2012)
65. W. Yan, N.A. Mortensen, M. Wubs, Phys. Rev. B **88**, 155414 (2013)
66. T.J.R. Hughes, *The Finite Element Method—Linear Static and Dynamic Finite Element Analysis* (Dower Publications Inc, Mineola, New York, 1987)
67. NanoPlasmonics Lab (2012), http://www.nanopl.org
68. L. Stella, P. Zhang, F.J. García-Vidal, A. Rubio, P. García-González, J. Phys. Chem. C **117**(17), 8941 (2013)
69. K. Andersen, K.L. Jensen, N.A. Mortensen, K.S. Thygesen, Phys. Rev. B **87**, 235433 (2013)

70. U. Kreibig, M. Vollmer, *Optical Properties of Metal Clusters* (Springer-Verlag, Berlin, 1995)
71. R. Esteban, A.G. Borisov, P. Nordlander, J. Aizpurua, Nat. Commun. **3**, 825 (2012)
72. J.B. Khurgin, G. Sun, Opt. Express **23**(24), 30739 (2015)
73. T. Christensen, From classical to quantum plasmonics in three and two dimensions. Ph.D. thesis, Technical University of Denmark (2015)
74. J.A. Scholl, A. Garcia-Etxarri, A.L. Koh, J.A. Dionne, Nano Lett. **13**(2), 564 (2013)
75. S. Raza, Probing plasmonic nanostructures with electron energy-loss spectroscopy. Ph.D. thesis, Technical University of Denmark (2014)
76. R. Esteban, A. Zugarramurdi, P. Zhang, P. Nordlander, F.J. García-Vidal, A.G. Borisov, J. Aizpurua, Faraday Discuss. **178**, 151 (2015)
77. U. Hohenester, Phys. Rev. B **91**(20), 205436 (2015)
78. S. Weiss, D. Botkin, D.F. Ogletree, M. Salmeron, D.S. Chemla, Phys. Stat. Sol. (b) **188**(1), 343 (1995)
79. M. Grifoni, P. Hänggi, Phys. Rep. **304**(5–6), 229 (1998)
80. M. Büttiker, H. Thomas, A. Prêtre, Phys. Lett. A **180**(4–5), 364 (1993)
81. U.D. Keil, T. Ha, J.R. Jensen, J.M. Hvam, Appl. Phys. Lett. **72**(23), 3074 (1998)
82. D. Jin, Q. Hu, D. Neuhauser, F. von Cube, Y. Yang, R. Sachan, T.S. Luk, D.C. Bell, N.X. Fang, Phys. Rev. Lett. **115**, 193901 (2015)
83. T.V. Teperik, P. Nordlander, J. Aizpurua, A.G. Borisov, Opt. Express **21**(22), 27306 (2013)
84. G. Toscano, S. Raza, S. Xiao, M. Wubs, A.P. Jauho, S.I. Bozhevolnyi, N.A. Mortensen, Opt. Lett. **37**(13), 2538 (2012)
85. R. Filter, C. Bösel, G. Toscano, F. Lederer, C. Rockstuhl, Opt. Lett. **39**(21), 6118 (2014)
86. C.L.C. Smith, N. Stenger, A. Kristensen, N.A. Mortensen, S.I. Bozhevolnyi, Nanoscale **7**(21), 9355 (2015)
87. E. Bermudez-Urena, C. Gonzalez-Ballestero, M. Geiselmann, R. Marty, I.P. Radko, T. Holmgaard, Y. Alaverdyan, E. Moreno, F.J. García-Vidal, S.I. Bozhevolnyi, R. Quidant, Nature Commun. **6**, 7883 (2015)
88. A.V. Krasavin, P. Ginzburg, G.A. Wurtz, A.V. Zayats, Nature Commun. **7**, 11497 (2016)
89. M.G. Blaber, M.D. Arnold, M.J. Ford, J. Phys. Chem. C **113**(8), 3041 (2009)

Chapter 13
Landau Damping—The Ultimate Limit of Field Confinement and Enhancement in Plasmonic Structures

Jacob B. Khurgin and Greg Sun

Abstract When either the dimensions of plasmonic structures, or the degree of field localization in them, become comparable to the mean free path of electrons, Landau damping becomes the dominant source of loss in plasmonics. Landau damping is in the heart of nonlocality in the plasmonic response and it is manifested as surface-collision (or Kreibig) damping inherent in nanoscale object. Ultimately this loss prevents further localization of the optical field and limits the attainable plasmonic enhancement, no matter what is the intrinsic quality of the materials used in plasmonic structures.

13.1 Introduction

Plasmonic structures are capable of supporting coupled oscillations of free electrons and electric fields, usually referred to as surface plasmon polaritons (SPP's). The SPP's come in different flavors, such as propagating SPP's on the metal-dielectric interface, gap and slab SPP's, localized SPP's in metal nanoparticles and their combinations, such as dimers and more complex nanoantennas [1–3]. The salient feature of all types of SPP-supporting nanostructures is the ability to achieve high degree of field confinement and enhancement, usually near the metal surface or in the narrow gap between the metal surfaces [4, 5]. As is the case for any oscillating mode, the degree to which the energy can be concentrated in the SPP mode depends on the quality factor $Q=\omega/\gamma$ where the damping rate γ is determined by the loss in the metal. The loss is inherent to the motion of free electrons, and has many underlying

---

J.B. Khurgin (✉)
Department of Electrical and Computer Engineering, Johns Hopkins University, 3400 N Charles St, Baltimore, MD 21218, USA
e-mail: jakek@jhu.edu

G. Sun
Department of Physics, University of Massachusetts at Boston, 100 Morrissey Blvd, Boston, MA 02125, USA
e-mail: Greg.sun@umb.edu

S.I. Bozhevolnyi et al. (eds.), *Quantum Plasmonics*,
Springer Series in Solid-State Sciences 185,
DOI 10.1007/978-3-319-45820-5_13

mechanisms, such as phonon and defect scattering, surface roughness, electron-electron interactions and others [6]. A number of concerted efforts are currently directed at the reduction of loss by either improving the purity and surface quality of the traditional plasmonic metals like Ag and Au [7] or by using alternative plasmonic materials, such as, for instance TiN [8, 9]. The path toward lower loss is an arduous one, but, it is reasonable to assume that eventually it will come to fruition and the field concertation in plasmonic structures will be increased beyond what is the state of the art today. Then it might be a right time to investigate of what will be the next, perhaps the ultimate limit of plasmonic enhancement in what one can deem to be "ideal" plasmonic structures incorporating "ideal" or "lossless" metal [6].

There already exists a fairly large body of work dealing with what happens when the concentration of electric field increases and the non-local effects are coming into play [10–14]. The nonlocality is manifested by the increase in the energy dissipation and spreading out (diffusion) of SPP mode. While phenomenological theories of nonlocality exist [12–14], it is important to stress that fundamentally nonlocality is the incarnation of Landau damping, i.e. direct excitation of electron hole pairs in the metal by the highly confined electric field of SPP's [15–17]. In this work we make the connection between Landau damping and surface collision damping also known as Kreibig damping. We develop a simple self-consistent model that shows the effect of Landau damping on the field enhancement, effective volume and linewidth of the SPP modes in various geometries, including propagating SPP's, localized SPP's in plasmonic nanoparticles, and the dimers. Plasmonic dimers [18–20] are not only very important plasmonic structures by themselves, but a dimer also can serve as a good model for all kinds of plasmonic nanoantennas.

Nanoantenna lets the SPP mode in the gap to be coupled to the propagating electro-magnetic waves, and the peak field inside the mode can be enhanced by orders of magnitude relative to the incident wave. Consequently all the linear and especially nonlinear processes get enhanced. In particular, the rate spontaneous emission of radiation is enhanced by the Purcell effect by as much as three orders of magnitude. Typical nanoantenna incorporates a dielectric gap between two metallic structures of various geometry and it is in this gap where the concentrated electric field is confined. Plasmonic dimer shown in Fig. 13.1a is a good example of nanoantenna. It consists of two metal spheres or radius a separated by a dielectric gap d_{gap} into which the object of the plasmonic enhancement (an atom, a molecule or a quantum dot) is placed. Because the relative simple shape of dimer simplifies mathematics and is relatively easy to fabricate, the dimer has been studied extensively and most of the conclusions obtained for the dimer are easily applicable to great many kinds of nanoantennas, (half-wave, Yagi, patch etc.) which in the end almost always incorporate two metal fixtures separated by a gap. Naturally, reduction of the gap size causes decrease in the SPP volume which is expected to be accompanied by the commensurate increase in the energy density of the SPP field and Purcell factor. However, once the gap size decreases to a few nanometers and less, both the field enhancement and Purcell factor cease to increase while the linewidth of the resonance eventually broadens to the degree where the SPP

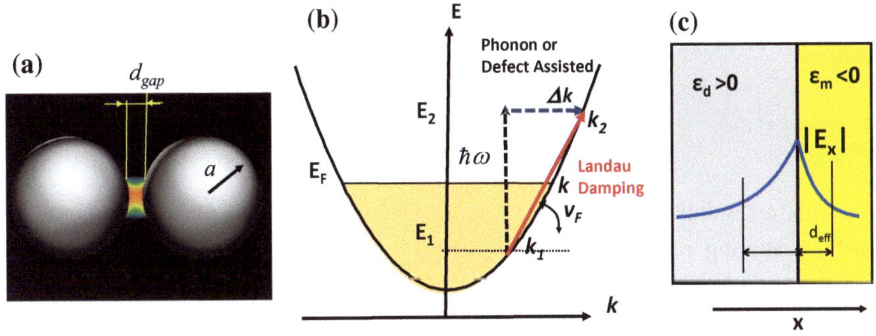

Fig. 13.1 **a** Plasmonic dimer, **b** Origin of Landau Damping—direct transition between two states with different wave-vectors $\boldsymbol{k}_1$and $\boldsymbol{k}$, **c** Propagating SPP on the metal-dielectric interface

resonances are no longer discernible [10–12]. Hence the size of the gap always needs to be optimized in order to attain maximum enhancement of a given process.

Using our self-consistent model we demonstrate that drastic changes of SPP properties occur in the gaps with the gap size as large as 1–2 nm, where they cannot be caused by the electron tunneling or other "quantum" phenomena. Thus it is the Landau damping presents the most practically-relevant limit to the achievable degree of plasmonic enhancement.

13.2 Spill Out and Nonlocality in the Hydrodynamic Model

The reason for the "saturation" and eventual decline of the enhancement in the gap has been studied extensively, yet at this point no consensus had been reached. There is definitely a "quantum limit" to the enhancement associated with the "wave-function spill-over" and tunneling between the metal electrodes [10]. The tunneling is expected to become significant when the gap size becomes comparable to a few times the decay length of the electron wave-function in the gap dielectric, $L_d \sim \hbar/\sqrt{2m\Phi}$, where the work function of the metal Φ is about 4–5 eV which brings the decay length to about 1Å and the onset of tunneling to a fraction of a nm. The electron spill out occurs due to inability of the electron density to undergo the change on the spatial scale smaller than the Thomas Fermi screening length $\lambda_{TF} \sim k_F^{-1}$ where k_F is a Fermi wave vector. The spatial extent of the spill out is then also on the order of a fraction of a nanometer. The quantum origin of electron spill out is obviously associated with the maximum wave vector of the occupied states in the conduction band being, but the spreading of the electron density can also be given semi-classical interpretation based on the non-locality (or spatial dispersion) of the dielectric constant of the metal, which, to the first order can be written as

$$\varepsilon(\omega,k)=\varepsilon_b(\omega)-\frac{\omega_p^2}{\omega^2+i\omega\gamma_b-\beta^2k^2}, \tag{13.1}$$

where $\varepsilon_b(\omega)$ is dielectric constant of bound electrons, ω_p is the plasma frequency, γ_b is the scattering rate in the bulk metal, $\beta^2=(3/5)v_F^2$, and v_F is Fermi velocity. According to (13.1) the nonlocality causes mostly the change in the real part of the dielectric function and one can obtain the equation for the current density in the metal

$$\beta^2\nabla(\nabla\cdot J)+(\omega^2+i\omega\gamma_b)J=i\omega_p^2\varepsilon_0\omega E \tag{13.2}$$

The first term in (13.2) is the "quantum pressure" or "convection term". If one expresses the current density as $\nabla\cdot J=i\omega en(r)$, where $n(r)$ is the induced electron density and e is the electron charge, then takes divergence of (13.2) and uses the Maxwell equation $\varepsilon_0\nabla\cdot E=en(r)$, the (13.2) takes form

$$\beta^2\nabla^2n(r)=(\omega_p^2-\omega^2-i\omega\gamma_b)n(r) \tag{13.3}$$

This means that the electron density at the metal surface is no longer described by the delta function as in local case, and for $\omega<<\omega_p$ the spatial extent of the electron density near the metal surface is on the order of $\beta/\omega_p\approx v_F/\omega_p\approx\lambda_{TF}$, i.e. just a few Å. This result agrees with the prediction based on maximum available wave-vector made above. At the same time, the spectra of the dimers start deviate from the predictions based on local theory by exhibiting blue shift and significant broadening when the size of the gap is still as large as a few nanometers. Clearly, neither tunneling nor spill out can be responsible for these results, and to account for these effects a diffusion term had been introduced in [11] by making a substitution $\beta^2\rightarrow\beta^2-i\omega D$ in (13.2), where the diffusion constant is $D\approx v_F^2\gamma_b^{-1}$. The diffusion term is dissipative and by comparing this term with the bulk energy dissipation term $i\omega\gamma_b$ in (13.2) one can see that the spatial extent of the field at which diffusion loss exceeds the bulk loss is roughly $l\approx v_F/\gamma_b$, i.e. precisely the mean free path, or a few nanometers. This is, of course, exactly the result expected from the phenomenological theory of Kreibig [21] in which the loss is associated with the electrons colliding with the metal surface with a frequency $\gamma_s\approx v_F/d$, where d is the characteristic dimension of the nanoparticle.

13.3 Landau Damping as the Cause of Nonlocality

The quantum mechanical origin of the surface collision damping and spectral broadening γ_s associated with it can be illustrated by first noting that direct transitions between two free electrons having different wave vectors k_1 and k_2 are prohibited by the momentum conservation as shown in Fig. 13.1b Collision with the metal wall provides recoil so that momentum conservation rule no longer applies and a photon (or, more properly, a surface plasmon polariton, SPP) with energy $\hbar\omega = E_2 - E_1$ can be absorbed. In that respect, a collision with the surface is considered to be no different from any other collision, with, say a phonon, or a defect. Obviously, if the electric field of the SPP is effectively confined within the distance d_{eff} from the surface, it is that distance rather than the nanoparticle size that enters the Kreibig's phenomenological expression. For example, for the SPP's propagating on the metal/dielectric interface (Fig. 13.1c) with electric field inside the metal

$$E(x,z) \sim E_0 \exp(-x/d_{eff}) \exp(ik_z z) \tag{13.4}$$

the surface collision damping becomes [22] $\gamma_s = 3/4 v_F/d_{eff}$. This increased loss near the surface of the metals is also known as anomalous skin effect [23].

Now, it would be tempting to offer a simple quantum mechanical interpretation of the surface damping using the Heisenberg's uncertainty principle. It can be argued that if the electric field is confined on the scale of d_{eff} then the energy of free electrons traversing the field are defined only within the uncertainty $\Delta E_s = \hbar\gamma_s \approx v_F/d_{eff}$, just as the energy of an electron undergoing, say collisions with phonons and defects, is defined within the uncertainty $\Delta E_b = \hbar\gamma_b$, where the bulk damping (broadening) $\gamma_b = 1/\tau_b$ and τ_b is the mean interval between those collisions. Therefore it is only natural that bulk and surface broadening mechanisms add up according to Matthiesen rule and the Drude expression for the effective dielectric constant of the SPP mode becomes

$$\varepsilon_{eff}(\omega, d_{eff}) = \varepsilon_b(\omega) - \frac{\omega_p^2}{\omega^2 + i\omega(\gamma_b + \gamma_s)} \tag{13.5}$$

While this expression is technically correct, it contains a deep flaw because according to it the effective dielectric constant of a given SPP mode depends only on its effective size and not on its shape. In other words, whether the overlap on the SPP mode with conduction electrons contains a sharp boundary (surface) or the mode shape is smooth, like the one in the dielectric waveguide, the damping associated with the finite electron's time-of-flight $\tau = d_{eff}/v_F$ will always ensue. To understand why this interpretation is erroneous one should go back to the momentum conservation (Fig. 13.1b) involved in the direct transition between two electronic states with energies E_1 and $E_2 = E_1 + \hbar\omega$ in the vicinity of the Fermi level. Linearizing the dispersion one obtains $\omega \approx v_F \cdot (k_2 - k_1)$ and $\Delta k = |k_2 - k_1| \geq k_0 = \omega/v_F$, where $k_0 = \omega/v_F$ is the so-called offset vector of Landau damping. In other words, if the magnitude of the wavevector $\boldsymbol{k}$ of longitudinal electromagnetic wave $E(r) \sim E_0 \exp(ik \cdot r)$ inside the metal exceeds k_0 then direct absorption by the free electrons becomes possible. This process is called *Landau damping*. At first look, it appears that Landau damping becomes a major factor only when the SPP mode dimensions approach $k_0^{-1} = v_F/\omega \approx 0.4$ nm for the wavelength of $\lambda = 600$ nm, but this first impression is wrong because the power spectrum of the electromagnetic wave insider the metal, $|E(k)|^2$ contains all kinds of wavevectors (spatial frequencies) and the spatial frequency components with $k \geq k_0$ all get absorbedvia diect transtions.

To evaluate the absorption, in place of the approximation (13.1) we shall use the full expression for the dielectric constant of the metal, derived by Lindhard [24]

$$\varepsilon(\omega, k) = \varepsilon_b + \frac{3\omega_p^2}{k^2 v_F^2}\left[1 - \frac{\omega}{2kv_F}\ln\frac{\omega + kv_F}{\omega - kv_F}\right] \tag{13.6}$$

Introducing normalized (to the onset of Landau Damping) wave-vector $q = k/k_0$ one can obtain separate expressions for the real and imaginary parts of ε. For the real part one can write $\varepsilon_r(\omega, q) = \varepsilon_r(\omega, 0) + \Delta\varepsilon_r(\omega, q)$, where $\varepsilon_r(\omega, 0) = \varepsilon_b - \omega_p^2/\omega^2$ is a long wavelength Drude dielectric constant and

$$\Delta\varepsilon_r = \frac{\omega_p^2}{\omega^2} g_r(q) = \frac{\omega_p^2}{\omega^2}\left(1 + \frac{3}{q^2} - \frac{3}{2q^3}\ln\left|\frac{1+q}{1-q}\right|\right) \tag{13.7}$$

is the dispersion of the non-local change of real part of the dielectric constant as plotted in Fig. 13.2a.

Characteristic feature of $\Delta\varepsilon_r$ is the change of sign near q = *1*, therefore, as one shall see soon, the change in the real part of the dielectric constant is relatively small and its effect mostly consists of relatively small shift of the resonant SPP frequency.

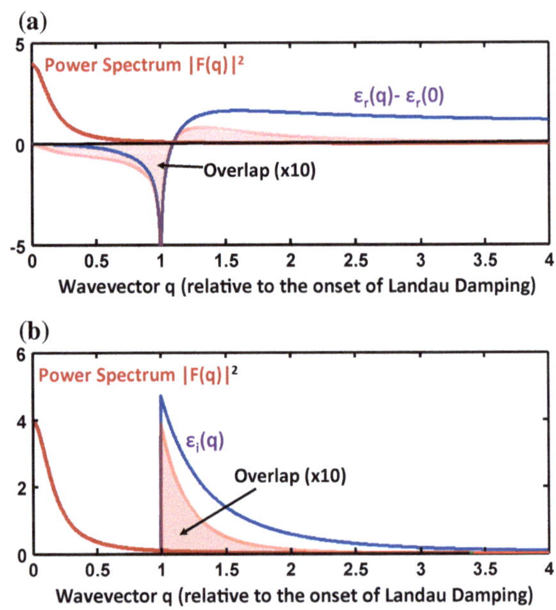

Fig. 13.2 Spatial dispersions of **a** real and **b** imaginary parts of metal dielectric constant and their overlaps with spatial power spectrum of the electric field in the metal

As for the imaginary part of dielectric constant, in the absence of scattering, for small wavevectors $k < \omega/v_F (q < 1)$ it remains zero, but for the larger wavevectors Landau damping gives rise to the imaginary part of dielectric constant, and, since logarithm of negative number has imaginary part of $i\pi$ one obtains

$$\varepsilon_i(\omega, |k| > \omega/v_F) = 3\pi\omega_p^2\omega/2k^3v_F^3 = \frac{3}{2}\pi\frac{\omega_p^2}{\omega^2 q^3}, \tag{13.8}$$

plotted in Fig. 13.2b. One can now evaluate its Fourier transform $F(k)$ and a power density spectrum $|F(k)|^2$ of the electric field $E(r)$ *inside the metal*. It is only logical then to assume that all the longitudinal field components with spatial frequencies with $k_x > \omega/v_F$ get Landau damped and contribute to the imaginary part of the dielectric constant. For each value of the wavevector $\boldsymbol{k}$ the power density of the longitudinal field is $|F_\parallel(k)|^2 = |F(k)\cdot k|^2/k^2$. One can then evaluate the effective dielectric constant $\varepsilon_{eff}(\omega)$ as a function of the size and shape of an SPP mode by evaluating the overlap of $|F_\parallel(k)|^2$ with $\varepsilon(\omega, k)$ as seen in Fig. 13.2a, b for real and imaginary parts of dielectric constant respectively. The imaginary part of the dielectric constant is then

$$\varepsilon_{eff,i} = \frac{3\pi\omega_p^2}{2\omega^2}\int\limits_{q>1}^{\infty} q^{-3}|F_\parallel(q)|^2 d^3q \Big/ \int\limits_0^{\infty} |F(q)|^2 d^3q \tag{13.9}$$

Now, according to the Drude formula $\varepsilon_{eff,i} = \omega_p^2\gamma_s/\omega^3$ and to conform to this expression one can introduce the surface collision damping as

$$\gamma_s = \frac{3\pi\omega}{2}\int\limits_{q>1}^{\infty} q^{-3}|F_\parallel(q)|^2 d^3q \Big/ \int\limits_0^{\infty} |F(q)|^2 d^3q \tag{13.10}$$

as well as the dimensionless quality factor of surface collision scattering,

$$Q_s = \frac{2}{3\pi}\int\limits_0^{\infty} |F(q)|^2 d^3q \Big/ \int\limits_{q>1}^{\infty} q^{-3}|F_\parallel(q)|^2 d^3q \tag{13.11}$$

One can now apply the result (13.10) to the aforementioned example of the SPP propagating at the metal/dielectric interface with electric field described by the (13.4) and the normalized power spectrum, accordingly

$$|F(k_x)|^2 = \frac{2}{\pi d_{eff}(k^2 + d_{eff}^{-2})}, \tag{13.12}$$

or, in normalized units

$$|F(q)|^2 = \frac{2v_F}{\pi d_{eff}\omega} \frac{1}{q^2 + (1/k_0 d_{eff})^2} \approx \frac{2v_F}{\pi d_{eff}\omega q^2} \tag{13.13}$$

where we have used the fact that as stated above, $d_{eff} >> k_0^{-1} \approx 0.4\,\text{nm}$. Substituting (13.13) into (13.10) one readily obtains $\gamma_s = 3/4 v_F/d_{eff}$ which is exactly the same as one obtained in [22] and is fully compatible with Kreibig interpretation of the surface scattering rate as the rate with which the electron confines inside the mode collides with the surface,

But now consider what happens if no sharp boundary is present and, let us say the optical filed is simply confined inside the metal into a smooth Gaussian mode,

$$E(x,z) \sim E_0 \exp(-x^2/2d_{eff}^2)\exp(ik_z z) \tag{13.14}$$

whose normalized longitudinal spectrum is also Gaussian

$$|F(q)|^2 = \frac{2}{\pi^{1/2}} \frac{\omega}{v_f} d_{eff} \exp\left[-(k_0 d_{eff} q)^2\right] \tag{13.15}$$

Substitution of (13.15) into (13.10) yields

$$\gamma_s \approx \frac{3\pi^{1/2}}{2} \frac{v_f}{d_{eff}} \exp\left[-(k_0 d_{eff})^2\right] \tag{13.16}$$

The Gaussian function (13.15) experiences dramatically faster decay than Lorentzian one (13.13) and therefore for $d_{eff} >> k_0^{-1} \approx 0.4\,\text{nm}$ the scattering rate is orders of magnitude lower than in the presence of sharp interface. Indeed the main "Kreibig-like" result $\gamma_s \sim v_f/d_{eff}$ only appears because the power spectrum of the longitudinal field $|F(k)|^2$ decays as k^{-2} for large wavevectors, which is precisely the power spectrum of the Heaviside function describing the sharp interface between the metal and surrounding dielectric. Therefore, the phenomenological model introduced by Kreibig according to which the damping is roughly the inverse of the transit time required the average electron inside the mode to reach the surface does have a deep underlying physical interpretation.

If we now extend the phenomenological Kreibig treatment to the real part of the dielectric constant and calculate the expected change in it as

$$\Delta\varepsilon_r(\omega) = \text{Re}\left[\frac{\omega_p^2}{\omega^2 + i\omega\gamma_b} - \frac{\omega_p^2}{\omega^2 + i\omega(\gamma_b + \gamma_s)}\right] \approx \gamma_s(2\gamma_b + \gamma_s)\frac{\omega_p^2}{\omega^4} \tag{13.17}$$

Calculating the same change using (13.7) leads to almost an identical to (13.18) result thanks to the almost complete cancellation of the contributions by the Fourier components below and above $q = 1$. Thus the relative change of the real part of the permittivity even when surface contribution equals the bulk contribution is only about $3(\gamma_b/\omega)^2$ i.e. a small fraction of per cent. Interestingly enough, one can interpret this small change as an appearance of the pole in the Drude formula (13.1) which can be written as a size-dependent permittivity

$$\varepsilon(\omega, d_{eff}) = \varepsilon_b(\omega) - \frac{\omega_p^2}{\omega^2 - \omega_0^2(d_{eff}) + i\omega\gamma_b}, \tag{13.18}$$

where the size-dependent resonant frequency $\omega_0 \sim v_F/d_{eff}$ is roughly the frequency at which the surface-enabled direct transition between two states near the Fermi energy can occur. At any rate, the main effect of the change of the real part of ε is the relatively insignificant shift of the resonant frequencies in various plasmonic structures.

The change of the imaginary part of permittivity on the other hand not only leads to additional damping, but also limits the extent to which the electric field can be concentrated in the plasmonic structures. We now consider three systems in which Landau damping serves as the ultimate limiting factor.

13.4 Limits of Confinement in Propagating SPP

This scheme, shown in Fig. 13.1c is the first and most basic plasmonic system. The wavevector of the SPP propagating on the boundary between dielectric with the frequency-independent relative permittivity ε_d and the metal with relative permittivity described by (13.5) can be found as

$$k_z(\omega) = k_D(\omega)\sqrt{\frac{\varepsilon_m(\omega)}{\varepsilon_d + \varepsilon_m(\omega)}} \tag{13.19}$$

where $k_D(\omega) = 2\pi\varepsilon_d^{1/2}\omega/c$ is the wavevector of the free propagating electromagnetic wave in the dielectric. The surface collision-induced damping rate is

$$\gamma_s = \frac{3}{4}\frac{v_F}{d_{eff}} f_x \tag{13.20}$$

where

$$d_{eff}^{-1} = \sqrt{k_z^2 - \varepsilon_m k_D^2} \tag{13.21}$$

and

$$f_x = \frac{|E_x|^2}{|E_z|^2 + |E_x|^2} = \frac{k_z^2}{k_z^2 + d_{eff}^{-2}} \tag{13.22}$$

is the fraction of the energy contained in the normal to the surface component of electric field. Next we introduce the frequency of SP resonance, $\omega_{SP} = \omega_p / \sqrt{(\varepsilon_b + \varepsilon_d)}$ and normalize the frequencies to it, $\tilde{\omega} = \omega / \omega_{SP}$. We also introduce effective index of SPP as $\beta = k_z / k_D$ and normalized penetration depth $\tilde{d} = d_{eff} k_D$ we obtain

$$\beta(\tilde{\omega}) = \sqrt{\frac{\varepsilon_b(\tilde{\omega}^2 - 1) - \varepsilon_d + i\varepsilon_b \tilde{\omega} Q_b^{-1} + i\varepsilon_b \tilde{\omega}^2 Q_s^{-1}}{(\varepsilon_b + \varepsilon_d)(\tilde{\omega}^2 - 1 + i\tilde{\omega} Q_b^{-1} + i\tilde{\omega}^2 Q_s^{-1})}}, \tag{13.23}$$

where the bulk quality factor $Q_b = \omega_{SP} / \gamma_b$ and the surface quality factor is $Q_s(\tilde{\omega}) = \omega_{SP} / \gamma_s = 4c / 3 n_d v_F f_x \tilde{d}^{-1}$ where, $n_d = \varepsilon_d^{1/2}$ and c is the speed of light in vacuum. Since according to (13.20) $Q_s(\tilde{\omega})$ itself is a function of $\beta(\tilde{\omega})$ (13.23) can be solved self-consistently using iterative methods.

Two important observations can be made from (13.23). First, the impact of surface damping would become noticeable when $\gamma_s \sim \gamma_b$, which will happen not far from the SP resonance, hence when $Q_s \sim Q_b$, and since in the vicinity of SP resonance $\tilde{d} \approx \beta^{-1}$ and $f_x \approx 1/2$ we obtain the value of effective index at which the surface dumping must be taken into account,

$$\beta_s \approx \frac{8c}{3\varepsilon_d^{1/2} v_F Q_b} \tag{13.24}$$

For the combination of Ag and GaN ($\varepsilon_d^{1/2} = 2.3$, $\gamma_b = 3.2 \times 10^{13} s^{-1}$, $\omega_{SP} = 4.5 \times 10^{15} s^{-1}$, $Q_b \approx 140$ [25]) we obtain $\beta_s \approx 1.6$ while for GaN–Au combination ($\gamma_b = 1.2 \times 10^{14} s^{-1}$; $Q_0 \approx 43$; [25]) we obtain $\beta_s \approx 5$. As expected, it is for a good metal like silver, that surface collision role becomes important early on, while for the less perfect metal like gold, the influence of surface collision does not become important until much later.

The second, and by far more important consequence is the ability to estimate the ultimate value of the effective index, and hence confinement, of SPP which could have been obtained in the hypothetical "ideal" metal with $\gamma_b = 0$, i.e. in the "ideal" metal that would have been free of defects, phonon scattering, electron-electron interaction, and residual interband absorption. Obviously, such metal does not, and probably cannot exist, however it is useful to see what kind of improvement can be achieved by reducing the loss in the metal. By inserting $\tilde{\omega} = 1$ and $Q_b^{-1} = 0$; into (13.23) we obtain a rather simple expression for the maximum effective index (real part) attainable with the "ideal" metal

$$\beta_{\max,r} \approx \left(\frac{\varepsilon_d^{1/2}}{\varepsilon_b + \varepsilon_d}\frac{4c}{3v_F}\right)^{1/3} < (c/n_d v_F)^{1/3} \tag{13.25}$$

The result is strikingly simple and depends only on the ratio of the speed of light in the dielectric and Fermi velocity and because of the surface collisions (Landau damping) one simply cannot reduce the wavelength of the SPP propagating on the metal-dielectric interface by more than a factor of about 4 or 5 relative to the plane wave propagating in dielectric, no matter how low is the loss in the bulk metal.

Turning now to the numerical calculations and following [22] we plot in the dispersion curves (13.23) for the SPP propagating on the boundary between the metal with silver-like dispersion ($\varepsilon_b = 4.1$, $\hbar\omega_{SP} = 9.3eV$;) and GaN $\varepsilon_d^{1/2} = 2.3$ resulting in SPP resonance near 415 nm. When it comes to the scattering constant in the metal we shall consider 5 different cases:

A. The best bulk silver with bulk damping constant $\gamma_b = 3.2 \times 10^{13}\ \mathrm{s}^{-1}$ and no surface collision damping taken into account $\gamma_s = 0$
B. The best bulk silver with bulk damping constant $\gamma_b = 3.2 \times 10^{13}\ \mathrm{s}^{-1}$ with surface collision damping taken into account
C. "Dirty silver" with bulk damping constant $\gamma_b = 1.2 \times 10^{14}\ \mathrm{s}^{-1}$ similar to that of gold and no surface collision damping taken into account $\gamma_s = 0$. The reason for using "dirty silver" instead of gold is that one cannot observe interface SP resonance in gold in combination with any dielectric due to high interband absorption, but to see how the surface collision damping affects metals with reasonably high bulk loss is important.
D. "Dirty silver" with bulk damping constant $\gamma_b = 1.2 \times 10^{14}\ \mathrm{s}^{-1}$ similar to that of gold with surface collision damping taken into account.
E. "Ideal metal" with no bulk damping $\gamma_b = 0$, with surface collision damping taken into account.

The results are shown in Fig. 13.3, next to the straight light line that describes propagation of plane electromagnetic wave in GaN. As expected, not taking into account surface collision damping for the best silver (curve A) may lead to very large propagation constant, with effective index exceeding 7, but once surface collisions damping have been included (curve B) the propagation constant is reduced almost two-fold. This simple fact explains while the effective indices of more than 4 have not been observed experimentally. If we now consider the more realistic silver, full of defects due to surface deposition process, whose damping rate is comparable to gold, the curves without (C) and with (D) surface collision damping, the latter's impact is less significant, although still prominent. But it is the curve (E) which is the most striking– if one starts with the best available silver and then hypothetically gets rid of all possible damping processes, then, even if the surface is atomically smooth the increase of the attainable propagation constant $\beta_{\max,r}$ will be only about 12 %, and the maximum effective index will not exceed roughly 4.4, just as predicted by (13.25). That means the minimum confinement

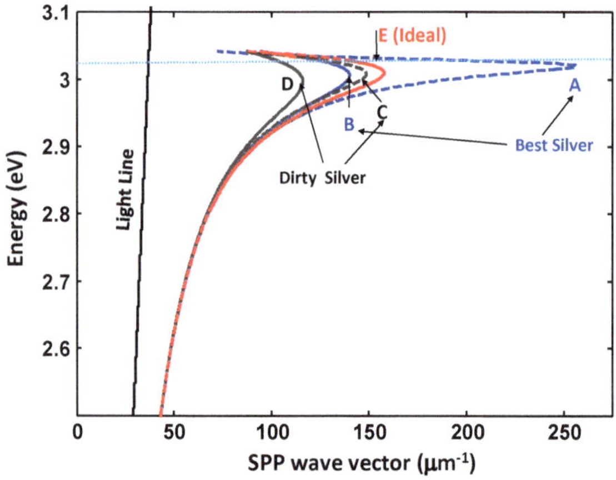

Fig. 13.3 Dispersion curves of SPP on the interface between Ga-N and Ag-like metal for the following cases: *A* The best bulk silver with bulk damping constant $\gamma_b = 3.2 \times 10^{13}\,\mathrm{s}^{-1}$ and no surface collision damping taken into account $\gamma_s = 0$ *B* Same with surface collision damping taken into account. *C* "Dirty silver" with bulk damping constant $\gamma_b = 1.2 \times 10^{14}\,\mathrm{s}^{-1}$ similar to that of gold and no surface collision damping taken into account $\gamma_s = 0$. *D* Same with bulk damping constant $\gamma_b = 1.2 \times 10^{14}\,\mathrm{s}^{-1}$ similar to that of gold with surface collision damping taken into account. (E) "Ideal metal" with no bulk damping $\gamma_b = 0$, with surface collision damping taken into account

depth in the normal direction $d_{\min,x}$ will not be less then roughly $\lambda_D/25n_d$ or 10 nm. It is reasonable to say then that all the imaging schemes involving SPP's—whether it is the "superlens" [26] or a "hyperlens [27]" will have their resolution limited by the surface collision (Landaudamping) to about a quarter of the wavelength in dielectric.

13.5 Landau (Surface Collision) Damping in Multipole Modes of Spherical Nanoparticles

We now turn our attention to the spherical nanoparticles and use our theory to estimate the impact of Landau damping on the properties of all the modes existing there. Consider the spherical metal nanoparticle of radius a embedded in a medium with dielectric constant $\varepsilon_d > 0$ shown in Fig. 13.4a. Specifically, we shall treat the nano-structure in which either Au or Ag nanospheres are embedded in the wide bandgap GaN. Each nanoparticle supports a large number of SPP modes (l, m) with resonant frequencies $\omega_l = \omega_p/[1 + (1 + 1/l)\varepsilon_d]^{1/2}$, ranging from the $\omega_1 = \omega_p/(1 + 2\varepsilon_d)^{1/2}$ for the lowest order radiating dipole mode to $\omega_{sp} = \omega_p/(1 + \varepsilon_d)^{1/2}$ for the highest order modes that behave as propagating SPs running in circles along the nanoparticle surface.

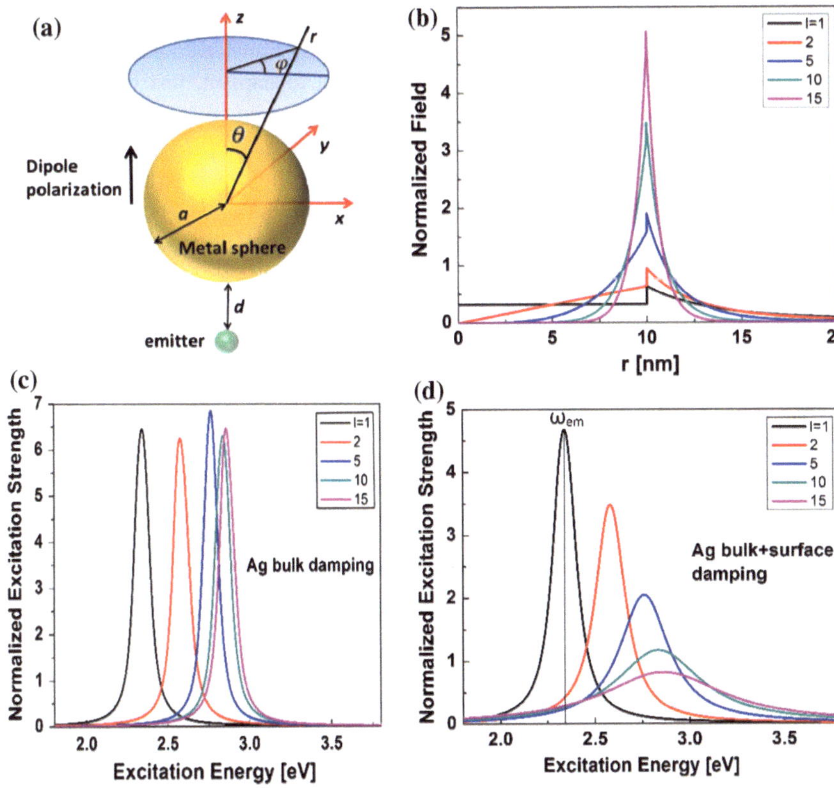

Fig. 13.4 **a** Geometry of luminesce enhancement by a spherical metal nanoparticle of radius a separated from the emitter by distance d. **b** Radial electric field of SPP modes of different order $l = 1, 2, 5, 10, 15$ versus distance from the center of nanoparticle of radius $a = 10$ nm. **c** Excitation spectra of those SPP modes for Ag nanoparticle of radius $a = 10$ nm without taking surface collision damping γ_s into account. **d** Same as **c** but with surface collision damping taken into account

If the polar axis z is associated with the polarization of the emitter as in Fig. 13.4a, only modes with $m = 0$ have non-vanishing field along this axis,

$$E_l = \begin{cases} E_{\max,l}\left(\frac{r}{a}\right)^{l-1}\left[-\frac{l}{l+1}P_l(\cos\theta)\hat{r} - \frac{1}{\sin\theta}\left[P_{l+1}(\cos\theta) - \cos\theta P_l(\cos\theta)\right]\widehat{\boldsymbol{\theta}}\right] & r < a \\ E_{\max,l}\left(\frac{a}{r}\right)^{l-1}\left[P_l(\cos\theta)\hat{r} - \frac{1}{\sin\theta}\left[P_{l+1}(\cos\theta) - \cos\theta P_l(\cos\theta)\right]\widehat{\boldsymbol{\theta}}\right] & r > a \end{cases} \tag{13.26}$$

where $\hat{r}$ and $\widehat{\boldsymbol{\theta}}$ are unit vectors in polar coordinates, and $P_l(\cos\theta)$ is a Legendre polynomial. The radial dependence of the electric field for different modes is shown in Fig. 13.4b along z axis ($\theta = 0$) where the energy of each mode has been normalized to $\hbar\omega_l$. Since the higher order mode occupies progressively smaller volume $V_{eff,l} \approx 4\pi a^3/(l+1)^2\varepsilon_d$ [28] the amplitudes of higher order modes increase with l, and if an emitter is placed close enough to the metal surface to interact with these dark high $-l$ modes, the energy will be very efficiently transferred into these modes and subsequently dissipated. But in order for that to happen, just having spatial overlap is not enough as the spectral overlap is also required, as shown in Fig. 13.4c for Ag nanosphere where each mode is represented by a Lorentzian with full width at half maximum (FWHM) equal to damping γ_l which for $l=1$ mode is equal to $\gamma_{l=1}=\gamma_{rad}+\gamma_b$ where γ_{rad} is the radiative decay rate of the nanoparticle proportional to the nanoparticle volume [28], and for all other modes $\gamma_{l>1}=\gamma_b$. All the modes have the same bulk damping rate γ_b, but the surface collisions are far more prevalent in the higher order modes. One can make a very rough estimate of these rates by evaluating the mean penetration depth of the radial field into the metal as

$$d_{eff,l}=\langle (a-r)E_l^2(r)r^2\rangle_r=\frac{a}{l+1} \quad (13.27)$$

which should result in surface collision damping rate of $\gamma_{s,l}\approx A(l+1)v_F/(2a)$ where according to [21] A is a constant of the order of unity. As a result, higher order modes are expected to be subject to stronger damping and their linewidths are expected to increase as shown in Fig. 13.4d.

The simple result (13.27) can be verified by evaluating the surface collision damping using the exact expression for the modal field (13.26) in the formula (13.10). A typical power spectrum of l-th mode is shown in Fig. 13.5a and its half-width is about l/a. It is clear that the damping rate can indeed be approximated as $\gamma_s^{(l)}=(v_F/2a)A^{(l)}$ for the wide range of the radii of the nanospheres and, as one can see from Fig. 13.5 the coefficient $A^{(l)}$ can be very well approximated as l and

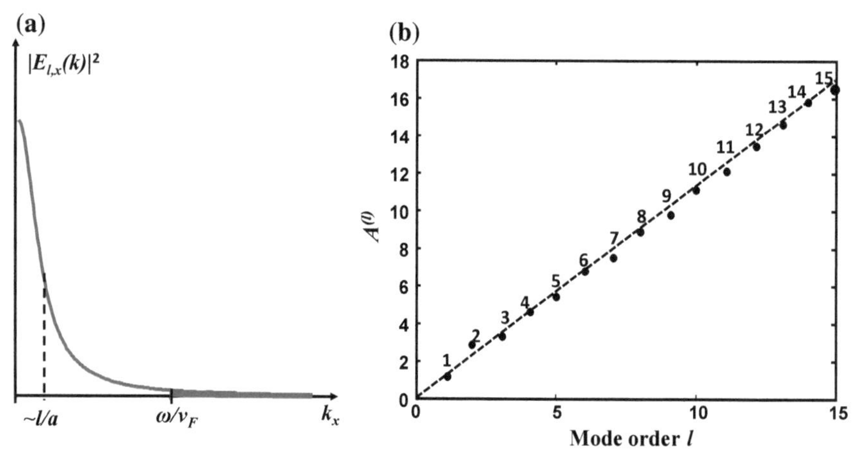

Fig. 13.5 **a** Power spectrum of the *l-th* order SPP mode of the nanosphere. The fraction of energy with wave vector larger than ω/v_F get absorbed by the metal (Landau damped). **b** Calculated values of damping coefficient $A^{(l)}$ for SPP modes from $l=1$ to $l=15$. Linear fit $A^{(l)}=l$ is shown by the dashed line

the surface collision damping as $\gamma_s^{(l)} \approx l v_F/2a$ which is only slightly different from the phenomenological estimate made from (13.27). The result of this surface damping adding on to the bulk damping rate leads to increase of the Lorentzian FWHM of various SPP modes as shown in Fig. 13.1d with the higher order ones being progressively more broadened.

This broadening makes an important impact on the enhancement of fluorescence of the emitters placed in the vicinity of the metal nanoparticles. The higher order $l > 1$ modes are multipoles hence they do not radiate efficiently. When the emitter is close to the nanoparticle surface only a fraction of its energy gets coupled into the dipole mode, from which it can be radiated leading to the fluorescence efficiency enhancement. The energy can also get coupled into the nonradiative higher order modes and dissipate exciting hot electrons and holes. This process of coupling into nonradiative modes is called "lifetime quenching". The resonant frequencies of higher order modes are blue-shifted relative to the dipole mode, hence one can attempt to engineer the structure to keep quenching to the minimum by avoiding the resonance with higher order modes. Unfortunately, dues to surface-collision broadening the linewidths of higher order modes are very broad, and for small nanoparticles the damping becomes enormous, cancelling the benefits of larger Purcell factors of the smaller particles. As a result, one can accelerate the decay but at the expense of the reduced efficiency. This is why, while being spectacularly successful in enhancement of Raman scattering, plasmonic structures so far have not emerged to efficient fluorescence sensors.

Next we shall see how the surface damping which has already been shown to put an ultimate limit on the field enhancement in propagating SPP's does the same for the plasmonic dimers, and by extension to all nanoantennas.

13.6 Impact of Landau (Surface Collision) Damping on Field Enhancement in Dimer

Let us try to construct a simple intuitive picture of why surface collision damping limits the degree to which the field can be squeezed. We can use the example of the propagating SPP considered above. Typically, the coupling rate between the propagating wave and the SPP mode is much lower than the damping rate of that mode. For this "overdamped" case, the higher is the damping rate the less energy gets coupled into the mode. In other words, the radiation tends to be coupled into the modes with higher quality factors, and, if the shape of the mode can be changed to reduce damping (while, of course, still satisfying Maxwell equations) then the shape with the lowest loss will describe the actually excited mode. This is, similar to using the variational method to solve quantum mechanical problems. Therefore, when the surface collisions cause increase in the damping of propagating SPP near the SP resonance the mode re-arranges itself and spreads out so that surface collision damping is reduced. Interestingly, if one considers gap SPP [29–31] the boundary conditions do not allow the mode there to spread out and the gap SPP can be squeezed to tiny size as the gap decreases—increased damping will only reduce propagation length and ability to couple the energy in and out of gap SPP mode, but will not change the shape of the mode.

The situation is different in the dimer shown in Fig. 13.1a where the field can be "squeezed" out of the gap—hence one would expect that as the gap gets smaller the shape of the mode will change and the degree of field enhancement in it will decrease. Before getting immersed in the numerical analysis let us use a simple coupled mode model [32] to paint a clear physical picture of what is going on. The gap mode can be represented as a coherent superposition of the individual modes of two nanospheres in the dimer $E_{\text{dim}}(r) = \sum_{k=1,2}\sum_{l=1}^{\infty} C_{k,l}E_{k,l}(r)$. Since the mode confinement increases with l, the more significant terms are there in this expansion the more confined gets the gap mode. The energy can couple directly into the dipole $l = 1$ modes of the each sphere, and from there into the $l > 1$ modes of the other sphere. This inter-sphere coupling increases as the gap decreases hence progressively higher and higher order modes gets mixed into the dimer mode and its effective confinement increases. But surface collision damping is bound to reduce the effective coupling of the high order modes, especially as the radius of the sphere gets small. As a result, confinement gets weaker and effective size of the mode is bound to increase.

Now, to solve the problem self-consistently we use the same iterative method as used for SPP. First we solve the Maxwell equations numerically using finite difference technique and assuming that the surface collision damping is not present in the expression for the dielectric permittivity (13.5) i.e. $\gamma_s^{(0)} = 0$. Once the field of the mode, $E_{\text{dim}}^{(1)}$, characterized by some effective volume $V_{eff}^{(1)}$ is calculated, we can find the new surface collision damping rate $\gamma_s^{(1)}$ by finding the spatial Fourier transform of $E_{\text{dim}}^{(1)}$ and then using (13.10). With this new damping rate in (13.5) the next iteration yields a new shape of the mode $E_{\text{dim}}^{(2)}$ that now is more spread out with larger effective spread $V_{eff}^{(2)} > V_{eff}^{(1)}$ and thus reduced surface damping rate $\gamma_s^{(2)} < \gamma_s^{(1)}$ serving as the input to the next iteration. According to Fig. 13.6a, b the process converges after only a few iterations yielding both γ_s and the shape of E_{dim} as well as the degree of field enhancement.

As shown in Fig. 13.6a for a dimer of radius $a = 2.5$ nm and gap size of $d_{\text{gap}} = 0.5$ nm the surface collision scattering rate after a few rather wild oscillations settles at a value of $\gamma_s = 2\times10^{15}s^{-1}$ which is almost two orders of magnitude larger than bulk damping rate $\gamma_b = 3\times10^{13}s^{-1}$. Ar the same time, according to Fig. 13.6b the effective volume of the mode also increases as the mode gets "pushed" out of the gap and settles at $V_{eff} = 1.8\,\text{nm}^{-3}$. According to the simple phenomenological theory described above one would expect to find $\gamma_s \sim v_F/V_{eff}^{1/3} \sim 1.8\times10^{15}s^{-1}$ indicating that results of Fig. 13.6 are indeed self-consistent.

Figure 13.7 shows dramatic impact of surface collision scattering on both the shape of the field in the SPP mode of the dimer and its magnitude. The field

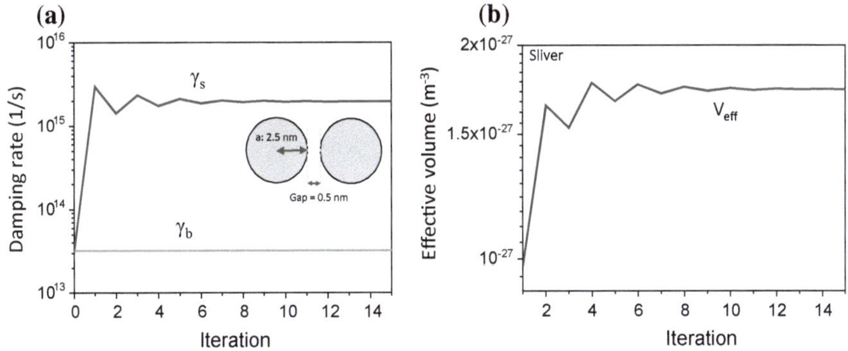

Fig. 13.6 Self consistent iterative calculation of the modes in silver dimer (radius $a = 2.5$ nm, $d_{gap} = 0.5$ nm) **a** Convergence of the surface collision damping constant γ_s. **b** Convergence of the effective volume V_{eff}

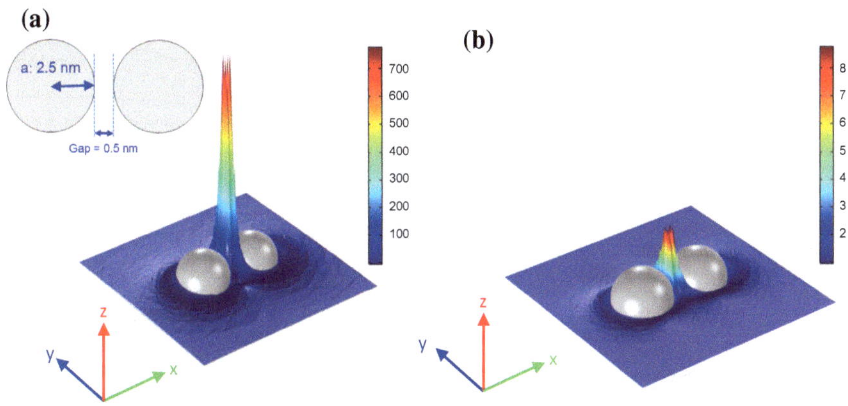

Fig. 13.7 Field enhancement in the silver dimer (radius $a = 2.5$ nm, $d_{gap} = 0.5$ nm) **a** without surface collision damping, **b** with surface collision damping

enhancement in the absence of surface collisions reaches 700 but once the surface collision damping is taken into account it gets reduced by nearly two orders of magnitude—which is more or less the ratio of γ_s/γ_b.

Even more dramatic ae the changes in the absorption spectrum of the dimer shown in Fig. 13.8 In the absence of surface collisions one can see a number of narrow resonances, but once the surface damping is factored in, only a single, very broad spectral feature remains, indicating that the plasmonic enhancement reaches its limit.

In Fig. 13.9a one can see how the surface collision damping grows as the gap gets narrow, while as can be seen from Fig. 13.9b the effective volume gets smaller.

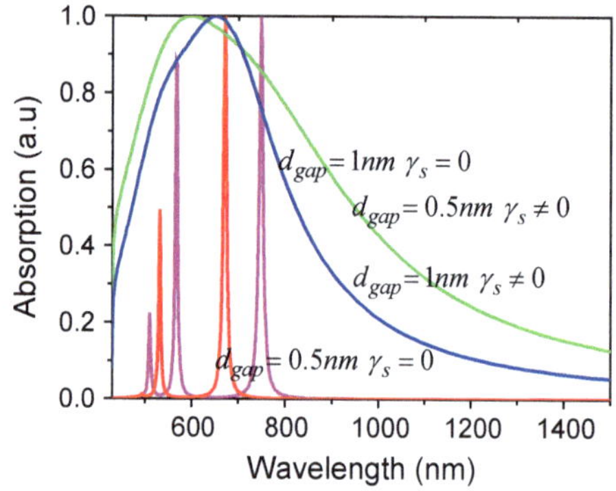

Fig. 13.8 Absorption spectra of silver dimers (radius a = 5 nm) for two different gaps with and without surface collision damping

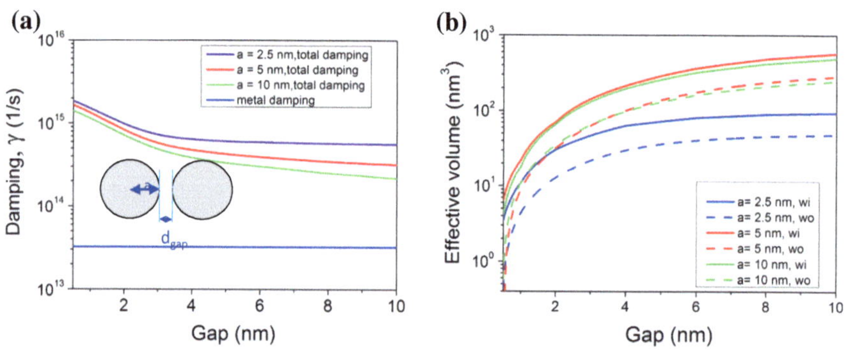

Fig. 13.9 **a** Surface collision damping γ_s in the silver dimer as a function of gap width for three different radii. **b** Effective mode volume as a function of gap width without (*dashed lines*) and with (*solid lines*) surface collisions

But note that while V_{eff} decreases very sharply (dashed curves) when the surface effect is disregarded, once γ_s is incorporated the decrease is not that significant. As expected, the mode gets "squeezed out" of the gap.

13.7 Conclusions

In this chapter we have analyzed the behavior of nanoscale metallic structures when characteristic dimensions of them become less than mean free path. We have shown that at that scale Landau damping becomes the dominant factor that causes increase of losses, broadening of the resonances and decrease of the field enhancement. We have also shown that many alternative description of this phenomena—diffusion in hydrodynamic model, phenomenological Kreibig model and others all have the same theoretical underpinning—Landau damping. For the structures with characteristic dimensions of less than a few nanometers the loss associated with bulk metal becomes irrelevant as Landau damping alone determines the degree of enhancement and confinement thus becoming an ultimate limit of plasmonic enhancement

References

1. M. Stockman, Nanoplasmonics: past, present, and glimpse into future. Opt. Express **19**, 22029–22106 (2011)
2. S.A. Maier, *Plasmonics: fundamentals and applications* (Springer, 2007)
3. W.L. Barnes, A. Dereux, T.W. Ebbesen, Surface plasmon subwavelength optics. Nature **424**, 824 (2003)
4. S. Kühn, U. Håkanson, L. Rogobete, V. Sandoghdar, Enhancement of single-molecule fluorescence using a gold nanoparticle as an optical nanoantenna. Phys. Rev. Lett. **97**, 017402 (2006)
5. L. Novotny, Single molecule fluorescence in inhomogeneous environments. Appl. Phys. Lett. **69**, 3806 (1996)
6. J.B. Khurgin, How to deal with the loss in plasmonics and metamaterials. Nat. Nanotechnol. **10**, 2–6 (2015)
7. Y. Wu, C. Zhang, N.M. Estakhri, Y. Zhao, J. Kim, M. Zhang, X.-X. Liu, Greg K. Pribil, A. Alù, C.K. Shih, X. Li, Intrinsic optical properties and enhanced plasmonic response of epitaxial silver. Adv. Mater. **26**, 6106–6110 (2014)
8. J.B. Khurgin, A. Boltasseva, Reflecting upon the losses in plasmonics and metamaterials. MRS Bull. **37**, 768–779 (2012)
9. A.J. Hoffman, L. Alekseyev, S.S. Howard, K.J. Franz, D. Wasserman, V.A. Podolskiy, E.E. Narimanov, D.L. Sivco, C. Gmachl, Negative refraction in semiconductor metamaterials. Nat. Mater. **6**, 946–950 (2007)
10. F.J. Garcıa de Abajo, Nonlocal effects in the plasmons of strongly interacting nanoparticles, dimers, and waveguides. J. Phys. Chem. C **112**, 17983–17987 (2008)
11. N. A. Mortensen, S. Raza, M. Wubs, T. Søndergaard, and S. I. Bozhevolnyi, A generalized nonlocal optical response theory for plasmonic nanostructures, Nat. Commun. **5** (2014)
12. S. Raza, S.I. Bozhevolnyi, M. Wubs, N.A. Mortensen, Nonlocal optical response in metallic nanostructures. J. Phys. Condens. Matter **27**, 1832014 (2015)
13. T. Christensen, W. Yan, S. Raza, S.A.P. Jauho, N.A. Mortensen, M. Wubs, Nonlocal response of metallic nanospheres probed by light, electrons, and atoms. ACS Nano **8**(2), 1745–1758 (2014)
14. A.V. Uskov, I.E. Protsenko, N.A. Mortensen, E.P. O'Reilly, Broadening of plasmonic resonance due to electron collisions with nanoparticle boundary: a quantum mechanical consideration. Plasmonics **9**, 185–192 (2014)
15. C. Yannouleas, R.A. Broglia, Landau damping and wall dissipation in large metal clusters, Ann. Phys. **217**, I 105–141 (1992)
16. R.A. Molina, D. Weinmann, R.A. Jalabert, Oscillatory size dependence of the surface plasmon linewidth in metallic nanoparticles. Phys. Rev. B **65**, 155427 (2002)
17. Z. Yan, S. Gao, Landau damping and lifetime oscillation of surface plasmons in metallic thin films studied in a jellium slab model. Surf. Sci. **602**, 460–464 (2008)

18. T. Atay, J.H. Song, A.V. Nurmikko, Strongly interacting plasmon nanoparticle pairs: from dipole–dipole interaction to conductively coupled regime. Nano Lett. **4**, 1627 (2004)
19. P. Nordlander, C. Oubre, E. Prodan, K. Li, M.I. Stockman, Plasmon hybridization in nanoparticle dimers. Nano Lett. **4**, 899 (2004)
20. M. Moskovits, D.H. Jeong, Engineering nanostructures for giant optical fields. Chem. Phys. Lett. **397**, 91 (2004)
21. U. Kreibig, M. Vollmer, *Optical properties of metal clusters* (Springer-Verlag, Berlin, 1995)
22. J.B. Khurgin, Ultimate limit of field confinement by surface plasmon polaritons. Faraday Discuss. **178**, 109–122 (2015)
23. R. G. Chambers, The anomalous skin effect. *Proceedings of the Royal Society of London. Series A, Mathematical and Physical Sciences* **215** 1123 (1952): 481–97
24. J. Lindhard, On the properties of a gas of charged particles. Mat. Fys. Medd. Dan. Vid. **28**(8), 1–57 (1954)
25. P.B. Johnson, R.W. Christy, Optical constants of noble metals. Phys. Rev. B **6**, 4370 (1972)
26. J.B. Pendry, Negative refraction makes a perfect lens. Phys. Rev. Lett. **85**, 3966 (2000)
27. Z. Jacob, L.V. Alekseyev, E. Narimanov, Optical hyperlens: far-field imaging beyond the diffraction limit. Opt. Express **14**, 8247–8256 (2006)
28. G. Sun, J.B. Khurgin, C.C. Yang, Impact of high-order surface Plasmon modes of metal nanoparticles on enhancement of optical emission. Appl. Phys. Lett. **95**, 171103 (2009)
29. K. Tanaka, M. Tanaka, T. Sugiyama, Simulation of practical nanometric optical circuits based on surface plasmon polariton gap waveguides. Opt. Express. **13,** 256–266 (2005)
30. I. V. Novikov and A. A. Maradudin, Channel polaritons. Phys. Rev. B. **66,** 035403 (2002)
31. S. I. Bozhevolnyi, V. S. Volkov, E. Devaux, and T. W. Ebbesen, Channel plasmon-polariton guiding by subwavelength metal grooves. Phys. Rev. Lett. **95,** 046802 (2005)
32. G. Sun, J. B. Khurgin, A. Bratkovsky, Coupled-mode theory of field enhancement in complex metal nanostructures. Phys. Rev. B. **84,** 045415 (2011)

Index

S.I. Bozhevolnyi et al. (eds.), *Quantum Plasmonics*,
Springer Series in Solid-State Sciences 185,
DOI 10.1007/978-3-319-45820-5